Edited by Editorial Board of UTokyo Engineering Course

# Linear Algebra II

## Advanced Topics for Applications

**UTokyo Engineering Course/Basic Mathematics**

---

*Linear Algebra I: Basic Concepts*
by Kazuo Murota and Masaaki Sugihara
ISBN: 978-981-125-702-5
ISBN: 978-981-125-797-1 (pbk)

UTokyo Engineering Course/ Basic Mathematics

Edited by Editorial Board of UTokyo Engineering Course

# Linear Algebra II

## Advanced Topics for Applications


**Kazuo Murota**

*The Institute of Statistical Mathematics, Japan;*
*Professor Emeritus at*
*The University of Tokyo, Japan & Kyoto University, Japan*
*& Tokyo Metropolitan University, Japan*

**Masaaki Sugihara**

*Professor Emeritus at The University of Tokyo, Japan*
*& Nagoya University, Japan*

*Published by*

World Scientific Publishing Co. Pte. Ltd.
5 Toh Tuck Link, Singapore 596224
*USA office:* 27 Warren Street, Suite 401-402, Hackensack, NJ 07601
*UK office:* 57 Shelton Street, Covent Garden, London WC2H 9HE

and

Maruzen Publishing Co., Ltd.
Kanda Jimbo-cho Bldg. 6F, Kanda Jimbo-cho 2-17
Chiyoda-ku, Tokyo 101-0051, Japan ('MP')

Library of Congress Control Number: 2022030856

**British Library Cataloguing-in-Publication Data**
A catalogue record for this book is available from the British Library.

**LINEAR ALGEBRA II**
**Advanced Topics for Applications**



ISBN 978-981-125-705-6 (hardcover)
ISBN 978-981-125-798-8 (paperback)
ISBN 978-981-125-706-3 (ebook for institutions)
ISBN 978-981-125-707-0 (ebook for individuals)

For any available supplementary material, please visit
https://www.worldscientific.com/worldscibooks/10.1142/12867#t=suppl

Desk Editor: Tan Rok Ting

Printed in Singapore

UTokyo Engineering Course

# About This Compilation

What is the purpose of engineering education at the University of Tokyo's Undergraduate and Graduate School of Engineering? This School was established 125 years ago, therefore we feel it is an appropriate time to ask this question again. More than a century has passed since Japan embarked on a path to introduce and negotiate Western knowledge and practices. Japan and the world are very different places now, and today our university stands as a world leading institute in engineering research and education. As such, it is our duty and mission to build a firm foundation of education that will support the creation and dissemination of engineering knowledge, practices and resources. Our School of Engineering must not only teach outstanding students from Japan but also those from throughout the world. Put another way, the engineering that we teach students is not only a responsibility of this School, but an imperative placed on us by society and the age in which we live. It is in this changed context, where we have gone from follower to leader, that we present this curriculum, The University of Tokyo (UTokyo) Engineering Course. The course is a reflection of the School's desire to engage with those outside the walls of the Ivory Tower, and to spread the best of engineering knowledge to the world outside our institution. At the same time, the course is also designed for the undergraduate and graduate students of the School. As such, the course contains the knowledge that should be learnt by our students, taught by our instructors and critically explored by all.

February 2012

Takehiko Kitamori
Dean, Undergraduate and Graduate Schools of Engineering
The University of Tokyo
(April 2010–March 2012)

**UTokyo Engineering Course**

# The Purpose of This Publication

Modern engineering is composed of the academic discipline of fundamental engineering and the academic discipline of integrated engineering that deals with specific systems and subjects. Interdisciplinary disciplines and multidisciplinary disciplines are amalgamations of multiple academic disciplines that result in new academic disciplines when the academic pursuit in question does not fit within one traditional fundamental discipline. Such interdisciplinary disciplines and multidisciplinary disciplines, once established, often develop into integrated engineering. Moreover, the movement toward interdisciplinarity and multidisciplinarity is well underway within both fundamental engineering and advanced research.

These circumstances are producing a variety of challenges in engineering. That is, the scope of research of integrated engineering is gradually growing larger, with economics, medicine and society converging into an enormously complex social system, which is resulting in the trend of connotative academic disciplines growing larger and becoming self-contained research fields, which, in turn, is resulting in a trend of neglect toward fundamental engineering. The challenge of fundamental engineering is how to connect engineering education that is built upon traditional disciplines with that of advanced engineering research in which interdisciplinarity and multidisciplinarity is continuing at a rapid pace. Truly this is an educational challenge shared by all the top engineering schools in the world. Without having a solid understanding of engineering, however, education related to learning state-of-the-art research methodologies will not hold up. This is the dichotomy of higher education in engineering; that is, higher education in engineering simply will not work out if either side of the equation is missing.

In the meantime, the internationalization of universities is going forward in routine fashion. In fact, here at the University of Tokyo (UTokyo), one quarter of the graduate students enrolled in engineering fields are of foreign nationality and the percentage of foreign undergraduate students is expected to increase more and more. On top of that, Japan is experiencing a reduction in the population of its youth. Therefore, the time is ripe to ramp up efforts to look outside of Japan in order to secure the human resources to sustain the future of advanced science and technology here in Japan. It is clear that the internationalization of engineering education is rapidly underway. As such, the need for a curriculum that is firmly rooted in engineering knowledge needs to be oriented toward both local and foreign students.

Due to these circumstances surrounding modern engineering, we at UTokyo's School of Engineering have systematically organized an engineering curriculum of fundamental engineering knowledge that will not be unduly influenced by the times, with the goal of firmly establishing a benchmark suitable for that of the top schools of engineering of science and technology for students to learn and teachers to teach. This engineering curriculum clarifies the disciplines and instruction policy of UTokyo's School of Engineering and is composed of three layers: Fundamental (sophomores (second semester) and juniors), Intermediate (seniors and graduates) and Advanced (graduates). Therefore, this engineering course is a policy for the thorough education of the engineering knowledge necessary for forming the foundation of our doctorate program as well. The following is an outline of the expected effect of this engineering course:

- Surveying the total outline of this engineering course will assist students in understanding which studies they should undertake for each field they are pursuing, and provide an overall image by which the students will know what fundamentals they should be studying in relation to their field.
- This course will build the foundation of education at UTokyo's School of Engineering and clarify the standard for what instructors should be teaching and what students need to know.
- As students progress in their major it may be necessary for them to go back and study a new fundamental course. Therefore the textbooks are designed with such considerations as well.

- By incorporating explanations from the viewpoint of engineering departments, the courses will make it possible for students to learn the fundamentals with a constant awareness of their application to engineering.

Takao Someya, Board Chair
Yukitoshi Motome, Shinobu Yoshimura, Executive Secretary
Editorial Board of UTokyo Engineering Course

## UTokyo Engineering Course

# Editorial Board

# Preface

In the University of Tokyo (UTokyo) Engineering Course there are two volumes on linear algebra, *Linear Algebra I: Basic Concepts* and *Linear Algebra II: Advanced Topics for Applications*. The objective of the first volume is to show the standard mathematical results in linear algebra from the engineering viewpoint. The objective of this second volume is to branch out from the first volume to illustrate useful specific topics pertaining to applications. These two volumes were originally written in Japanese and published from Maruzen Publishing in 2013 and 2015. The present English version is their almost faithful translations by the authors.

Linear algebra is primarily concerned with systems of equations and eigenvalue problems for matrices and vectors with real or complex entries. In contrast, the topics treated in this volume, *Linear Algebra II*, can be regarded as variations, extensions or generalizations of these central issues, incorporating additional features relevant in applications. The contents of the chapters of this volume are described below.

Chapter 1, "Matrices and Graphs," presents a methodology that focuses solely on structural information while disregarding numerical information. In this book, the graph-theoretic method is utilized mainly in the theory of nonnegative matrices, which deals with eigenvectors of nonnegative matrices and has an important application to stationary distributions of Markov chains. However, the method itself has engineering versatility, offering effective methods in other applications, *e.g.*, decompositions of large-scale systems and preprocessing of large-scale numerical computations.

Chapter 2, "Nonnegative Matrices," features the sign (positive or negative) of real numbers, which is a characteristic property of real numbers not shared by complex numbers. The theory of nonnegative matrices forms a basis of probabilistic methods for the obvious reason that the probabilities are represented by nonnegative numbers. The theory is also related to monotonicity, which is one of the most fundamental concepts in understanding system characteristics in terms of input–output relations. In an electric network, for example, the output current increases with the increase of the applied (or input) voltage. This kind of monotonicity corresponds to the concept of M-matrices, which can be viewed as a variant of nonnegative matrices.

Chapter 3, "Systems of Linear Inequalities," deals with inequalities rather than equality relations (equations). Inequalities are often used in expressing constraints in optimization, and the theory of systems of linear inequalities provides a theoretical basis for the method of linear programming. Mathematically, this involves an intriguing fact known as duality.

Chapter 4, "Integer Matrices," deals with matrices with entries of integers (as opposed to real numbers). Mathematically, the divisibility relation (division) is highlighted as the central issue. In engineering applications, integers are naturally used in counting the number of things and people, but they are also used, *e.g.*, as stoichiometric coefficients in describing chemical reactions. In addition, network structures have a close relation with integer matrices.

Chapter 5, "Polynomial Matrices," deals with matrices with entries of polynomial functions (as opposed to real or complex constants). Also in this case, the divisibility relation (division) plays the primary role, and the underlying mathematical structure is quite similar to that of integer matrices. In engineering applications, polynomial matrices often appear through the Laplace transformation, and the theory of polynomial matrices is useful in numerical solution of differential equations (simulations of dynamic systems) and in control theory. Further theoretical results about polynomial matrices, which are beyond the scope of this chapter, are used extensively as modern tools in control theory.

Chapter 6, "Generalized Inverses," shows that it is possible to define, for rectangular or square singular matrices, a concept that is similar to the inverse matrices. Generalized inverses are a convenient tool widely used to cope with inconsistency in the degrees of freedom of variables and equations, typically in statistics and structural engineering.

Chapter 7, "Group Representation Theory," can be viewed as a theory of linear algebra that incorporates an additional aspect of group-theoretic symmetry. Matrices possessing group-theoretic symmetry can be decomposed into block-diagonal forms in a systematic manner. The framework of group representation theory enables us to mathematically formulate our institution that systems are endowed with symmetry (in some sense or other) and to take advantage of these characteristic properties in our analysis. The methodology of exploiting symmetry is used successfully in many fields of science and engineering, including physics, chemistry, structural engineering, and optimization.

In this book, every effort has been made to develop theories with sufficient mathematical rigor, while relating them to engineering applications. The book takes the approach of developing theories through concrete matrix representations (normal forms) rather than abstract concepts. It aims at self-contained presentations of mathematical arguments, so that the reader can follow them without referring to other sources. The chapters are, for the most part, independent of each other, and can be read in any order according to the reader's interest. The main objective of this book is to present the mathematical aspects of linear algebraic methods for engineering that will potentially be effective in various application areas. Although applicabilities of those mathematical methods are indicated briefly, no attempt has been made to provide actual case studies that demonstrate the effectiveness of the respective methodologies.

In writing this book, we received lot of help from many people. We would like to express appreciation to all. In particular, Naonori Kakimura, who took on the painstaking task of thoroughly reading and checking the manuscript for the Japanese version, Yoshiko Ikebe, who read through the English translation and gave suggestions for improvements in English writing, Kensuke Aishima, who helped us in the numerical computation for example problems, and Mikio Furuta, Ken Hayami, Yoshihiro Kanno, Yusuke Kobayashi, Yusuke Kuroki, Kazuhisa Makino, Satoko Moriguchi, Mizuyo Takamatsu, Akihisa Tamura, Ken'ichiro Tanaka, and the graduate students of the University of Tokyo, who provided us with a variety of feedback.

Kazuo Murota
Masaaki Sugihara

# Contents

# List of Figures

# Chapter 1

# Matrices and Graphs

In this chapter we present a method of representing the structure of matrices in terms of graphs. Here, by "structure" we mean a structure of combinatorial nature, no more than the distinction between zeros and nonzeros of the matrix entries, disregarding numerical or quantitative information. The methodology of focusing on the structure is an effective means for qualitative analysis of large-scale systems and efficient large-scale numerical computations. The method of this chapter will also be used in Chap. 2 for the theory of nonnegative matrices.

## 1.1 Matrices and Directed Graphs

### 1.1.1 *Graph Representation*

A *graph* is a discrete structure composed of *vertices* and *edges*,[1] such as the one shown in Fig. 1.1. We speak of a *directed graph* or *digraph*, when we want to emphasize that the edges have directions. Graphs are easier to understand when they are illustrated as a drawing, but the way they are drawn is not important, because the relevant information is which edges connect which vertices. The set $V$ of the vertices of a graph $G$ is referred to as the *vertex set* of $G$, and the set $E$ of the edges as the *edge set* of $G$. A graph with vertex set $V$ and edge set $E$ is often denoted as $G = (V, E)$.

Consider a square matrix $A = (a_{ij})$ of order $n$ with row and column sets (sets of row and column indices) $\{1, \ldots, n\}$, where a one-to-one correspondence between the row and column indices is assumed. Such is the case, for example, with matrices $A$ in the eigenvalue problem $A\boldsymbol{x} = \alpha\boldsymbol{x}$. This is also

[1]Vertices are also called *points* or *nodes*, and edges are also called *arcs* or *branches*. For more information on graphs, see, *e.g.*, [26–28, 32].

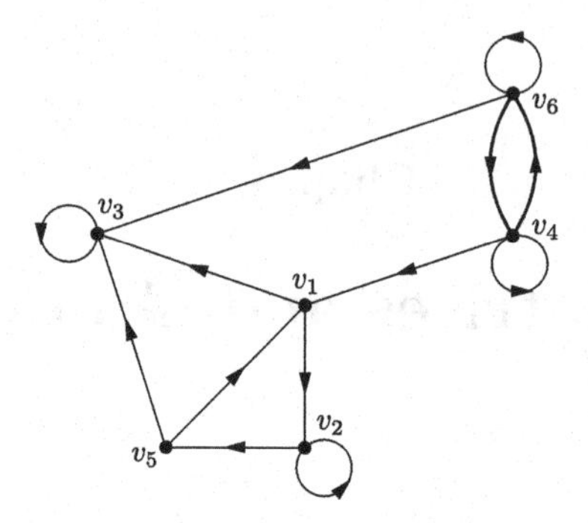


Fig. 1.1 Graph representation of a matrix (Example 1.1).

true for the transition probability matrices of Markov chains, which will be treated in Sec. 2.3.2. Note that the concept of "diagonal entries" is meaningful in a matrix if (and only if) a one-to-one correspondence between the row and column indices is presupposed (see also Remark 1.2).

The structure of a matrix $A$ is represented by a graph $G(A) = (V, E)$ that has vertex set

$$V = \{v_1, \ldots, v_n\} \tag{1.1}$$

and edge set

$$E = \{(v_i, v_j) \mid a_{ij} \neq 0\}. \tag{1.2}$$

That is, $G(A)$ is a graph that has an edge $(v_i, v_j)$ from vertex $v_i$ to vertex $v_j$ for each nonzero entry at the $(i, j)$ position of the matrix $A$. Specifically, if a diagonal entry $a_{ii}$ is nonzero, then $G(A)$ possesses a self-loop at vertex $v_i$, where a *self-loop* is an edge whose initial and end vertices are the same. Note that the definition (1.2) presupposes a one-to-one correspondence between the row and column indices.

**Example 1.1.** Suppose that the following matrix is given:

$$A = \begin{array}{c} \\ 1 \\ 2 \\ 3 \\ 4 \\ 5 \\ 6 \end{array} \begin{array}{c} \begin{array}{cccccc} 1 & 2 & 3 & 4 & 5 & 6 \end{array} \\ \left[ \begin{array}{cccccc} & \alpha_{12} & \alpha_{13} & & & \\ & \alpha_{22} & & & \alpha_{25} & \\ & & \alpha_{33} & & & \\ \alpha_{41} & & & \alpha_{44} & & \alpha_{46} \\ \alpha_{51} & & \alpha_{53} & & & \\ & & \alpha_{63} & \alpha_{64} & & \alpha_{66} \end{array} \right] \end{array}. \tag{1.3}$$

Here the blank entries represent zeros and $\alpha_{12}, \alpha_{13}, \alpha_{22}, \alpha_{25}, \ldots, \alpha_{66}$ are nonzero entries. The structure of this matrix is represented by the graph of Fig. 1.1. The vertex set of this graph is $V = \{v_1, v_2, v_3, v_4, v_5, v_6\}$, since the row set and the column set of this matrix are both $\{1, 2, 3, 4, 5, 6\}$. The edge set $E$ is given as

$$\begin{aligned} E = \{&(v_1, v_2), (v_1, v_3), (v_2, v_2), (v_2, v_5), (v_3, v_3), (v_4, v_1), (v_4, v_4), (v_4, v_6), \\ &(v_5, v_1), (v_5, v_3), (v_6, v_3), (v_6, v_4), (v_6, v_6)\} \end{aligned}$$

with edges corresponding to the nonzero entries $\alpha_{12}, \alpha_{13}, \alpha_{22}, \alpha_{25}, \ldots, \alpha_{66}$. Self-loops exist at vertices $v_2$, $v_3$, $v_4$, and $v_6$, corresponding to the nonzero diagonal entries $a_{22}$, $a_{33}$, $a_{44}$, and $a_{66}$. ■

The structure of a matrix $A$ can often be studied successfully by examining the structure of the associated graph $G(A)$. Specific properties such as irreducibility and periodicity are treated subsequently in Secs. 1.1.2, 1.1.3, and 1.1.4.

**Remark 1.1.** In this book we have adopted the convention of drawing an edge $(v_i, v_j)$ from vertex $v_i$ to vertex $v_j$ for each $a_{ij}$ $(\neq 0)$. Another convention, also often used, is to draw an edge in the reverse direction, thereby defining

$$E = \{(v_j, v_i) \mid a_{ij} \neq 0\}.$$

One can choose whichever convention that is more convenient in applications. ■

**Remark 1.2.** In this subsection, we defined a graph representation for square matrices in which a one-to-one correspondence can naturally be assumed between the row and column indices. In applications, we equally often encounter another type of matrices, square or rectangular, in which no natural one-to-one correspondence is given between the row and column indices. For example, in a system of linear equations $A\boldsymbol{x} = \boldsymbol{b}$, the ordering of the equations and that of the variables are (generally) unrelated, and therefore no natural one-to-one correspondence exists between the row and column indices of the coefficient matrix $A$. For such matrices, a different type of graph representation, introduced in Sec. 1.2, is appropriate. ■

### 1.1.2 *Decomposition into Strongly Connected Components*

#### 1.1.2.1 *A simple example*

Suppose that the graph in Fig. 1.1 represents the state transition of some sort of a system. The graph shows that the system has six states

$\{v_1, v_2, v_3, v_4, v_5, v_6\}$ and that the system can possibly change its state from $v_i$ to $v_j$ when edge $(v_i, v_j)$ exists. From this graph, we can see the following:

- If the system is in state $v_6$, the system can possibly reach an arbitrary state (eventually at some point in time).
- If the system is in state $v_3$, the only possible future state for the system is $v_3$.
- If the system is in state $v_1$, it is impossible for the system to reach state $v_6$.

In terms of reachability, the vertex set $V$ of the graph is classified into three blocks of

$$V_1 = \{v_4, v_6\}, \qquad V_2 = \{v_1, v_2, v_5\}, \qquad V_3 = \{v_3\}$$

as shown in Fig. 1.2. Here the edge directions are restricted to

$$V_1 \to V_2, \qquad V_2 \to V_3, \qquad V_1 \to V_3 \tag{1.4}$$

and hence

$$\text{if } k > l, \text{ there is no edge from } V_k \text{ to } V_l. \tag{1.5}$$

If the vertices are ordered as

$$v_4, v_6 \mid v_1, v_2, v_5 \mid v_3 \tag{1.6}$$

compatibly with the ordering of $V_1 = \{v_4, v_6\}$, $V_2 = \{v_1, v_2, v_5\}$, and $V_3 = \{v_3\}$, then, for an arbitrary edge $(v_i, v_j)$ connecting two vertices of different blocks, the initial vertex $v_i$ always appears before the end vertex $v_j$.

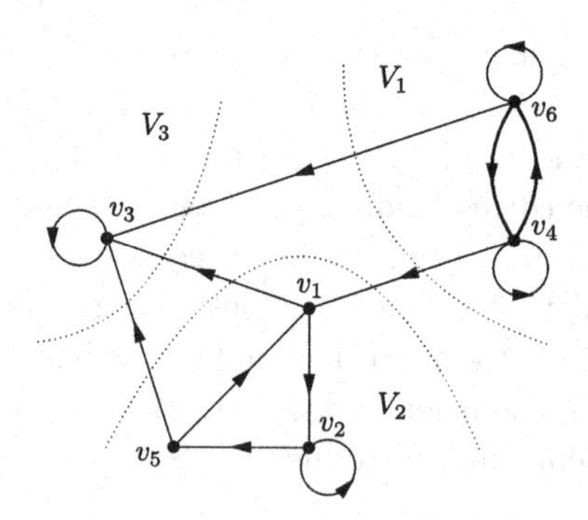


Fig. 1.2 Strongly-connected component decomposition of the graph in Fig. 1.1.

The decomposition illustrated above is an example of the decomposition into strongly-connected components. This decomposition is defined for general directed graphs as follows.

1.1.2.2 *Definition*

Let $G = (V, E)$ be a directed graph. A set of consecutive edges, consistently directed, is called a *directed path.* By grouping those vertices which can be mutually reached from each other by a directed path, we can partition the vertex set $V$ into blocks, or subsets of $V$. Denoting the blocks by $V_1, \ldots, V_p$ $(p \geqq 1)$ we have the resulting partition of $V$ as

$$\bigcup_{k=1}^{p} V_k = V; \qquad V_k \neq \emptyset \quad (k = 1, \ldots, p). \tag{1.7}$$

This is the so-called *strongly-connected component decomposition,* or the *decomposition into strongly-connected components.* Hereafter, we will call this the *SCC-decomposition.* Each $V_k$ is called a *strongly-connected component,* or a *strong component.* Moreover, a hierarchical structure, *i.e.*, an ordering, can be introduced among the strong components by defining a binary relation (order relation) $V_k \succeq V_l$ as[2]

$$V_k \succeq V_l \iff \text{a directed path exists from some } u \in V_k \text{ to some } v \in V_l. \tag{1.8}$$

It follows, in particular, that

$$\text{an edge } (u, v) \in E \text{ exists for some } u \in V_k,\ v \in V_l \implies V_k \succeq V_l. \tag{1.9}$$

In some cases, the SCC-decomposition consists of a single component (with $p = 1$). Such a graph is said to be *strongly connected.*

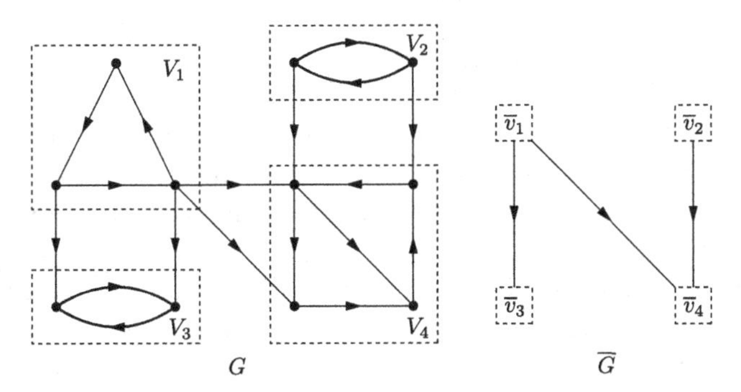


Fig. 1.3 Strongly-connected component decomposition of the graph $G$ and the reduced graph $\overline{G}$.

[2]For any vertex $v$ we may understand (by convention) that a directed path of length zero exists from $v$ to $v$. Then $V_k \succeq V_k$ is true, even when $V_k$ consists of a single vertex.

For the graph in Fig. 1.2 we have

$$V_1 \succeq V_2 \succeq V_3 \tag{1.10}$$

by (1.4). In this particular example, the components $V_k$ are linearly ordered. However, this is not always the case. For instance, the graph $G$ in Fig. 1.3 (left) has four strongly-connected components $V_1$, $V_2$, $V_3$, and $V_4$ with the order relation given by

$$V_1 \succeq V_3, \qquad V_1 \succeq V_4, \qquad V_2 \succeq V_4. \tag{1.11}$$

There is no ordering between $V_1$ and $V_2$ (neither $V_1 \succeq V_2$ nor $V_2 \succeq V_1$). In this way, the binary relation $\succeq$ defined in (1.8) is a partial order.[3]

To represent the partial order in the SCC-decomposition, it is convenient to construct a *reduced graph* $\overline{G}$ by contracting each of the strongly-connected components $V_1, \ldots, V_p$ to a single vertex (see Fig. 1.3). Denote by $\overline{v}_k$ the vertex corresponding to a strongly-connected component $V_k$. Then the vertex set $\overline{V}$ of the reduced graph $\overline{G}$ is given as

$$\overline{V} = \{\overline{v}_1, \ldots, \overline{v}_p\}$$

and the edge set $\overline{E}$ of $\overline{G}$ is defined by

$$(\overline{v}_k, \overline{v}_l) \in \overline{E} \iff (u, v) \in E \text{ for some } u \in V_k,\ v \in V_l;\ k \neq l.$$

Since the reduced graph $\overline{G} = (\overline{V}, \overline{E})$ is free from directed cycles, the binary relation $\succeq$ between the vertices defined as

$$\overline{v}_k \succeq \overline{v}_l \iff \text{a directed path exists from } \overline{v}_k \text{ to } \overline{v}_l \tag{1.12}$$

is a partial order. With the understanding that a vertex $\overline{v}_k$ of the reduced graph $\overline{G}$ corresponds to the strongly-connected component $V_k$ of the original graph $G$, the partial order in (1.12) coincides with the partial order between strongly-connected components defined in (1.8). Otherwise stated, the partial order between strongly-connected components in (1.8) is represented by the reduced graph $\overline{G}$.

The partial order in the SCC-decomposition of a graph can be interpreted as a hierarchical structure intrinsic to the engineering system represented by the graph. As such, the partial order often has significant meanings in applications. For example, if an edge represents an information communication channel, the information sharing structure is expressed

[3] A *partial order* is defined as a binary relation that satisfies the reflexive, antisymmetric, and transitive laws. See Remark 1.3.

by the partial order among strongly-connected components. The SCC-decomposition can be found efficiently in $O(|V|+|E|)$ time,[4] which renders it applicable to large-scale systems.

**Remark 1.3.** A mathematical formal definition of the strongly-connected component decomposition (SCC-decomposition) of a graph $G = (V, E)$ is given here.[5]

First, define a binary relation $\geq$ on the vertex set $V$ as

$$u \geq v \iff \text{a directed path exists from } u \text{ to } v.$$

Then we have:

- Reflexive law: $v \geq v$ for all $v \in V$,
- Transitive law: $(\ u \geq v,\ v \geq w \Rightarrow u \geq w\ )$ for all $u, v, w \in V$.

That is, the binary relation $\geq$ is a preorder. (In general, a binary relation that satisfies the reflexive and transitive laws is called a *preorder* or *quasi-order.*)

Next, using the preorder $\geq$, define another binary relation $\sim$ on the vertex set $V$ as

$$u \sim v \iff u \geq v \text{ and } v \geq u.$$

Then we have:

- Reflexive law: $v \sim v$ for all $v \in V$,
- Symmetric law: $(\ u \sim v \Rightarrow v \sim u\ )$ for all $u, v \in V$,
- Transitive law: $(\ u \sim v,\ v \sim w \Rightarrow u \sim w\ )$ for all $u, v, w \in V$.

That is, the binary relation $\sim$ is an equivalence relation. (In general, a binary relation that satisfies the reflexive, symmetric, and transitive laws is called an *equivalence relation.*)

A partition of the set $V$ is induced from this equivalence relation $\sim$. For each $v \in V$ the *equivalence class* containing $v$ is a subset of $V$ defined as

$$[v] = \{u \in V \mid u \sim v\}.$$

These sets satisfy:

- $v \in [v]$ for all $v \in V$.
- $(\ u \in [v] \Rightarrow [u] = [v]\ )$ for all $u, v \in V$.

---

[4]Here $|V|$ and $|E|$ denote the size (number of elements) of sets $V$ and $E$, respectively, and $O(|V|+|E|)$ means being bounded by $c(|V|+|E|)$ with some constant $c$.

[5]The argument here will be useful in learning basic concepts concerning *binary relation.*

- $( [u] \cap [v] \neq \emptyset \Rightarrow [u] = [v] )$ for all $u, v \in V$.

An equivalence class $[v]$ is called a *strongly-connected component* and the partition of $V$ determined by the equivalence classes is called the *decomposition into strongly-connected components* (*SCC-decomposition*). In general, the set of equivalence classes is called a *quotient set*, and is often denoted as $V/\sim$. That is,

$$V/\sim \;=\; \{ [v] \mid v \in V \}.$$

Finally, define a binary relation $\succeq$ on the quotient set $V/\sim$ as

$$[u] \succeq [v] \iff u \geq v. \tag{1.13}$$

This definition is *well-defined* in the sense that the truth value of the right-hand side of (1.13) is determined independently of the choice of the representatives $u$ and $v$ of the respective equivalence classes $[u]$ and $[v]$. To see this, suppose that $[u] = [u']$, which means that we consider two different representatives $u$ and $u'$ for the same equivalence class, and similarly suppose that $[v] = [v']$. Then $u \geq v$ is true if and only if $u' \geq v'$ is true. This binary relation $\succeq$ has the following properties.

- Reflexive law: $[v] \succeq [v]$ for all $[v] \in V/\sim$.
- Antisymmetric law:
  $( [u] \succeq [v],\ [v] \succeq [u] \Rightarrow [u] = [v] )$ for all $[u], [v] \in V/\sim$.
- Transitive law:
  $( [u] \succeq [v],\ [v] \succeq [w] \Rightarrow [u] \succeq [w] )$ for all $[u], [v], [w] \in V/\sim$.

That is, the binary relation $\succeq$ is a partial order. (In general, a binary relation that satisfies the reflexive, antisymmetric, and transitive laws is called a *partial order*, and a set on which a partial order is defined is called a *partially ordered set*.) In this manner, a partial order $\succeq$ is defined on the set of the strongly-connected components. ■

### 1.1.3 *Block-Triangularization*

Given a square matrix, the SCC-decomposition can be utilized effectively to decompose the matrix into a block-triangular form by rearranging its rows and columns. We shall explain this by using the matrix in Example 1.1.

**Example 1.2.** We consider the matrix

$$
A = \begin{array}{c} \\ 1 \\ 2 \\ 3 \\ 4 \\ 5 \\ 6 \end{array}
\begin{array}{c} \begin{array}{cccccc} 1 & 2 & 3 & 4 & 5 & 6 \end{array} \\
\left[\begin{array}{cccccc}
 & \alpha_{12} & \alpha_{13} & & & \\
 & \alpha_{22} & & & \alpha_{25} & \\
 & & \alpha_{33} & & & \\
\alpha_{41} & & & \alpha_{44} & & \alpha_{46} \\
\alpha_{51} & & \alpha_{53} & & & \\
 & & \alpha_{63} & \alpha_{64} & & \alpha_{66}
\end{array}\right] \end{array} \tag{1.14}
$$

in (1.3) again. The row set and the column set of this matrix are $N = \{1, 2, 3, 4, 5, 6\}$. When the rows and columns are arranged naturally in the order of $(1, 2, 3, 4, 5, 6)$, the matrix $A$ takes the form of (1.14). If the row and column indices are rearranged in accordance with the SCC-decomposition (1.6), the matrix is transformed or decomposed into a form of *block-triangular matrix*[6]

$$
B = \begin{array}{c} \\ 4 \\ 6 \\ 1 \\ 2 \\ 5 \\ 3 \end{array}
\begin{array}{c} \begin{array}{cccccc} 4 & 6 & 1 & 2 & 5 & 3 \end{array} \\
\left[\begin{array}{cc|ccc|c}
\alpha_{44} & \alpha_{46} & \alpha_{41} & & & \\
\alpha_{64} & \alpha_{66} & & & & \alpha_{63} \\
\hline
 & & & \alpha_{12} & & \alpha_{13} \\
 & & & \alpha_{22} & \alpha_{25} & \\
 & & \alpha_{51} & & & \alpha_{53} \\
\hline
 & & & & & \alpha_{33}
\end{array}\right] \end{array} \tag{1.15}
$$

with three diagonal blocks

$$
B_1 = \begin{bmatrix} \alpha_{44} & \alpha_{46} \\ \alpha_{64} & \alpha_{66} \end{bmatrix}, \quad
B_2 = \begin{bmatrix} & \alpha_{12} & \\ & \alpha_{22} & \alpha_{25} \\ \alpha_{51} & & \end{bmatrix}, \quad
B_3 = \begin{bmatrix} \alpha_{33} \end{bmatrix}. \tag{1.16}
$$

[6] In this book,a block-triangular matrix means a *block-upper triangular matrix*, unless otherwise stated.

Matrices $A$ and $B$ are mutually related as $B = P^{\top} AP$ through a permutation matrix[7]

$$P = \begin{array}{c} \\ 1 \\ 2 \\ 3 \\ 4 \\ 5 \\ 6 \end{array} \begin{array}{c} \begin{array}{cccccc} 1 & 2 & 3 & 4 & 5 & 6 \end{array} \\ \begin{bmatrix} 0 & 0 & 1 & 0 & 0 & 0 \\ 0 & 0 & 0 & 1 & 0 & 0 \\ 0 & 0 & 0 & 0 & 0 & 1 \\ 1 & 0 & 0 & 0 & 0 & 0 \\ 0 & 0 & 0 & 0 & 1 & 0 \\ 0 & 1 & 0 & 0 & 0 & 0 \end{bmatrix} \end{array}. \tag{1.17}$$

■

The method of *block-triangularization*, as illustrated above, is useful in a variety of situations in linear algebra. For example, the eigenvalue problem for the matrix $A$ above can be "decomposed" into the eigenvalue problems for the three diagonal blocks $B_1$, $B_2$, and $B_3$.

Let us consider the relationship between the block-triangularization and the SCC-decomposition. In the matrix $A$ of (1.14) in Example 1.2, the set of row (and column) indices $N = \{1, 2, 3, 4, 5, 6\}$ can be partitioned into three subsets

$$N_1 = \{4, 6\}, \qquad N_2 = \{1, 2, 5\}, \qquad N_3 = \{3\}, \tag{1.18}$$

for which we have

$$a_{ij} = 0 \qquad (i \in N_k,\ j \in N_l,\ k > l). \tag{1.19}$$

This corresponds to the fact that $P^{\top} AP$ with the permutation matrix $P$ of (1.17) is in a block-triangular form. On the other hand, the condition (1.19) itself is guaranteed by the property (1.5) of the SCC-decomposition of the associated graph.

For a square matrix $A$ in general, finding a permutation matrix $P$ such that the transformed matrix

$$P^{\top} AP \tag{1.20}$$

takes a block-triangular form is equivalent to partitioning the row (column) set $N$ of the matrix $A$ into (two or more) subsets $N_1, N_2, \ldots, N_p$ that satisfy the condition in (1.19). For the graph $G(A)$ associated with matrix $A$, we

---

[7] A *permutation matrix* is a square matrix with entries of 0 or 1, having exactly one 1 in each row and each column. Multiplying a matrix $A$ with a permutation matrix from the left corresponds to rearranging the rows of $A$, and multiplying from the right corresponds to rearranging the columns of $A$.

consider its SCC-decomposition (1.7), in which we index (or number) the components so that the condition

$$V_k \succeq V_l \implies k < l \tag{1.21}$$

is satisfied, and define the corresponding subsets $N_k$ by

$$N_k = \{i \mid v_i \in V_k\}.$$

Then the condition (1.19) is satisfied by the matrix $A$ as a consequence of the property (1.9) of the SCC-decomposition. Accordingly, the matrix $A$ takes the following form:

$$A = \begin{bmatrix} A_{11} & A_{12} & \cdots & A_{1p} \\ O & A_{22} & \cdots & A_{2p} \\ \vdots & \ddots & \ddots & \vdots \\ O & \cdots & O & A_{pp} \end{bmatrix}, \tag{1.22}$$

where the permutation matrix $P$ is suppressed, with the understanding that the rows and columns of matrix $A$ are rearranged appropriately in advance. In this way, a block-triangularization is obtained from the SCC-decomposition. Moreover, it is also evident that an arbitrary block-triangularization can be constructed in this way from the SCC-decomposition by ordering the components as in (1.21).

**Remark 1.4.** In the examples above, the subsets $V_k$ were actually numbered compatibly with the condition in (1.21). For the graph in Fig. 1.2, we had

$$V_1 \succeq V_2 \succeq V_3$$

in (1.10), and for the graph $G$ of Fig. 1.3 we had

$$V_1 \succeq V_3, \qquad V_1 \succeq V_4, \qquad V_2 \succeq V_4$$

in (1.11). In general, the numbering of $V_k$ satisfying the condition (1.21) is not uniquely determined. In the graph $G$ of Fig. 1.3, for example, we may adopt an alternative numbering $(V_1', V_2', V_3', V_4') = (V_2, V_1, V_4, V_3)$, which results in

$$V_2' \succeq V_4', \qquad V_2' \succeq V_3', \qquad V_1' \succeq V_3',$$

where the condition (1.21) is still maintained. For this graph, the following five different numberings are possible:

$$\begin{array}{lll} (V_1, V_2, V_3, V_4), & (V_1, V_2, V_4, V_3), & (V_1, V_3, V_2, V_4), \\ (V_2, V_1, V_3, V_4), & (V_2, V_1, V_4, V_3). & \end{array}$$

■

A square matrix of order $n$ cannot always be decomposed to a non-trivial block-triangular form (possessing two or more diagonal blocks) by rearranging its rows and columns. A matrix that cannot be decomposed is called an *irreducible matrix*, and a matrix that can be decomposed is called a *reducible matrix*. For instance, the matrix $A$ in (1.14) is reducible with the expression in (1.15).

Reducibility of a square matrix $A = (a_{ij})$ can also be defined without referring to the rearrangement of rows and columns. That is, we can define $A$ to be *reducible* if[8]

$$\text{there exists } N' \ (\emptyset \subsetneqq N' \subsetneqq N) \text{ such that } a_{ij} = 0 \quad (i \in N',\ j \in N \setminus N'), \tag{1.23}$$

where $N$ denotes the set of row (and column) indices of $A$. A square matrix is defined to be *irreducible*, if it is not reducible.

A square matrix $A$ is irreducible if and only if the associated graph $G(A)$ is strongly connected. When a matrix $A$ is reducible, it can be brought into the form of (1.22) with $p \geqq 2$ through the SCC-decomposition of $G(A)$. Then each diagonal block in (1.22) is an irreducible matrix, called an *irreducible component* of the matrix $A$.

**Remark 1.5.** We mention already at this point that, in discussing nonnegative matrices in Chap. 2, the definition of irreducibility will be modified in the case of $n = 1$. See Sec. 2.1.2 for details. ■

### 1.1.4 *Period*

Irreducible matrices can be decomposed further on the basis of period.[9]

Let us begin with the definition of period. In this section, $A$ is assumed to be an irreducible square matrix of order $n$ with $n \geqq 2$. The graph associated with $A$ is denoted by $G(A) = (V, E)$, where

$$V = \{v_1, \ldots, v_n\}, \qquad E = \{(v_i, v_j) \mid a_{ij} \neq 0\}$$

as in (1.1) and (1.2). By the assumed irreducibility of matrix $A$, the graph $G(A)$ is strongly connected and, for every pair of vertices $v_i$ and $v_j$, there exists a directed path from $v_i$ to $v_j$. Hence, for each $v_i$, there exists a directed cycle that passes through the vertex $v_i$. The *period* of an irreducible matrix $A$, denoted by $\sigma$, is defined as the greatest common divisor of the

[8]The notation $N \setminus N'$ means the set of elements of $N$ that do not belong to $N'$. That is, $N \setminus N' = \{j \mid j \in N, j \notin N'\}$.

[9]The concept of period plays an important role in Secs. 2.1 to 2.3.

lengths of all the directed cycles[10] in $G(A)$. An irreducible square matrix with period one is said to be *primitive*. The period is also referred to as the *index of primitivity*.

**Example 1.3.** Consider a matrix

$$A = \begin{array}{c} \\ 1 \\ 2 \\ 3 \\ 4 \\ 5 \\ 6 \\ 7 \\ 8 \end{array} \begin{array}{c} \begin{array}{cccccccc} 1 & 2 & 3 & 4 & 5 & 6 & 7 & 8 \end{array} \\ \begin{bmatrix} & \alpha_{12} & & \alpha_{14} & \alpha_{15} & & & \\ & & \alpha_{23} & & & & & \\ & & & & & \alpha_{36} & & \\ & & \alpha_{43} & & & & \alpha_{47} & \\ & & \alpha_{53} & & & & \alpha_{57} & \\ & & & & \alpha_{65} & & & \alpha_{68} \\ \alpha_{71} & & & & & \alpha_{76} & & \\ & & & & & & \alpha_{87} & \end{bmatrix} \end{array}. \tag{1.24}$$

Here, $\alpha_{12}, \alpha_{14}, \ldots, \alpha_{87}$ represent nonzero entries and the blank entries are zeros. The associated graph $G(A)$, shown in Fig. 1.4, is strongly connected. The lengths of the directed cycles in $G(A)$ are, for example:

| | | |
|---|---|---|
| Directed cycle | $v_1 \to v_4 \to v_7 \to v_1$ | has length 3, |
| Directed cycle | $v_5 \to v_7 \to v_6 \to v_5$ | has length 3, |
| Directed cycle | $v_1 \to v_5 \to v_3 \to v_6 \to v_8 \to v_7 \to v_1$ | has length 6. |

It can be verified that the lengths of directed cycles are all multiples of 3. Therefore, the period $\sigma$ is given by $\sigma = 3$. ■

For each pair of vertices $u, v \in V$, the length of any directed path from $u$ to $v$ is a constant modulo $\sigma$. For example, for $u = v_1$ and $v = v_5$ in Fig. 1.4 the path lengths are as follows:

| | | |
|---|---|---|
| Directed path | $v_1 \to v_5$ | has length 1, |
| Directed path | $v_1 \to v_4 \to v_7 \to v_6 \to v_5$ | has length 4, |
| Directed path | $v_1 \to v_4 \to v_3 \to v_6 \to v_5$ | has length 4. |

It can be verified that every directed path from $v_1$ to $v_5$ has a path length that is a multiple of 3 plus 1.

The observation made above for a particular example is generally true for an irreducible matrix $A$, and the length of any directed path from a

[10]Here we may restrict ourselves to simple directed cycles (cycles that pass through the same vertex only once), although we have to include self-loops (if any). A directed cycle that passes through the same vertex more than once decomposes into a union of simple directed cycles, and consequently, the greatest common divisor of the cycle lengths remains the same even if non-simple directed cycles are included in consideration.

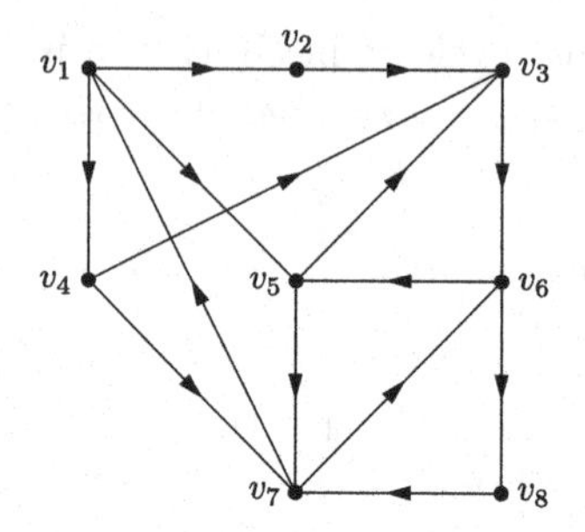


Fig. 1.4 A strongly-connected graph of period 3.

vertex to another vertex is a constant modulo $\sigma$, independently of the choice of a directed path. This can be proved as follows.

Let $P_1$ and $P_2$ be directed paths from vertex $u$ to vertex $v$ with the lengths denoted by $\ell(P_1)$ and $\ell(P_2)$, respectively. Since the graph $G(A)$ is strongly connected, a directed path exists from $v$ to $u$. Let $Q$ be such a directed path, and denote the length of $Q$ by $\ell(Q)$. By connecting $P_i$ and $Q$ for $i = 1, 2$, we can obtain two directed cycles (not necessarily simple), the lengths of which are both multiples of $\sigma$, *i.e.*,

$$\ell(P_1) + \ell(Q) = 0 \qquad \bmod \sigma, \qquad \ell(P_2) + \ell(Q) = 0 \qquad \bmod \sigma,$$

from which follows

$$\ell(P_1) = \ell(P_2) \qquad \bmod \sigma.$$

This shows that the lengths of directed paths from $u$ to $v$ are a constant modulo $\sigma$.

Now fix a vertex $u_0 \in V$ and, for each vertex $v \in V$, denote by $\lambda(v)$ the length modulo $\sigma$ of a directed path from $u_0$ to $v$. We may arbitrarily take a directed path $P$ from $u_0$ to $v$ and set

$$\lambda(v) = \ell(P) \qquad \bmod \sigma,$$

where $0 \leqq \lambda(v) \leqq \sigma - 1$. Note here that, by the argument above, the value of $\ell(P) \mod \sigma$ on the right-hand side is determined independently of the chosen $P$. The vertices are classified into $\sigma$ subsets according to the value of $\lambda(v)$:

$$V_k = \{v \in V \mid \lambda(v) = k\} \qquad (0 \leqq k \leqq \sigma - 1),$$

which determine a partition $\{V_0, V_1, \ldots, V_{\sigma-1}\}$ of the vertex set $V$. Each edge goes from a certain block $V_k$ to the next numbered block $V_{k+1}$. That is, if $u \in V_k$ for an edge $(u, v)$, then we have $v \in V_{k+1}$, where $V_\sigma = V_0$ by convention.

The decomposition of $G(A)$ above can be recast for the matrix $A$ itself as follows. Define $N_k$ by

$$N_k = \{i \mid \lambda(v_i) = k\} \qquad (0 \leqq k \leqq \sigma - 1)$$

and put $N_\sigma = N_0$, where $N_k$ corresponds to $V_k$. For any nonzero entry $a_{ij} \neq 0$, if $i \in N_k$, then $j \in N_{k+1}$. Therefore, by rearranging the row (and column) indices of the matrix $A$ in accordance with these $N_0, N_1, \ldots, N_{\sigma-1}$, we obtain a matrix that explicitly demonstrates its structure. For example,

$$A = \begin{bmatrix} O & A_{12} & O \\ O & O & A_{23} \\ A_{31} & O & O \end{bmatrix} \tag{1.25}$$

when $\sigma = 3$, and

$$A = \begin{bmatrix} O & A_{12} & O & O \\ O & O & A_{23} & O \\ O & O & O & A_{34} \\ A_{41} & O & O & O \end{bmatrix} \tag{1.26}$$

when $\sigma = 4$.

**Example 1.4.** In Fig. 1.4 (Example 1.3), let us choose vertex $v_1$ as the starting vertex $u_0$. Then

$$V_0 = \{v_1, v_6\}, \qquad V_1 = \{v_2, v_4, v_5, v_8\}, \qquad V_2 = \{v_3, v_7\},$$

which shows that the period $\sigma$ is 3. Accordingly, we have

$$N_0 = \{1, 6\}, \qquad N_1 = \{2, 4, 5, 8\}, \qquad N_2 = \{3, 7\}.$$

By rearranging the rows and columns, the matrix $A$ in (1.24) can be transformed to

$$\begin{array}{c} \\ 1 \\ 6 \\ 2 \\ 4 \\ 5 \\ 8 \\ 3 \\ 7 \end{array} \begin{array}{c} \begin{array}{cc|cccc|cc} 1 & 6 & 2 & 4 & 5 & 8 & 3 & 7 \end{array} \\ \left[\begin{array}{cc|cccc|cc} & & \alpha_{12} & \alpha_{14} & \alpha_{15} & & & \\ & & & & \alpha_{65} & \alpha_{68} & & \\ \hline & & & & & & \alpha_{23} & \\ & & & & & & \alpha_{43} & \alpha_{47} \\ & & & & & & \alpha_{53} & \alpha_{57} \\ & & & & & & & \alpha_{87} \\ \hline & \alpha_{36} & & & & & & \\ \alpha_{71} & \alpha_{76} & & & & & & \end{array}\right] \end{array}, \tag{1.27}$$

which is of the form of (1.25). ■

## 1.2 Matrices and Bipartite Graphs

In Sec. 1.1 we have dealt with graph representations of square matrices in which a one-to-one correspondence exists between the row and column indices. For rectangular matrices and also for square matrices in which no natural one-to-one correspondence exists between the row and column indices, bipartite graphs offer a more suitable graph representation.

### 1.2.1 *Graph Representation*

The structure of an $m \times n$ matrix $A = (a_{ij})$ can be represented by an undirected graph[11] $G$ having the union $U \cup V$ of

$$U = \{u_1, \ldots, u_m\}, \qquad V = \{v_1, \ldots, v_n\}$$

as its vertex set, and

$$E = \{(u_i, v_j) \mid a_{ij} \neq 0\} \tag{1.28}$$

as its edge set. The vertex $u_i$ corresponds to row index $i$ and the vertex $v_j$ corresponds to column index $j$. For a nonzero $(i, j)$ entry of matrix $A$, an edge $(u_i, v_j)$ connects vertex $u_i$ to vertex $v_j$. The vertex set of graph $G$ is divided into two parts, $U$ and $V$, and each edge connects a vertex in $U$ to a vertex in $V$ (and no edge connects two vertices in the same part $U$ or $V$). Such a graph is called a *bipartite graph* and is denoted as $G = (U, V; E)$.

**Example 1.5.** Recall the matrix $A$ in (1.3) of Example 1.1, but suppose here that there is no natural one-to-one correspondence between the row and column indices of $A$. Accordingly, we denote the row set as $\{1', 2', 3', 4', 5', 6'\}$ and the column set as $\{1, 2, 3, 4, 5, 6\}$ to obtain

$$A = \begin{array}{c} \\ 1' \\ 2' \\ 3' \\ 4' \\ 5' \\ 6' \end{array} \begin{array}{c} \begin{array}{cccccc} 1 & 2 & 3 & 4 & 5 & 6 \end{array} \\ \begin{bmatrix} & \alpha_{12} & \alpha_{13} & & & \\ & \alpha_{22} & & & \alpha_{25} & \\ & & \alpha_{33} & & & \\ \alpha_{41} & & & \alpha_{44} & & \alpha_{46} \\ \alpha_{51} & & \alpha_{53} & & & \\ & & \alpha_{63} & \alpha_{64} & & \alpha_{66} \end{bmatrix} \end{array}. \tag{1.29}$$

[11]An *undirected graph* is a graph wherein no direction is attached to the edges. Since no direction is considered for an edge, it is completely legitimate to write $(v_j, u_i)$ for $(u_i, v_j)$ in (1.28). Mathematically, an undirected edge between $u_i$ and $v_j$ is an unordered pair of two vertices $u_i$ and $v_j$, and as such it is more precise to write $\{u_i, v_j\}$ instead of $(u_i, v_j)$. In this book, however, we write $(u_i, v_j)$ for consistency with our notation in (1.2).

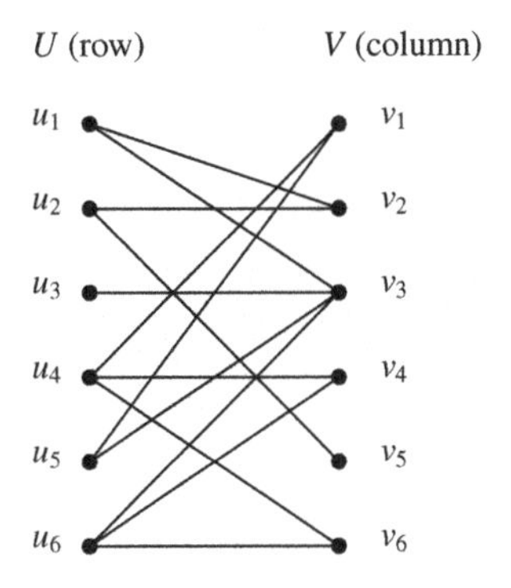


Fig. 1.5 Bipartite graph representation of a matrix (Example 1.5).

The structure of this matrix is represented by the bipartite graph of Fig. 1.5. This graph has the vertex set composed of two parts,

$$U = \{u_1, u_2, u_3, u_4, u_5, u_6\}, \qquad V = \{v_1, v_2, v_3, v_4, v_5, v_6\},$$

and the edge set

$$E = \{(u_1, v_2), (u_1, v_3), (u_2, v_2), (u_2, v_5), (u_3, v_3), (u_4, v_1), (u_4, v_4), (u_4, v_6), (u_5, v_1), (u_5, v_3), (u_6, v_3), (u_6, v_4), (u_6, v_6)\}.$$

■

### 1.2.2 *Block-Triangularization*

In Sec. 1.1.3, we have considered block-triangularization for square matrices $A$ through a transformation of the form of $P^{\top}AP$, where the row and column indices of $A$ are assumed to have a one-to-one correspondence. Here we consider another kind of block-triangularization, which is applicable to matrices for which the row and column indices are not assumed to have a one-to-one correspondence. For such matrices it is natural to rearrange the row and column indices independently, and hence to consider *block-triangularization* through a transformation of the form

$$P^{\top}AQ$$

using two permutation matrices $P$ and $Q$.

For example, the block-triangularization of the matrix $A$ of (1.29) is given by

$$
P^{\top} A Q = \begin{array}{c} \\ 4' \\ 6' \\ 2' \\ 1' \\ 5' \\ 3' \end{array}
\begin{array}{c} \begin{array}{cccccc} 4 & 6 & 5 & 2 & 1 & 3 \end{array} \\
\left[ \begin{array}{cc|c|c|c|c}
\alpha_{44} & \alpha_{46} & & & \alpha_{41} & \\
\alpha_{64} & \alpha_{66} & & & & \alpha_{63} \\ \hline
 & & \alpha_{25} & \alpha_{22} & & \\ \hline
 & & & \alpha_{12} & & \alpha_{13} \\ \hline
 & & & & \alpha_{51} & \alpha_{53} \\ \hline
 & & & & & \alpha_{33}
\end{array} \right] \end{array}, \tag{1.30}
$$

which should be compared with (1.15) in Example 1.2. The diagonal blocks of $P^{\top} AQ$ above are

$$
B_1 = \begin{bmatrix} \alpha_{44} & \alpha_{46} \\ \alpha_{64} & \alpha_{66} \end{bmatrix}, \quad B_2 = [\alpha_{25}], \quad B_3 = [\alpha_{12}], \quad B_4 = [\alpha_{51}], \quad B_5 = [\alpha_{33}]
$$

with their row and column sets given by

$$
\{4', 6'\} \times \{4, 6\}, \quad \{2'\} \times \{5\}, \quad \{1'\} \times \{2\}, \quad \{5'\} \times \{1\}, \quad \{3'\} \times \{3\},
$$

respectively. These row and column sets will be denoted as $R_k \times C_k$ for $k = 1, \ldots, 5$.

Furthermore, a *partial order* $\succeq$ can be defined on the set of diagonal blocks $B_1, \ldots, B_5$ on the basis of the existence or nonexistence of nonzero entries in the upper triangular part. For example, we have the following:

- $B_1 \succeq B_4$ because $\alpha_{41}$ exists in the block of $R_1 \times C_4$.
- $B_2 \succeq B_3$ because $\alpha_{22}$ exists in the block of $R_2 \times C_3$.
- $B_3 \succeq B_5$ because $\alpha_{13}$ exists in the block of $R_3 \times C_5$.
- No direct order relation exists between $B_1$ and $B_2$ because nonzero entries do not exist in the block of $R_1 \times C_2$.
- $B_2 \succeq B_5$ follows from $B_2 \succeq B_3$ and $B_3 \succeq B_5$ by the transitive law of partial order $\succeq$, although nonzero entries do not exist in the block of $R_2 \times C_5$.

In this way, the partial order is given as follows:

$$
B_1 \succeq B_4 \succeq B_5, \qquad B_2 \succeq B_3 \succeq B_5.
$$

In accordance with the block-triangular form in (1.30), the bipartite graph in Fig. 1.5 is decomposed as in Fig. 1.6.

In the above, we have illustrated the block-triangularization for a particular instance of a (structurally) nonsingular square matrix. Such a block-triangularization can also be defined for (structurally) singular square

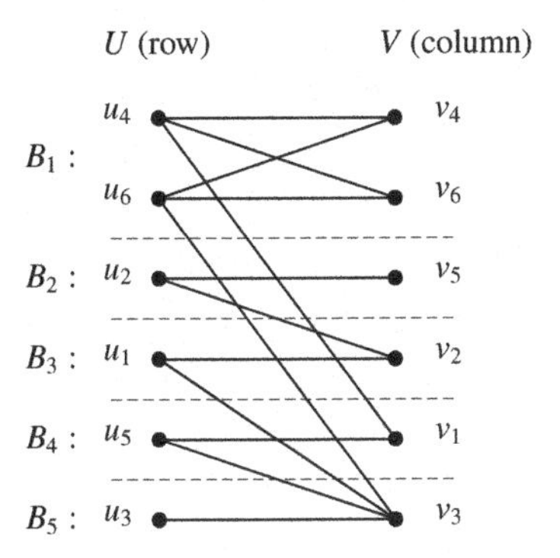


Fig. 1.6 DM-decomposition of the bipartite graph in Example 1.5.

matrices and rectangular matrices in general, and is called the *Dulmage–Mendelsohn decomposition* (or *DM-decomposition*). It is known that the DM-decomposition is uniquely determined and it can be constructed by an efficient graph-theoretic algorithm. For more details, see [16, Chap. 6], [29, Sec. 2.2], and [30, Sec. 2.4].

**Remark 1.6.** In a bipartite graph $G = (U, V; E)$, a subset $M$ of the edge set $E$ is called a *matching* if all the *end-vertices* of the edges belonging to $M$ are different. A matching $M$ with the maximum number of edges is called a *maximum matching.* For example, in the bipartite graph of Fig. 1.5,

$$M = \{(u_4, v_4), (u_6, v_6), (u_2, v_5), (u_1, v_2), (u_5, v_1), (u_3, v_3)\}$$

(consisting of the horizontal edges in Fig. 1.6) is a maximum matching with $|M| = 6$, where $|M|$ denotes the number of edges contained in $M$. For a subset $X$ of the vertex set $U$, the set of the vertices in $V$ connected with some vertex in $X$ is denoted as $\Gamma(X)$. For example, in the bipartite graph in Fig. 1.5, we have $\Gamma(X) = \{v_2, v_3, v_5\}$ for $X = \{u_1, u_2\}$. If the bipartite graph represents a matrix $A$, the subset $\Gamma(X)$ is expressed as

$$\Gamma(X) = \{v_j \mid \exists i : u_i \in X, a_{ij} \neq 0\}.$$

The size of the maximum matching is known to be equal to the minimum value of $|\Gamma(X)| - |X| + |U|$ over all subsets $X \subseteqq U$. That is

$$\max_M\{|M| \mid M \text{ is a matching}\} = \min_X\{|\Gamma(X)| - |X| + |U|\}. \tag{1.31}$$

It is worth mentioning that the DM-decomposition is constructed based on the relation (1.31).

The relation (1.31) above also implies the following well-known fact in combinatorics (to be used in Sec. 2.5.2).

**Hall's Theorem**: *For a square matrix $A$ of order $n$, there exists a permutation $\sigma$ of $\{1, \ldots, n\}$ such that $a_{i\sigma(i)} \neq 0$ for all $i = 1, \ldots, n$ if and only if $|\Gamma(X)| \geqq |X|$ for all subsets $X$ of $\{u_1, \ldots, u_n\}$.* ■

# Chapter 2

# Nonnegative Matrices

This chapter presents the Perron–Frobenius theorem, which is a fundamental theorem for eigenvalues and eigenvectors of nonnegative matrices. As a primary application of this theorem, the theory of stochastic matrices (transition probability matrices of Markov chains, in particular) is developed, where effective use is made of the graph representation of matrix structures introduced in Chap. 1. We also explain the basic properties of M-matrices and Birkhoff's theorem for doubly stochastic matrices.

## 2.1 Nonnegative Matrices

### 2.1.1 *Definition*

We say that a matrix is a *nonnegative matrix* if all its entries are nonnegative real numbers. That is, an $m \times n$ matrix $A = (a_{ij})$ is called a nonnegative matrix if

$$a_{ij} \geqq 0 \qquad (i = 1, \ldots, m;\ j = 1, \ldots, n). \tag{2.1}$$

Similarly, a matrix $A$ is called a *positive matrix* if

$$a_{ij} > 0 \qquad (i = 1, \ldots, m;\ j = 1, \ldots, n). \tag{2.2}$$

We write

$$A \geqq O, \qquad A > O$$

to mean that $A$ is a nonnegative matrix and a positive matrix, respectively. For two matrices $A$ and $B$, we write

$$A \geqq B, \qquad A > B$$

to mean that $A - B$ is a nonnegative matrix and a positive matrix, respectively. We also write

$$B \leqq A, \qquad B < A$$

for the same meanings.

A vector whose components are all nonnegative real numbers is called a *nonnegative vector.* A vector whose components are all positive real numbers is called a *positive vector.* The notations $\geqq$, $\leqq$, $>$, and $<$ defined above are also used for vectors.

A typical example of a nonnegative vector is a vector of probabilities, and a typical example of a nonnegative matrix is a transition probability matrix of a Markov chain (see Sec. 2.3).

### 2.1.2 *Irreducibility*

The concept of irreducibility of matrices has been introduced in Sec. 1.1.3. A square matrix $A$ of order $n$ is irreducible if the associated directed graph $G(A)$ is strongly connected. A square matrix is reducible if it is not irreducible. This definition implies, in particular, that every matrix of order 1 is irreducible. But in dealing with nonnegative matrices, it is convenient to treat the case of $n = 1$ as an exceptional case by redefining:

When $n = 1$, $A$ is *irreducible* if $A \neq O$, and *reducible* if $A = O$. (2.3)

This modified definition is adopted in many books. We also adopt this convention in discussing nonnegative matrices in this chapter, while retaining the definition for $n \geqq 2$ introduced in Sec. 1.1.3.

### 2.1.3 *Powers and Graphs*

Powers of a nonnegative matrix can be studied successfully with the aid of the graph representation explained in Sec. 1.1. Let $A = (a_{ij})$ be a nonnegative matrix of order $n$, and denote the associated directed graph by $G(A) = (V, E)$. Recall from (1.1) and (1.2) that

$$V = \{v_1, \ldots, v_n\}, \qquad E = \{(v_i, v_j) \mid a_{ij} \neq 0\}.$$

For a nonnegative integer $m$, we denote the $(i, j)$ entry of $A^m$ by $a_{ij}^{(m)}$, *i.e.*,

$$A^m = (a_{ij}^{(m)}),$$

where $A^0 = I$ (unit matrix). Obviously, the matrix $A^m$ is nonnegative, and we are interested in which entries are strictly positive.

For example, the $(1, 2)$ entry of $A^3$ can be expressed as

$$a_{12}^{(3)} = \sum_{k=1}^{n} \sum_{l=1}^{n} a_{1k} a_{kl} a_{l2}, \tag{2.4}$$

and each term on the right-hand side is nonnegative. Therefore, $a_{12}^{(3)} > 0$ if and only if there exist $k$ and $l$ such that

$$a_{1k} > 0, \quad a_{kl} > 0, \quad a_{l2} > 0. \tag{2.5}$$

The condition in (2.5) is equivalent to the existence of a directed path $v_1 \to v_k \to v_l \to v_2$ of length 3 from vertex $v_1$ to vertex $v_2$ in the directed graph $G(A)$. A key fact to be noted here is that no terms cancel each other in the summation on the right-hand side of (2.4) by virtue of the nonnegativity of the terms.

For a nonnegative integer $m$, in general, the $(i, j)$ entry of $A^m$ can be written as

$$a_{ij}^{(m)} = \sum_{k_1=1}^{n} \sum_{k_2=1}^{n} \cdots \sum_{k_{m-1}=1}^{n} a_{ik_1} a_{k_1k_2} \cdots a_{k_{m-1}j}. \tag{2.6}$$

Thus, by the same argument as above, we can show the following proposition.

**Proposition 2.1.** *Let $A$ be a nonnegative matrix of order $n$, and $m$ be a nonnegative integer. Then $a_{ij}^{(m)} > 0$ if and only if there exists a directed path of length $m$ from the vertex $v_i$ corresponding to $i$ to the vertex $v_j$ corresponding to $j$ in the associated graph $G(A)$.*

For an irreducible nonnegative matrix $A$, the graph $G(A)$ is strongly connected, and therefore, for any pair $(i, j)$, there exists a directed path from vertex $v_i$ to vertex $v_j$. Denoting the length of such a directed path by $m$, we have $a_{ij}^{(m)} > 0$.

**Proposition 2.2.** *Let $A$ be an irreducible nonnegative matrix of order $n$. For any $(i, j)$, there exists an $m$ (where $0 \leqq m \leqq n-1$) such that $a_{ij}^{(m)} > 0$.*

**Remark 2.1.** It should be noted in Proposition 2.2 that $m$ is dependent on $(i, j)$. For any $(i, j)$, a nonnegative integer $m = m_{ij}$ can be selected such that $a_{ij}^{(m)} > 0$, but it is not necessarily true that there exists a single $m$ for which $a_{ij}^{(m)} > 0$ holds for all $(i, j)$. In other words, there does not always exist an $m$ such that $A^m > O$. For example, the matrix

$$A = \begin{bmatrix} 0 & 1 \\ 1 & 0 \end{bmatrix}$$

is an irreducible nonnegative matrix, but we have

$$A^{2k+1} = \begin{bmatrix} 0 & 1 \\ 1 & 0 \end{bmatrix}, \qquad A^{2k} = \begin{bmatrix} 1 & 0 \\ 0 & 1 \end{bmatrix}.$$

It is known that, for a general irreducible $A$, there exists an $m$ such that $A^m > O$ if and only if $A$ is primitive (*i.e.*, of period 1) (see pp. 239–240 of [15]). ∎

The next proposition states that if $A$ is an irreducible nonnegative matrix, all entries of $(I + A)^m$ are positive for all sufficiently large $m$.

**Proposition 2.3.** *Let $A$ be an irreducible nonnegative matrix of order $n$. If $m \geqq n - 1$, then*

$$(I + A)^m > O.$$

**Proof.** With *binomial coefficients* $\binom{m}{k} = \dfrac{m!}{k!(m-k)!}$ we have an expansion

$$(I + A)^m = \sum_{k=0}^{m} \binom{m}{k} A^k.$$

By nonnegativity of the matrix entries it follows that $(I + A)^m > O$ if and only if, for each $(i, j)$, there exists $k$ such that $0 \leqq k \leqq m$ and $a_{ij}^{(k)} > 0$. By Proposition 2.2, this holds if $m \geqq n - 1$. □

## 2.2 Perron–Frobenius Theorem

### 2.2.1 *Theorems*

This section presents the *Perron–Frobenius theorem*, which is a fundamental theorem for eigenvalues of nonnegative matrices.[1]

**Theorem 2.1 (Perron–Frobenius Theorem).** *An irreducible nonnegative matrix $A$ has a positive eigenvalue $\alpha$ with the following properties.*

(1) *$\alpha$ admits a positive eigenvector (with all components being positive).*
(2) *The absolute values of the other eigenvalues of $A$ do not exceed $\alpha$.*
(3) *$\alpha$ is a simple eigenvalue of $A$, that is, a simple root of the characteristic equation of $A$.*

**Proof.** The proof is provided in Sec. 2.2.2. □

From (2) of Theorem 2.1, $\alpha$ is equal to the spectral radius $\rho(A)$ of $A$, where the *spectral radius* is the largest absolute value of an eigenvalue.

[1] O. Perron proved the theorem for positive matrices, and G. Frobenius, independently of Perron, proved the theorem in the general case.

Thus, the positive real number $\alpha$ in Theorem 2.1 above is determined uniquely as $\alpha = \rho(A)$. The associated eigenvector is also uniquely determined, up to a scalar factor, since $\alpha$ is a simple eigenvalue by (3). As $\alpha > 0$ and $\alpha = \rho(A)$, we have $\rho(A) > 0$.

For nonnegative matrices that are not necessarily irreducible, the Perron–Frobenius theorem (Theorem 2.1) is extended as follows.

**Theorem 2.2.** *A square nonnegative matrix $A$ has a nonnegative eigenvalue $\alpha$ with the following properties.*

(1) *$\alpha$ admits an eigenvector that is a nonnegative vector.*
(2) *The absolute values of the other eigenvalues of $A$ do not exceed $\alpha$.*

**Proof.** The proof is provided in Sec. 2.2.3. □

By (2) of Theorem 2.2, $\alpha$ is equal to the spectral radius $\rho(A)$ of $A$. Whereas $\alpha = \rho(A)$ is a simple eigenvalue when $A$ is irreducible, this is not the case in general. An example of this is the unit matrix of order 2 or higher.

The eigenvalue described in Theorem 2.2 is called the *Perron–Frobenius root* of a nonnegative matrix $A$. Denoting this by $\lambda_{\mathrm{PF}}(A)$ we have

$$\lambda_{\mathrm{PF}}(A) = \rho(A). \tag{2.7}$$

The following is an important theorem that shows the monotonicity of the Perron–Frobenius root $\lambda_{\mathrm{PF}}(A)$.

**Theorem 2.3.** *Let $A$ be an irreducible nonnegative matrix and $B$ be a nonnegative matrix. If $A \geqq B$, then $\lambda_{\mathrm{PF}}(A) \geqq \lambda_{\mathrm{PF}}(B)$. The equality $\lambda_{\mathrm{PF}}(A) = \lambda_{\mathrm{PF}}(B)$ holds only if $A = B$.*

**Proof.** This follows from Proposition 2.7 in Sec. 2.2.2, together with (2.7) above. □

**Remark 2.2.** As stated in Sec. 2.1.2, when $n \geqq 2$, a matrix $A$ is irreducible if and only if the associated graph $G(A)$ is strongly connected. Recall, however, that a matrix $A$ of order 1 ($n = 1$) is treated as an exceptional case where $A$ is irreducible if $A \neq O$, and reducible if $A = O$. ■

**Remark 2.3.** If a real number $\alpha$ has the property (1) in Theorem 2.1, it necessarily has the property (2). That is, if $A\boldsymbol{x} = \alpha\boldsymbol{x}$ for a positive vector $\boldsymbol{x}$, then it follows that $\alpha = \rho(A)$ (Proposition 2.5 in Sec. 2.2.2). ■

**Remark 2.4.** An irreducible nonnegative matrix may possibly have several positive eigenvalues. For example, the matrix

$$A = \begin{bmatrix} 2 & 1 \\ 1 & 2 \end{bmatrix}$$

has eigenvalues $\alpha_1 = 3$ and $\alpha_2 = 1$, and the eigenvectors are given as

$$\begin{bmatrix} 2 & 1 \\ 1 & 2 \end{bmatrix}\begin{bmatrix} 1 \\ 1 \end{bmatrix} = 3 \cdot \begin{bmatrix} 1 \\ 1 \end{bmatrix}, \qquad \begin{bmatrix} 2 & 1 \\ 1 & 2 \end{bmatrix}\begin{bmatrix} 1 \\ -1 \end{bmatrix} = 1 \cdot \begin{bmatrix} 1 \\ -1 \end{bmatrix}.$$

The eigenvector $\begin{bmatrix} 1 & 1 \end{bmatrix}^\top$ associated with the eigenvalue $\alpha_1 = 3$ is a positive vector, but the eigenvector $\begin{bmatrix} 1 & -1 \end{bmatrix}^\top$ associated with $\alpha_2 = 1$ is not positive. In this case, the eigenvalue $\alpha_1$ serves as the $\alpha$ in Theorem 2.1, but $\alpha_2$ does not. ■

**Remark 2.5.** Let us examine the eigenvalues whose absolute values are equal to $\rho(A)$, *i.e.*, those eigenvalues $\lambda$ with

$$|\lambda| = \rho(A). \tag{2.8}$$

The Perron–Frobenius root $\lambda_{\mathrm{PF}}(A)$ is certainly one of such $\lambda$, but there can be other eigenvalues that have this property. For example, the matrix

$$A = \begin{bmatrix} 0 & 1 & 0 \\ 0 & 0 & 1 \\ 1 & 0 & 0 \end{bmatrix}$$

has three eigenvalues 1, $\exp(2\pi\mathrm{i}/3)$, and $\exp(4\pi\mathrm{i}/3)$, and all of these three satisfy the condition (2.8). Generally, it is known that if an irreducible nonnegative matrix $A$ has a period $\sigma$, there exist exactly $\sigma$ different eigenvalues satisfying the condition (2.8), and these are given by

$$\lambda^{(k)} = \rho(A) \cdot \zeta^k \qquad (k = 0, 1, \ldots, \sigma - 1) \tag{2.9}$$

with $\zeta = \exp(2\pi\mathrm{i}/\sigma)$.

The associated eigenvectors can also be constructed easily. For example, when the period is $\sigma = 4$, the matrix $A$ takes the following form (see (1.26)):

$$A = \begin{bmatrix} O & A_{12} & O & O \\ O & O & A_{23} & O \\ O & O & O & A_{34} \\ A_{41} & O & O & O \end{bmatrix}$$

and similarly for general values of $\sigma$. If $\boldsymbol{x} = (\boldsymbol{x}_1, \boldsymbol{x}_2, \ldots, \boldsymbol{x}_\sigma)$ denotes[2] the eigenvector $\boldsymbol{x}$ associated with the Perron–Frobenius root $\lambda_{\mathrm{PF}}(A)$ that is

[2] To be strict, this should be written as $\boldsymbol{x} = (\boldsymbol{x}_1^\top, \boldsymbol{x}_2^\top, \ldots, \boldsymbol{x}_\sigma^\top)^\top$, but the transpositions are omitted for simplicity.

partitioned compatibly with the matrix $A$, then the eigenvector associated with $\lambda^{(k)}$ is given by

$$\boldsymbol{x}^{(k)} = (\boldsymbol{x}_1,\ \zeta^k \boldsymbol{x}_2,\ \zeta^{2k}\boldsymbol{x}_3, \ldots,\ \zeta^{(\sigma-1)k}\boldsymbol{x}_\sigma). \qquad \blacksquare$$

### 2.2.2 *Proof for Irreducible Matrices*

This section provides a proof of Theorem 2.1 (the Perron–Frobenius theorem). This proof relies crucially on the proposition below.

**Proposition 2.4.** *An irreducible nonnegative matrix $A$ has a positive eigenvector associated with a positive eigenvalue, that is, there exist $\alpha > 0$ and $\boldsymbol{x} > \boldsymbol{0}$ such that $A\boldsymbol{x} = \alpha \boldsymbol{x}$.*

**Proof.** We consider relaxing the equality condition $A\boldsymbol{z} - \lambda \boldsymbol{z} = \boldsymbol{0}$ for an eigenvector to a weaker inequality condition $A\boldsymbol{z} - \lambda\boldsymbol{z} \geqq \boldsymbol{0}$, and define

$$\mu(\boldsymbol{z}) = \max\{\lambda \mid A\boldsymbol{z} - \lambda\boldsymbol{z} \geqq \boldsymbol{0}\}$$

for nonnegative vectors $\boldsymbol{z}$ ($\neq \boldsymbol{0}$). We have $\mu(\boldsymbol{z}) \geqq 0$, since the inequality in the maximization is satisfied by $\lambda = 0$. An alternative expression

$$\mu(\boldsymbol{z}) = \min_{i: z_i > 0} \frac{(A\boldsymbol{z})_i}{z_i}$$

shows that $\mu(\boldsymbol{z})$ is finite-valued.

Next, we consider varying $\boldsymbol{z}$ to maximize $\mu(\boldsymbol{z})$. With this in mind we define

$$\alpha = \sup_{\boldsymbol{z} \geqq \boldsymbol{0},\ \boldsymbol{z} \neq \boldsymbol{0}} \mu(\boldsymbol{z}).$$

We note the following facts:

(1) If we denote the maximum component of $\boldsymbol{z} \geqq \boldsymbol{0}$ ($\boldsymbol{z} \neq \boldsymbol{0}$) by $z_{i^*}$, we have

$$\mu(\boldsymbol{z}) = \min_{i: z_i > 0} \frac{(A\boldsymbol{z})_i}{z_i} \leq \frac{\sum_{j=1}^n a_{i^*j} z_j}{z_{i^*}} \leq \max_{1 \leq i \leq n} \sum_{j=1}^n a_{ij}.$$

This shows that $\alpha$ is finite-valued.

(2) Since $\mu(c\boldsymbol{z}) = \mu(\boldsymbol{z})$ for any positive real number $c$, we have an alternative expression

$$\alpha = \sup_{\boldsymbol{z} \geqq \boldsymbol{0},\ \|\boldsymbol{z}\| = 1} \mu(\boldsymbol{z}).$$

By the definition of supremum, there exists a sequence $\boldsymbol{x}_1, \boldsymbol{x}_2, \ldots, \boldsymbol{x}_k, \ldots$ in $S = \{\boldsymbol{z} \mid \boldsymbol{z} \geqq \boldsymbol{0},\ \|\boldsymbol{z}\| = 1\}$ that satisfies $\mu(\boldsymbol{x}_k) \to \alpha$.

Since $S$ is a bounded closed set (compact set),[3] the sequence $(\boldsymbol{x}_k)$ contains a convergent subsequence $\boldsymbol{x}_{k(1)}, \boldsymbol{x}_{k(2)}, \ldots, \boldsymbol{x}_{k(l)}, \ldots$ with its limit $\boldsymbol{x} = \lim_{l\to\infty} \boldsymbol{x}_{k(l)}$ belonging to $S$. By letting $l \to \infty$ in the inequality

$$A\boldsymbol{x}_{k(l)} - \mu(\boldsymbol{x}_{k(l)})\boldsymbol{x}_{k(l)} \geqq \boldsymbol{0},$$

we obtain $A\boldsymbol{x} - \alpha\boldsymbol{x} \geqq \boldsymbol{0}$. Therefore, we have

$$A\boldsymbol{x} - \alpha\boldsymbol{x} \geqq \boldsymbol{0}, \qquad \boldsymbol{x} \geqq \boldsymbol{0}, \qquad \boldsymbol{x} \neq \boldsymbol{0}.$$

(3) We have $\alpha > 0$. This is because each row sum of $A$ is positive by the irreducibility of $A$, and $\mu(\boldsymbol{1})$ for $\boldsymbol{1} = (1, 1, \ldots, 1)^\top$ is equal to the minimum value of the row sums of $A$; hence $\alpha \geqq \mu(\boldsymbol{1}) > 0$.

For the $\alpha$ and $\boldsymbol{x}$ above, we now show that

$$A\boldsymbol{x} - \alpha\boldsymbol{x} = \boldsymbol{0}, \qquad \boldsymbol{x} > \boldsymbol{0}.$$

From the irreducibility of matrix $A$ and Proposition 2.3 in Sec. 2.1.3, we have $(I + A)^{n-1} > O$, and therefore $\boldsymbol{y} = (I + A)^{n-1}\boldsymbol{x}$ is a positive vector, *i.e.*, $\boldsymbol{y} > \boldsymbol{0}$. We also have

$$A\boldsymbol{y} - \alpha\boldsymbol{y} = (I + A)^{n-1}(A\boldsymbol{x} - \alpha\boldsymbol{x}).$$

To show $A\boldsymbol{x} - \alpha\boldsymbol{x} = \boldsymbol{0}$ by contradiction, suppose that $A\boldsymbol{x} - \alpha\boldsymbol{x} \neq \boldsymbol{0}$. Then $A\boldsymbol{y} - \alpha\boldsymbol{y} > \boldsymbol{0}$, which implies $\mu(\boldsymbol{y}) > \alpha$. But this contradicts the definition of $\alpha$. Therefore, $A\boldsymbol{x} - \alpha\boldsymbol{x} = \boldsymbol{0}$ must be true. Finally, we note $\boldsymbol{y} = (I + A)^{n-1}\boldsymbol{x} = (1 + \alpha)^{n-1}\boldsymbol{x}$ and $\boldsymbol{y} > \boldsymbol{0}$, to conclude $\boldsymbol{x} > \boldsymbol{0}$. □

**Proposition 2.5.** *If $A\boldsymbol{x} = \alpha\boldsymbol{x}$, $A \geqq O$, and $\boldsymbol{x} > \boldsymbol{0}$, then $\alpha = \rho(A)$.*

**Proof.** Obviously, $\alpha \leqq \rho(A)$. Let us show the reverse inequality. Let

$$X = \mathrm{diag}\,(x_1, x_2, \ldots, x_n), \qquad B = X^{-1}AX.$$

The $(i, j)$ entry $b_{ij}$ of the matrix $B$ is given by $b_{ij} = a_{ij}x_j/x_i \geqq 0$, and the sum of the $i$th row of $B$ is

$$\sum_{j=1}^{n} b_{ij} = \frac{1}{x_i}\sum_{j=1}^{n} a_{ij}x_j = \frac{1}{x_i}(\alpha x_i) = \alpha.$$

[3] Generally, a sequence in a compact set contains a convergent subsequence. Note that, in $\mathbb{R}^n$, being a compact set is equivalent to being a bounded closed set. We also mention that the function $\mu(\boldsymbol{z})$ is not necessarily a continuous function, which fact motivates us to give the present proof using a convergent subsequence.

Then, from a general fact stated in Proposition 2.6 below, we obtain

$$\rho(B) \leqq \max_{1 \leqq i \leqq n} \sum_{j=1}^{n} |b_{ij}| = \max_{1 \leqq i \leqq n} \sum_{j=1}^{n} b_{ij} = \alpha.$$

On the other hand, $\rho(B) = \rho(A)$ by $B = X^{-1}AX$. Therefore, $\rho(A) \leqq \alpha$. □

**Proposition 2.6.** *For a square matrix $C = (c_{ij})$ of order $n$ with entries of complex numbers, we have*

$$\rho(C) \leqq \max_{1 \leqq i \leqq n} \sum_{j=1}^{n} |c_{ij}|. \tag{2.10}$$

**Proof.** Consider an eigenvalue $\lambda$ and the associated eigenvector $\boldsymbol{z}$. Let $z_k$ denote the component of $\boldsymbol{z}$ with the largest absolute value. Then $z_k \neq 0$, and the $k$th equation of $\lambda \boldsymbol{z} = C\boldsymbol{z}$ reads

$$\lambda z_k = \sum_{j=1}^{n} c_{kj} z_j.$$

From this, we obtain

$$|\lambda| = \left| \sum_{j=1}^{n} c_{kj} \frac{z_j}{z_k} \right| \leqq \sum_{j=1}^{n} |c_{kj}| \frac{|z_j|}{|z_k|} \leqq \sum_{j=1}^{n} |c_{kj}|.$$

This inequality holds for each eigenvalue $\lambda$, and therefore, the inequality in (2.10) follows. □

Finally, let us show that $\alpha = \rho(A)$ is a simple eigenvalue. We first present a fundamental fact.

**Proposition 2.7.** *Let $A$ be an irreducible nonnegative matrix and $B$ be a nonnegative matrix. If $A \geqq B$, then $\rho(A) \geqq \rho(B)$. The equality $\rho(A) = \rho(B)$ holds only if $A = B$.*

**Proof.** By Propositions 2.4 and 2.5 applied to $A^\top$, there exists $\boldsymbol{w} > \boldsymbol{0}$ such that

$$A^\top \boldsymbol{w} = \rho(A^\top)\boldsymbol{w} = \rho(A)\boldsymbol{w}. \tag{2.11}$$

Take any eigenvalue $\lambda$ of $B$ and the associated eigenvector $\boldsymbol{y} = (y_1, y_2, \ldots, y_n)^\top$, and set $\boldsymbol{v} = (|y_1|, |y_2|, \ldots, |y_n|)^\top$. The inner product of (2.11) and $\boldsymbol{v}$ gives

$$\boldsymbol{w}^\top A\boldsymbol{v} = \rho(A)\boldsymbol{w}^\top \boldsymbol{v}. \tag{2.12}$$

From $A \geqq B$, $\boldsymbol{w} > \mathbf{0}$, and $\boldsymbol{v} \geqq \mathbf{0}$, it follows that

$$\boldsymbol{w}^\top A\boldsymbol{v} \geqq \boldsymbol{w}^\top B\boldsymbol{v}. \tag{2.13}$$

On the other hand, $B\boldsymbol{y} = \lambda\boldsymbol{y}$ implies $B\boldsymbol{v} \geqq |\lambda|\boldsymbol{v}$, which yields

$$\boldsymbol{w}^\top B\boldsymbol{v} \geqq |\lambda|\boldsymbol{w}^\top \boldsymbol{v}. \tag{2.14}$$

By (2.12), (2.13), and (2.14), we obtain

$$\rho(A)\boldsymbol{w}^\top \boldsymbol{v} = \boldsymbol{w}^\top A\boldsymbol{v} \geqq \boldsymbol{w}^\top B\boldsymbol{v} \geqq |\lambda|\boldsymbol{w}^\top \boldsymbol{v}. \tag{2.15}$$

Since $\boldsymbol{w}^\top \boldsymbol{v} > 0$, this shows $\rho(A) \geqq |\lambda|$. Therefore we have $\rho(A) \geqq \rho(B)$, since $\lambda$ is an arbitrary eigenvalue of $B$.

Let us consider the case where the equality $\rho(A) = \rho(B)$ holds. For an eigenvalue $\lambda$ with $|\lambda| = \rho(B)$, all the inequalities in (2.15) are satisfied in equalities. Therefore,

$$\boldsymbol{w}^\top (A\boldsymbol{v} - B\boldsymbol{v}) = 0, \qquad \boldsymbol{w}^\top (B\boldsymbol{v} - |\lambda|\boldsymbol{v}) = 0.$$

Here, we have $\boldsymbol{w} > \mathbf{0}$, $A\boldsymbol{v} - B\boldsymbol{v} \geqq \mathbf{0}$, and $B\boldsymbol{v} - |\lambda|\boldsymbol{v} \geqq \mathbf{0}$, and hence

$$A\boldsymbol{v} = B\boldsymbol{v} = \rho(A)\boldsymbol{v}.$$

Since $(I+A)^{n-1} > O$ from the irreducibility of matrix $A$ and Proposition 2.3 in Sec. 2.1.3, and since $\boldsymbol{v} \geqq \mathbf{0}$ and $\boldsymbol{v} \neq \mathbf{0}$, we obtain

$$\mathbf{0} < (I + A)^{n-1}\boldsymbol{v} = (1 + \rho(A))^{n-1}\boldsymbol{v},$$

which shows $\boldsymbol{v} > \mathbf{0}$. From this and $(A - B)\boldsymbol{v} = \mathbf{0}$ we can conclude that $A = B$. □

**Proposition 2.8.** *For an irreducible nonnegative matrix $A$, $\alpha = \rho(A)$ is a simple eigenvalue of $A$.*

**Proof.** Consider the characteristic polynomial $\phi(t) = \det(tI - A)$. Let $B_i$ denote the principal submatrix of $A$, of order $n-1$, with the $i$th row and the $i$th column deleted from $A$, and put $\phi_i(t) = \det(tI - B_i)$. The derivative of $\phi(t)$ is given (see Remark 2.6) by

$$\phi'(t) = \sum_{i=1}^{n} \phi_i(t). \tag{2.16}$$

Let $C_i$ denote the matrix of order $n$ obtained from $A$ by changing all entries in the $i$th row or in the $i$th column of $A$ to zero. By Proposition 2.7, we have $\rho(C_i) < \rho(A)$, whereas it is obvious that $\rho(B_i) = \rho(C_i)$. Let $\tau_i$ denote the maximum real root of the equation $\phi_i(t) = 0$, where we define $\tau_i = 0$ if no real root exists. Then we have $\tau_i \leqq \rho(B_i)$. This is obvious if $\tau_i = 0$, and otherwise, $\tau_i$ is an eigenvalue of $B_i$ and is bounded by the spectral radius $\rho(B_i)$. Thus, we have

$$\tau_i \leqq \rho(B_i) = \rho(C_i) < \rho(A).$$

Since there exists no real root of $\phi_i(t)$ greater than $\tau_i$ and $\phi_i(t) \to +\infty$ as $t \to +\infty$, we have $\phi_i(t) > 0$ for every $t > \tau_i$. In particular, for $t = \rho(A)$ we have $\phi_i(\rho(A)) > 0$ for all $i$. With this in (2.16) we see that $\phi'(\rho(A)) > 0$. Therefore, $\rho(A)$ is a simple root of $\phi(t) = 0$, that is, $\rho(A)$ is a simple eigenvalue of $A$. □

The proof of Theorem 2.1 (Perron–Frobenius theorem) is thus completed.

**Remark 2.6.** Consider a matrix $A(t)$ of order $n$ that includes a parameter $t$, and let $\det A(t)$ denote its determinant. The derivative of $\det A(t)$ with respect to $t$ is given by the following formula:

$$\frac{\mathrm{d}}{\mathrm{d}t} \det A(t) = \sum_{j=1}^{n} \det A^{(j)}(t),$$

where $A^{(j)}(t)$ denotes the matrix obtained from $A(t)$ by replacing each entry in the $j$th column by its derivative with respect to $t$. This formula can be derived easily using the multi-linearity of a determinant. For example,

$$\frac{\mathrm{d}}{\mathrm{d}t} \det \begin{bmatrix} a_{11}(t) & a_{12}(t) \\ a_{21}(t) & a_{22}(t) \end{bmatrix} = \det \begin{bmatrix} a'_{11}(t) & a_{12}(t) \\ a'_{21}(t) & a_{22}(t) \end{bmatrix} + \det \begin{bmatrix} a_{11}(t) & a'_{12}(t) \\ a_{21}(t) & a'_{22}(t) \end{bmatrix}$$

in the case of $n = 2$. ■

### 2.2.3 *Proof for Reducible Matrices*

Theorem 2.2 (for reducible matrices) is proved here on the basis of Theorem 2.1 (for irreducible matrices). The following two different proofs are shown.

(1) The zero entries of the given matrix $A$ are replaced by a positive number $\varepsilon > 0$ to form an irreducible matrix $A_\varepsilon$, to which Theorem 2.1 is applied. Then we consider the limit as $\varepsilon \to 0$.
(2) The block-triangularization method of Sec. 1.1.3 is used to decompose the matrix $A$ into irreducible components. We first concentrate on an irreducible component, or a diagonal block, that determines the spectral radius of $A$. Theorem 2.1 is applied to that diagonal block to ensure the existence of a nonnegative eigenvector for that diagonal block. From this eigenvector, a nonnegative eigenvector of the whole matrix $A$ is constructed according to a recurrence relation.

The first method of proof is more popular in linear algebra textbooks. While this renders a shorter argument, it requires knowledge of limit operations and continuity. The second method, describing the overall structure explicitly in terms of the combination of irreducible components, is more likely to provide a better understanding of engineering systems.

#### 2.2.3.1 *Proof by limit operation*

Replace all the zero entries of matrix $A$ by a positive number $\varepsilon > 0$, and denote the resulting matrix by $A_\varepsilon$. This matrix $A_\varepsilon$ is irreducible (because all of its entries are positive), and the Perron–Frobenius theorem (Theorem 2.1) is applicable. Thus, $\alpha_\varepsilon = \rho(A_\varepsilon)$ is a simple eigenvalue of $A_\varepsilon$, and furthermore, there exists a positive vector $\boldsymbol{x}_\varepsilon > \mathbf{0}$ satisfying

$$A_\varepsilon \boldsymbol{x}_\varepsilon = \alpha_\varepsilon \boldsymbol{x}_\varepsilon. \tag{2.17}$$

The vector $\boldsymbol{x}_\varepsilon$ is uniquely determined under normalization $\|\boldsymbol{x}_\varepsilon\| = 1$.

Next, let us consider the limit as $\varepsilon \to 0$. Obviously, we have

$$\lim_{\varepsilon \to 0} A_\varepsilon = A.$$

Eigenvalues are continuous functions of the matrix entries, and hence

$$\lim_{\varepsilon \to 0} \rho(A_\varepsilon) = \rho(A).$$

Thus, if we set $\alpha = \rho(A)$, then $\lim_{\varepsilon \to 0} \alpha_\varepsilon = \alpha$. Since the vector $\boldsymbol{x}_\varepsilon$ is normalized to $\|\boldsymbol{x}_\varepsilon\| = 1$ and the unit sphere is a bounded closed set (compact set), there exists a decreasing sequence $\varepsilon_1 > \varepsilon_2 > \cdots > \varepsilon_k > \cdots$ of $\varepsilon$ converging to 0 such that the limit

$$\lim_{k \to \infty} \boldsymbol{x}_{\varepsilon_k} = \boldsymbol{x}$$

exists. Then $\boldsymbol{x} \geqq \mathbf{0}$, and the relation $A\boldsymbol{x} = \alpha \boldsymbol{x}$ holds.

#### 2.2.3.2 *Proof by block-triangularization*

With a rearrangement of rows and columns in advance, it may be assumed that the given matrix is in the form of

$$A = \begin{bmatrix} A_{11} & A_{12} & \cdots & A_{1p} \\ O & A_{22} & \cdots & A_{2p} \\ \vdots & \ddots & \ddots & \vdots \\ O & \cdots & O & A_{pp} \end{bmatrix}. \tag{2.18}$$

Here, each of the diagonal blocks $A_{11}, \ldots, A_{pp}$ is either an irreducible nonnegative matrix or a zero matrix of order 1.

For the spectral radius we have the obvious relation

$$\rho(A) = \max_{1 \leqq k \leqq p} \rho(A_{kk}),$$

which is a consequence of the fact that the eigenvalues of $A$ are the collection of the eigenvalues of $A_{11}, \ldots, A_{pp}$. Let $q$ denote the smallest index $k$ which attains the maximum on the right-hand side. Then

$$\rho(A_{kk}) < \rho(A_{qq}) = \rho(A) \qquad (k = 1, \ldots, q-1). \tag{2.19}$$

First, let us consider the (rather exceptional) case of $\rho(A) = 0$. In this case the diagonal blocks $A_{11}, \ldots, A_{pp}$ are all zero matrices of order 1, and in particular, the first column of $A$ is $\mathbf{0}$. Therefore, $A\boldsymbol{x} = \alpha\boldsymbol{x}$ for $\alpha = 0$ and $\boldsymbol{x} = (1, 0, \ldots, 0)^\top \geqq \mathbf{0}$.

Next, we consider the case of $\rho(A) > 0$. Set $\alpha = \rho(A)$. Then $\alpha = \rho(A_{qq})$. By the Perron–Frobenius theorem (Theorem 2.1) applied to $A_{qq}$, there exists a positive vector $\boldsymbol{x}_q > \mathbf{0}$ satisfying

$$A_{qq}\boldsymbol{x}_q = \alpha\boldsymbol{x}_q. \tag{2.20}$$

Using this $\boldsymbol{x}_q$, we determine vectors $\boldsymbol{x}_{q-1}, \boldsymbol{x}_{q-2}, \ldots, \boldsymbol{x}_1$ by solving the equation

$$A_{kk}\boldsymbol{x}_k + \sum_{l=k+1}^{q} A_{kl}\boldsymbol{x}_l = \alpha\boldsymbol{x}_k \tag{2.21}$$

for $k = q-1, q-2, \ldots, 1$ (in this order). As explained below, these vectors $\boldsymbol{x}_{q-1}, \boldsymbol{x}_{q-2}, \ldots, \boldsymbol{x}_1$ are uniquely determined and are nonnegative. Putting $\boldsymbol{x}_k = \mathbf{0}$ for $k = q+1, q+2, \ldots, p$, we define a vector $\boldsymbol{x}$ as $\boldsymbol{x} = (\boldsymbol{x}_1, \boldsymbol{x}_2, \ldots, \boldsymbol{x}_p)$ (or more precisely, $\boldsymbol{x} = (\boldsymbol{x}_1^\top, \boldsymbol{x}_2^\top, \ldots, \boldsymbol{x}_p^\top)^\top$). This $\boldsymbol{x}$ is a nonnegative (nonzero) vector, which satisfies $A\boldsymbol{x} = \alpha\boldsymbol{x}$ by (2.20) and (2.21).

It remains to prove that (2.21) determines nonnegative vectors $\boldsymbol{x}_k$ ($1 \leqq k \leqq q-1$). We show $\boldsymbol{x}_k \geqq \mathbf{0}$ by induction on $k$ in the order of $k = q-1, q-2, \ldots, 1$. The equation (2.21) can be rewritten as

$$(\alpha I - A_{kk})\boldsymbol{x}_k = \sum_{l=k+1}^{q} A_{kl}\boldsymbol{x}_l.$$

Since $\alpha = \rho(A_{qq}) > \rho(A_{kk})$ and $A_{kk} \geqq O$, the matrix $\alpha I - A_{kk}$ is a nonsingular M-matrix (see Sec. 2.4.3), and therefore, $(\alpha I - A_{kk})^{-1}$ is a nonnegative matrix. On the other hand, the right-hand side $\sum_{l=k+1}^{q} A_{kl}\boldsymbol{x}_l$ above is a nonnegative vector by the induction hypothesis. Therefore, $\boldsymbol{x}_k \geqq \mathbf{0}$. We have completed the proof of Theorem 2.2.

## 2.3 Stochastic Matrices

### 2.3.1 *Definition*

A nonnegative matrix is called a *stochastic matrix* if each row sum is equal to 1. That is, an $m \times n$ nonnegative matrix $P = (p_{ij})$ is a stochastic matrix if

$$\sum_{j=1}^{n} p_{ij} = 1 \qquad (i = 1, \ldots, m). \tag{2.22}$$

Using the vector $\mathbf{1} = (1, 1, \ldots, 1)^\top$, this condition can be written as

$$P\mathbf{1} = \mathbf{1}. \tag{2.23}$$

If $P$ is a square stochastic matrix, it follows from (2.23) that $\boldsymbol{x} = \mathbf{1}$ is an eigenvector of $P$ associated with the eigenvalue $\lambda = 1$. On the other hand, we know from Proposition 2.6 that the spectral radius $\rho(P)$ is 1 or less. Thus,

$$\rho(P) = 1. \tag{2.24}$$

Then Theorem 2.2 (extended version of the Perron–Frobenius theorem) shows that there exists a nonnegative left eigenvector associated with the eigenvalue 1. That is, there exists a nonnegative row vector $\boldsymbol{y} \neq \mathbf{0}$ such that

$$\boldsymbol{y}P = \boldsymbol{y}. \tag{2.25}$$

### 2.3.2 *Markov Chains*

#### 2.3.2.1 *Random walk*

Consider a person who moves randomly (such as a drunk person) along a line with $n$ locations.

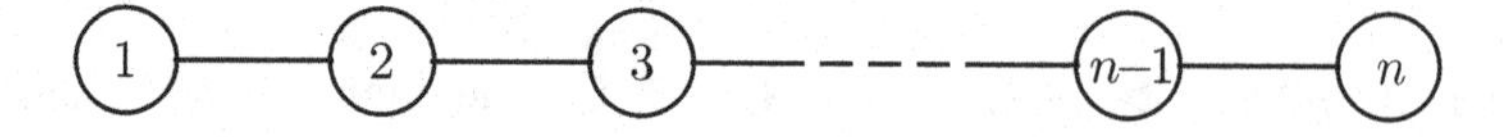


From the current location, the person moves to one of the two adjacent locations with equal probability, except in the following cases. From location 1, the person always moves to location 2, and from location $n$, the person either moves to location $n - 1$ or stays at location $n$ with equal probability. The $n \times n$ matrix $P = (p_{ij})$ consisting of the probabilities $p_{ij}$ of movement from location $i$ to location $j$ takes the form of

$$P = \begin{bmatrix} & 1 & & & \\ 0.5 & & 0.5 & & \\ & 0.5 & & 0.5 & \\ & & 0.5 & & 0.5 \\ & & & 0.5 & 0.5 \end{bmatrix}$$

(the case of $n = 5$ is shown). This matrix $P$ is a stochastic matrix.

#### 2.3.2.2 *Transition probability matrix*

As in the random walk above, suppose that we have a finite set of states $\{1, 2, \ldots, n\}$ in a given system, and the system changes its state stochastically at discrete time $k = 0, 1, 2, \ldots$. We assume[4] that the probability $p_{ij}$ of the system changing its state from state $i$ to state $j$ is independent of the time $k$. The matrix $P = (p_{ij})$ consisting of these probabilities is called the *transition probability matrix* or *transition matrix*, and its properties determine the characteristics of the system.

As the probability $p_{ij}$ is a nonnegative real number, we have

$$p_{ij} \geqq 0 \qquad (i, j = 1, \ldots, n). \tag{2.26}$$

In addition, if the system is in state $i$ at a certain time instance, it will be in state $j$ for some $j = 1, \ldots, n$ at the next time. This gives

$$\sum_{j=1}^{n} p_{ij} = 1 \qquad (i = 1, \ldots, n). \tag{2.27}$$

Thus, the matrix $P = (p_{ij})$ is a stochastic matrix.

Let $p_i^{(k)}$ denote the probability of the system being in state $i$ at time $k$, and define a row vector $\boldsymbol{p}^{(k)}$ consisting of these probabilities, *i.e.*,

$$\boldsymbol{p}^{(k)} = (p_1^{(k)}, \ldots, p_n^{(k)}).$$

Obviously, we have

$$p_i^{(k)} \geqq 0 \qquad (i = 1, \ldots, n), \tag{2.28}$$

$$\sum_{i=1}^{n} p_i^{(k)} = 1. \tag{2.29}$$

[4] Such a system is called a *Markov chain*. For more details, see [34, 35].

In general, a vector is called a *probability vector* or *stochastic vector*, if each of its components is a nonnegative real number and the sum of its components is equal to 1. Thus, $\boldsymbol{p}^{(k)}$ is a probability vector for each $k$.

The probability of a state transition is expressed by the matrix $P = (p_{ij})$ as

$$p_j^{(k+1)} = \sum_{i=1}^{n} p_i^{(k)} p_{ij} \qquad (j = 1, \ldots, n;\ k = 0, 1, 2, \ldots). \tag{2.30}$$

On the right-hand side of this relation, $p_i^{(k)} p_{ij}$ indicates the probability of the system in state $i$ at time $k$ changing to state $j$. Equation (2.30) can be interpreted as saying that the probability $p_j^{(k+1)}$ of the system being in state $j$ at time $k+1$ is equal to the sum of the probabilities $p_i^{(k)} p_{ij}$ over $i = 1, \ldots, n$. The relation (2.30) is written in matrix-vector form as

$$\boldsymbol{p}^{(k+1)} = \boldsymbol{p}^{(k)} P \qquad (k = 0, 1, 2, \ldots). \tag{2.31}$$

From this recurrence relation, we obtain

$$\boldsymbol{p}^{(k)} = \boldsymbol{p}^{(0)} P^k \qquad (k = 0, 1, 2, \ldots). \tag{2.32}$$

If there exists a limit of $\boldsymbol{p}^{(k)}$ as time $k \to \infty$, the limit is called a *limit distribution* and expressed as

$$\boldsymbol{p}^{(\infty)} = \lim_{k \to \infty} \boldsymbol{p}^{(k)}. \tag{2.33}$$

If a limit distribution exists, letting $k \to \infty$ in (2.31) results in

$$\boldsymbol{p}^{(\infty)} = \boldsymbol{p}^{(\infty)} P. \tag{2.34}$$

A probability distribution that remains invariant under a state transition is called a *stationary distribution.* That is, a stationary distribution is a probability distribution expressed by a probability vector $\boldsymbol{\pi}$ satisfying

$$\boldsymbol{\pi} = \boldsymbol{\pi} P. \tag{2.35}$$

Equation (2.34) shows that a limit distribution, if any, is a stationary distribution.

In this context it is natural to ask the following questions.

- (Existence and uniqueness of stationary distributions) Does a stationary distribution exist for a given system? Is it uniquely determined?
- (Existence of limit distribution and dependence on initial distribution) Does a limit distribution exist for a given initial distribution $\boldsymbol{p}^{(0)}$? How does the limit distribution depend on the initial distribution?

### 2.3.2.3 *Stationary distributions*

To study stationary distributions, let us first look at a simple example, and then consider the general case.

**Example 2.1.** In the random walk shown at the beginning of Sec. 2.3.2, the vector

$$\boldsymbol{\pi} = \left( \frac{1}{2n-1}, \frac{2}{2n-1}, \ldots, \frac{2}{2n-1}, \frac{2}{2n-1} \right)$$

gives a unique stationary distribution. The defining identity $\boldsymbol{\pi} = \boldsymbol{\pi} P$ for a stationary distribution is written componentwise as

$$\begin{aligned}
\pi_1 &= \frac{1}{2}\pi_2, \\
\pi_2 &= \pi_1 + \frac{1}{2}\pi_3, \\
\pi_i &= \frac{1}{2}\pi_{i-1} + \frac{1}{2}\pi_{i+1} \qquad (i = 3, \ldots, n-1), \\
\pi_n &= \frac{1}{2}\pi_{n-1} + \frac{1}{2}\pi_n.
\end{aligned}$$

The vector $\boldsymbol{\pi}$ above is determined from these relations together with the normalization condition $\sum_{i=1}^{n} \pi_i = 1$. ■

**Example 2.2.** For the transition probability matrix $P = \begin{bmatrix} 1 & 0 \\ 0 & 1 \end{bmatrix}$, for example, every probability vector is a stationary distribution. Thus, a stationary distribution is not always determined uniquely. ■

As mentioned before, the extended version of the Perron–Frobenius theorem (Theorem 2.2) implies that there exists a nonnegative row vector $\boldsymbol{y} \neq \mathbf{0}$ that satisfies (2.25). When $\boldsymbol{y}$ is normalized to $\boldsymbol{\pi} = \boldsymbol{y}/c$ with $c = \sum_{i=1}^{n} y_i$, the resulting vector $\boldsymbol{\pi}$ is a probability vector and satisfies the condition $\boldsymbol{\pi} = \boldsymbol{\pi} P$ for a stationary distribution. Hence, a stationary distribution always exists.

As shown in Example 2.2, a stationary distribution is not uniquely determined, in general. However, if $P$ is irreducible, it follows from the Perron–Frobenius theorem (Theorem 2.1) that the stationary distribution $\boldsymbol{\pi}$ is uniquely determined, and furthermore, $\boldsymbol{\pi}$ is a positive vector. Example 2.1 above falls into this case.

Let us state these important facts as a theorem.

**Theorem 2.4.**

(1) *For any transition probability matrix $P$, there exists a stationary distribution $\pi$.*
(2) *If a transition probability matrix $P$ is irreducible, the stationary distribution $\pi$ is uniquely determined, and all of its components $\pi_i$ are positive.*

**Remark 2.7.** Let us consider the stationary distribution for a reducible $P$. To simplify notation, we assume that the given stochastic matrix $P$ is block-triangularized (decomposed into strongly connected components) in the following particular form

$$P = \begin{bmatrix} P_{11} & O & P_{13} & P_{14} \\ O & P_{22} & O & P_{24} \\ O & O & P_{33} & O \\ O & O & O & P_{44} \end{bmatrix}. \tag{2.36}$$

Here, the diagonal blocks $P_{11}$, $P_{22}$, $P_{33}$, and $P_{44}$ are irreducible nonnegative matrices, and $P_{13}$, $P_{14}$, and $P_{24}$ are (nonzero) nonnegative matrices. The graph expressing the partial order among the irreducible blocks of $P$ is shown in Fig. 2.1 (see Sec. 1.1.3 for details on such graphs). Note that the minimal blocks[5] $P_{33}$ and $P_{44}$ are stochastic matrices, with each row sum being equal to 1.

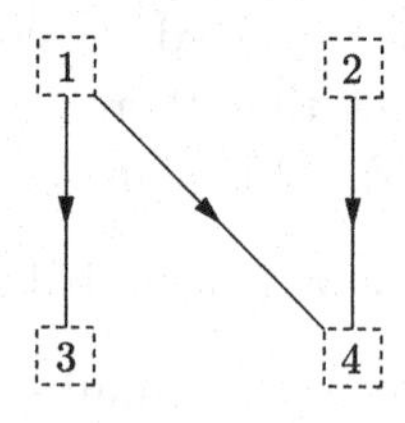


Fig. 2.1 Partial order associated with $P$ in (2.36).

Let $\pi$ be any stationary distribution of $P$, and express $\pi$ as $\pi =$

[5]In a partially ordered set, in general, an element is called *minimal* if there exists no other element that is smaller than that element.

$(\pi_1, \pi_2, \pi_3, \pi_4)$. Then we can write the condition $\pi = \pi P$ as

$$\pi_1 = \pi_1 P_{11}, \tag{2.37}$$

$$\pi_2 = \pi_2 P_{22}, \tag{2.38}$$

$$\pi_3 = \pi_3 P_{33} + \pi_1 P_{13}, \tag{2.39}$$

$$\pi_4 = \pi_4 P_{44} + \pi_1 P_{14} + \pi_2 P_{24}. \tag{2.40}$$

Since $\rho(P_{11}) < 1$ and $\rho(P_{22}) < 1$ by Theorem 2.3 and (2.7), it follows from (2.37) and (2.38) that $\pi_1 = \mathbf{0}$ and $\pi_2 = \mathbf{0}$, where $\mathbf{0}$ is a row zero vector. By substituting $\pi_1 = \mathbf{0}$ and $\pi_2 = \mathbf{0}$ into (2.39) and (2.40), we obtain

$$\pi_3 = \pi_3 P_{33}, \qquad \pi_4 = \pi_4 P_{44}.$$

Here $P_{33}$ and $P_{44}$ are irreducible stochastic matrices, and Theorem 2.4 shows that there uniquely exist stationary distributions $\pi^{(3)}$ and $\pi^{(4)}$ for $P_{33}$ and $P_{44}$, respectively. In addition, the eigenvalue 1 is a simple eigenvalue by Theorem 2.1(3). Therefore, we must have

$$\pi_3 = \alpha_3 \pi^{(3)}, \qquad \pi_4 = \alpha_4 \pi^{(4)}$$

for some $\alpha_3 \geqq 0$ and $\alpha_4 \geqq 0$. Thus, the stationary distribution of the stochastic matrix $P$ is represented in the form

$$\pi = \left(\mathbf{0}, \mathbf{0}, \alpha_3 \pi^{(3)}, \alpha_4 \pi^{(4)}\right). \tag{2.41}$$

We have $\alpha_3 \geqq 0$, $\alpha_4 \geqq 0$, and $\alpha_3 + \alpha_4 = 1$, in accordance with the fact that the sum of probabilities is equal to 1.

Also in the general case, the stationary distribution $\pi$ can be expressed using block-triangularization (decomposition into strongly connected components). Assume that the irreducible blocks are indexed by $1, 2, \ldots, p$ and the blocks $q+1, q+2, \ldots, p$ are minimal blocks ($q \geqq 0$). For $k = q+1, q+2, \ldots, p$, let $\pi^{(k)}$ denote the (uniquely determined) stationary distribution associated with the minimal irreducible block $k$. Then we have

$$\pi = (\mathbf{0}, \ldots, \mathbf{0}, \alpha_{q+1} \pi^{(q+1)}, \ldots, \alpha_p \pi^{(p)}),$$

where $\alpha_k \geqq 0$ for each $k$ and $\displaystyle\sum_{k=q+1}^{p} \alpha_k = 1$. Thus the stationary distribution is expressed with reference to the partial order in the decomposition into strongly connected components. ■

2.3.2.4 *Limit distributions*

Next, we consider limit distributions. Recall $\boldsymbol{p}^{(\infty)} = \lim_{k\to\infty} \boldsymbol{p}^{(k)}$ from (2.33). As already stated, we have the following questions.

- Does a limit distribution $\boldsymbol{p}^{(\infty)}$ exist for a given initial distribution $\boldsymbol{p}^{(0)}$?
- How does the limit distribution $\boldsymbol{p}^{(\infty)}$ depend on the initial distribution $\boldsymbol{p}^{(0)}$?

As we have seen in (2.34), if a limit distribution exists, then it is a stationary distribution. Conversely, for any stationary distribution $\boldsymbol{\pi}$, the choice of initial distribution $\boldsymbol{p}^{(0)} = \boldsymbol{\pi}$ obviously leads to the limit distribution $\boldsymbol{p}^{(\infty)} = \boldsymbol{\pi}$. We note, however, that not every initial distribution $\boldsymbol{p}^{(0)}$ has a limit distribution $\boldsymbol{p}^{(\infty)}$, which is demonstrated in the following example.

**Example 2.3.** Consider the transition probability matrix

$$P = \begin{bmatrix} 0 & 1 & 0 \\ 0 & 0 & 1 \\ 1 & 0 & 0 \end{bmatrix}.$$

For $\boldsymbol{p}^{(0)} = (a, b, c)$ we have

$$\boldsymbol{p}^{(3k)} = (a, b, c), \qquad \boldsymbol{p}^{(3k+1)} = (c, a, b), \qquad \boldsymbol{p}^{(3k+2)} = (b, c, a).$$

Therefore, if $(a, b, c) \neq (1/3, 1/3, 1/3)$, the limit distribution $\boldsymbol{p}^{(\infty)}$ does not exist. This $P$ is irreducible, but not primitive (with period 3). ■

A limit distribution does not always exist, as shown in the above example. However, if $P$ is irreducible and primitive, a limit distribution exists for any initial distribution, and is uniquely determined independently of the initial distribution. The next theorem states this.

**Theorem 2.5.** *Suppose that a transition probability matrix $P$ is irreducible and primitive, and denote its stationary distribution*[6] *by $\boldsymbol{\pi}$.*

(1) *We have*

$$\lim_{k\to\infty} P^k = \mathbf{1}\boldsymbol{\pi} = \begin{bmatrix} \pi_1 & \pi_2 & \cdots & \pi_n \\ \pi_1 & \pi_2 & \cdots & \pi_n \\ \vdots & \vdots & & \vdots \\ \pi_1 & \pi_2 & \cdots & \pi_n \end{bmatrix}. \tag{2.42}$$

[6] By Theorem 2.4, the stationary distribution is uniquely determined when $P$ is irreducible.

(2) *For any initial distribution* $\boldsymbol{p}^{(0)}$, *the limit distribution* $\boldsymbol{p}^{(\infty)}$ *exists, and is given by* $\boldsymbol{p}^{(\infty)} = \boldsymbol{\pi}$. *In particular, the limit distribution does not depend on the initial distribution.*

**Proof.** (1) As shown in (2.24), we have $\rho(P) = 1$. Since $P$ is irreducible, it has 1 as a simple eigenvalue (Theorem 2.1). Since $P$ is primitive, it has no other eigenvalues of absolute value 1 (Remark 2.5). By the Jordan normal form of $P$, there exists a nonsingular matrix $S$ and a matrix $J$ with $\rho(J) < 1$ such that

$$SPS^{-1} = \begin{bmatrix} 1 & \mathbf{0}^\top \\ \mathbf{0} & J \end{bmatrix}.$$

Since the eigenvalue 1 is a simple eigenvalue, $\boldsymbol{\pi} = \boldsymbol{\pi}P$, $P\mathbf{1} = \mathbf{1}$, and $\boldsymbol{\pi}\mathbf{1} = 1$, we can assume that the first row of $S$ is equal to $\boldsymbol{\pi}$ and the first column of $S^{-1}$ is equal to $\mathbf{1}$, namely,

$$S = \begin{bmatrix} \boldsymbol{\pi} \\ \hline * \end{bmatrix}, \qquad S^{-1} = \begin{bmatrix} \mathbf{1} \,|\, * \end{bmatrix}.$$

Note also that $\lim_{k\to\infty} J^k = O$ as a consequence of $\rho(J) < 1$. Therefore we have

$$\lim_{k\to\infty} P^k = S^{-1} \left[ \lim_{k\to\infty} (SPS^{-1})^k \right] S = \begin{bmatrix} \mathbf{1} \,|\, * \end{bmatrix} \begin{bmatrix} 1 & \mathbf{0}^\top \\ \mathbf{0} & O \end{bmatrix} \begin{bmatrix} \boldsymbol{\pi} \\ \hline * \end{bmatrix} = \mathbf{1}\boldsymbol{\pi}.$$

(2) For any $\boldsymbol{p}^{(0)}$, we have $\boldsymbol{p}^{(0)}\mathbf{1} = 1$. By using (2.42) we obtain

$$\boldsymbol{p}^{(\infty)} = \lim_{k\to\infty} \boldsymbol{p}^{(k)} = \lim_{k\to\infty} \boldsymbol{p}^{(0)} P^k = \boldsymbol{p}^{(0)} \lim_{k\to\infty} P^k = \boldsymbol{p}^{(0)}\mathbf{1}\boldsymbol{\pi} = \boldsymbol{\pi}.$$

□

**Remark 2.8.** In Example 2.3 we have seen that the transition probability matrix

$$P = \begin{bmatrix} 0 & 1 & 0 \\ 0 & 0 & 1 \\ 1 & 0 & 0 \end{bmatrix},$$

having period 3, does not have a limit distribution. We can explain this phenomenon with reference to the eigenvectors of $P$, which we investigated in Remark 2.5 in Sec. 2.2.1. The eigenvalues of $P$ are given by 1, $\zeta = \exp(2\pi\mathrm{i}/3)$ and $\overline{\zeta} = \exp(-2\pi\mathrm{i}/3)$. With the use of

$$S = \begin{bmatrix} 1 & 1 & 1 \\ 1 & \overline{\zeta} & \zeta \\ 1 & \zeta & \overline{\zeta} \end{bmatrix}, \qquad S^{-1} = \frac{1}{3} \begin{bmatrix} 1 & 1 & 1 \\ 1 & \zeta & \overline{\zeta} \\ 1 & \overline{\zeta} & \zeta \end{bmatrix},$$

we can obtain diagonal forms

$$SPS^{-1} = \begin{bmatrix} 1 & 0 & 0 \\ 0 & \zeta & 0 \\ 0 & 0 & \overline{\zeta} \end{bmatrix}, \qquad (SPS^{-1})^k = \begin{bmatrix} 1 & 0 & 0 \\ 0 & \zeta^k & 0 \\ 0 & 0 & \overline{\zeta}^k \end{bmatrix}.$$

Since

$$\boldsymbol{p}^{(k)} = \boldsymbol{p}^{(0)} P^k = (\boldsymbol{p}^{(0)} S^{-1})(SPS^{-1})^k S = (\boldsymbol{p}^{(0)} S^{-1}) \begin{bmatrix} 1 & 0 & 0 \\ 0 & \zeta^k & 0 \\ 0 & 0 & \overline{\zeta}^k \end{bmatrix} S,$$

$\boldsymbol{p}^{(k)}$ has a limit as $k \to \infty$ if and only if the second and third components of $\boldsymbol{p}^{(0)} S^{-1}$ both vanish. For $\boldsymbol{p}^{(0)} = (a, b, c)$ we have

$$\boldsymbol{p}^{(0)} S^{-1} = \frac{1}{3} \left( a + b + c, \ a + b\zeta + c\overline{\zeta}, \ a + b\overline{\zeta} + c\zeta \right),$$

and hence $\boldsymbol{p}^{(k)}$ has a limit if and only if $a + b\zeta + c\overline{\zeta} = a + b\overline{\zeta} + c\zeta = 0$. This condition is equivalent to $(a, b, c) = (1/3, 1/3, 1/3)$, in agreement with Example 2.3. ■

## 2.4 M-Matrices

### 2.4.1 *Definition*

A nonsingular real matrix is called an *M-matrix*,[7] if all the diagonal entries are positive, all the off-diagonal entries are nonpositive (0 or negative), and its inverse is a nonnegative matrix. That is, a nonsingular matrix $B = (b_{ij})$ of order $n$ is an M-matrix if

$$b_{ii} > 0 \qquad (i = 1, \ldots, n), \tag{2.43}$$

$$b_{ij} \leqq 0 \qquad (i \neq j; \ i, j = 1, \ldots, n), \tag{2.44}$$

$$B^{-1} \geqq O. \tag{2.45}$$

The concept of M-matrices is useful in studying nonnegativity and/or monotonicity of solutions of systems of equations. M-matrices play important roles in numerical analysis, control engineering, economics, and so on. We present such examples in the next section, and discuss mathematical properties of M-matrices in Sec. 2.4.3.

[7]More generally, singular M-matrices can also be defined, but in this book we restrict ourselves to nonsingular M-matrices.

### 2.4.2 *Examples*

**Example 2.4.** Let us consider *Poisson's equation*

$$-\triangle u = f,$$

which is a typical differential equation. Here, the symbol $\triangle$ denotes the *Laplacian*, which is given, for a one-dimensional problem, by

$$\triangle = \frac{\mathrm{d}^2}{\mathrm{d}x^2}.$$

This differential equation may be solved numerically through discretization, *e.g.*, by replacing differentiation by finite difference. In the one-dimensional problem, for example, if the interval in question is divided into equal subintervals, we obtain a system of linear equations with a coefficient matrix such as

$$B = \begin{bmatrix} 2 & -1 & & \\ -1 & 2 & -1 & \\ & -1 & 2 & -1 \\ & & -1 & 2 \end{bmatrix}; \tag{2.46}$$

this particular matrix arises from the case where the interval is divided into four parts and the Dirichlet boundary condition is imposed. The matrix $B$ in (2.46) is an M-matrix. Indeed, the diagonal entries of $B$ are positive, the off-diagonal entries are nonpositive, and its inverse

$$B^{-1} = \frac{1}{5}\begin{bmatrix} 4 & 3 & 2 & 1 \\ 3 & 6 & 4 & 2 \\ 2 & 4 & 6 & 3 \\ 1 & 2 & 3 & 4 \end{bmatrix}$$

is a nonnegative matrix. As the matrix $B$ is an approximation of a differential operator, its inverse $B^{-1}$ corresponds to an integral operator. As such, the matrix $B^{-1}$ is naturally expected to be a nonnegative matrix. ■

**Example 2.5.** M-matrices arise naturally from electric circuits composed of linear resistors of positive resistances. As an example, let us consider the electric circuits shown in Fig. 2.2 consisting of five branches (linear resistors) and four nodes. The *conductance* (reciprocal of resistance) of each branch is denoted by $g_j > 0$ $(j = 1, \ldots, 5)$, and the (electric) potential at node $k$ is denoted by $p_k$ $(k = 0, 1, 2, 3)$, where the node 0 is assumed to be grounded, *i.e.*, $p_0 = 0$. Also, the current flowing into the circuit from the outside at

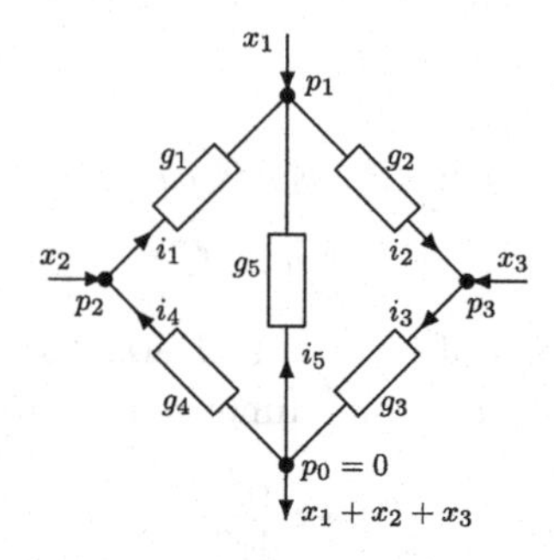


Fig. 2.2 A circuit consisting of resistors.

node $k$ is denoted by $x_k$ $(k = 1, 2, 3)$, which means that the current flowing out at node 0 is $x_1 + x_2 + x_3$. The branch voltage vector $\boldsymbol{v} = (v_1, \ldots, v_5)^\top$ and the node potential vector $\boldsymbol{p} = (p_1, p_2, p_3)^\top$ are related as

$$\boldsymbol{v} = N^\top \boldsymbol{p}$$

with

$$N = \begin{bmatrix} -1 & 1 & 0 & 0 & -1 \\ 1 & 0 & 0 & -1 & 0 \\ 0 & -1 & 1 & 0 & 0 \end{bmatrix}. \tag{2.47}$$

By Ohm's law, the branch current vector $\boldsymbol{i} = (i_1, \ldots, i_5)^\top$ can be expressed as

$$\boldsymbol{i} = G\boldsymbol{v},$$

where $G = \operatorname{diag}(g_1, \ldots, g_5)$ is a diagonal matrix having conductances $g_j$ as its diagonal entries. In addition, the conservation law of current is expressed as

$$N\boldsymbol{i} = \boldsymbol{x}.$$

From the above arguments, the relation between the node potential vector $\boldsymbol{p}$ and the inflow current vector $\boldsymbol{x} = (x_1, x_2, x_3)^\top$ is obtained as

$$NGN^\top \boldsymbol{p} = \boldsymbol{x}.$$

The coefficient matrix

$$B = NGN^\top = \begin{bmatrix} g_1 + g_2 + g_5 & -g_1 & -g_2 \\ -g_1 & g_1 + g_4 & 0 \\ -g_2 & 0 & g_2 + g_3 \end{bmatrix} \tag{2.48}$$

is called the *node-conductance matrix*, and is an M-matrix. Nonnegativity of the inverse $B^{-1}$ corresponds to the natural monotonicity that an increase of the current flowing into a node results in an increase (non-decrease) of potentials at all nodes. ■

**Example 2.6.** M-matrices also appear in the field of economics. Suppose that we are interested in the economy of a certain region or country for a specific time period (usually, one year), and want to consider the quantity of the production of various industrial sectors (for example, agriculture, forestry, and fisheries; mining, electrical machinery, and so on) used as inputs by other industrial sectors (including the sectors themselves).

Industrial sectors are referred to, in an abstract way, as "industry 1", "industry 2", ..., and "industry $n$." The total production (output) of industry $i$ is denoted by $x_i$, and the amount of this production used by industry $j$ as input is denoted by $x_{ij}$. Non-industrial use called final demand (including household consumption, government demand, and so on) is collectively denoted by $c_i$. This gives us the following equations:

$$\begin{aligned} x_{11} + x_{12} + \cdots + x_{1j} + \cdots + x_{1n} + c_1 &= x_1 \quad \text{(production of industry 1)}, \\ x_{21} + x_{22} + \cdots + x_{2j} + \cdots + x_{2n} + c_2 &= x_2 \quad \text{(production of industry 2)}, \\ &\vdots \\ x_{n1} + x_{n2} + \cdots + x_{nj} + \cdots + x_{nn} + c_n &= x_n \quad \text{(production of industry } n). \end{aligned} \tag{2.49}$$

We can view these equations as expressing how the outputs are used as inputs, and also, conversely, as expressing how the outputs are produced from inputs.

In taking the latter interpretation that industry $j$ uses inputs of amounts $x_{1j}, x_{2j}, \ldots, x_{nj}$ to produce its output of amount $x_j$, it would be natural to assume, at least as a first-order approximation, that $x_{1j}, x_{2j}, \ldots, x_{nj}$ are proportional to $x_j$ as

$$x_{1j} = a_{1j}x_j, \qquad x_{2j} = a_{2j}x_j, \quad \ldots, \quad x_{nj} = a_{nj}x_j. \tag{2.50}$$

This represents the model that the production volume is doubled when each of the input amounts is doubled. Naturally, the constants of proportionality $a_{ij}$ are nonnegative numbers; they are called the *input coefficients*.

Substituting the relationship of (2.50) into (2.49), we obtain

$$\begin{aligned} (1 - a_{11})x_1 - a_{12}x_2 - \cdots - a_{1n}x_n &= c_1, \\ -a_{21}x_1 + (1 - a_{22})x_2 - \cdots - a_{2n}x_n &= c_2, \\ &\vdots \\ -a_{n1}x_1 - a_{n2}x_2 - \cdots + (1 - a_{nn})x_n &= c_n. \end{aligned} \tag{2.51}$$

This system of equations is the fundamental equations for the *input-output analysis* initiated by W. Leontief. The matrix $A = (a_{ij})$ is called the *input coefficient matrix*. In using this model in practice, the concrete values of the input coefficients $a_{ij}$ are estimated from the data $x_{ij}$ and $x_i$ for a specific year as

$$\widehat{a}_{ij} = x_{ij}/x_i.$$

The definitions of the $c_i$ and $x_i$ appearing in (2.51) mean that they should naturally be nonnegative. If there exist some $c_i > 0$ and $x_i \geqq 0$ $(i = 1, \ldots, n)$ that satisfy the equations in (2.51), then the condition (c) in Theorem 2.6 in the next section is satisfied, and consequently, the coefficient matrix $I - A$ is an M-matrix. ■

### 2.4.3 *Mathematical Properties*

Several necessary and sufficient conditions for a matrix to be an M-matrix are known. In particular, the condition (e) in the theorem below allows us to say that an M-matrix is a matrix represented as $sI - A$ for a nonnegative matrix $A$ and a real number $s > \rho(A)$, where $\rho(A)$ denotes the spectral radius of the matrix $A$.

**Theorem 2.6.** *The following conditions* (a) *to* (g) *are equivalent for an* $n \times n$ *matrix* $B = (b_{ij})$ *with the sign pattern of* (2.43) *and* (2.44), *i.e., with positive diagonal entries and nonpositive off-diagonal entries:*

(a) *$B$ is nonsingular and $B^{-1} \geqq O$.*
(b) *For any (n-dimensional) nonnegative vector $\mathbf{c}$, there exists a nonnegative vector $\boldsymbol{x}$ that satisfies $B\boldsymbol{x} = \mathbf{c}$.*
(c) *For some positive vector $\mathbf{c}$, there exists a nonnegative vector $\boldsymbol{x}$ that satisfies $B\boldsymbol{x} = \mathbf{c}$.*
(d) *For some positive vector $\mathbf{c}$, there exists a positive vector $\boldsymbol{x}$ that satisfies $B\boldsymbol{x} = \mathbf{c}$.*
(e) *For $s = \max\limits_{1 \leqq i \leqq n} b_{ii}$ and $A = sI - B$, we have $\rho(A) < s$.*
(f) *All the leading principal minors[8] of $B$ are positive.*
(g) *$B$ admits an LU decomposition[9] $B = LU$ such that*

- *$L$ is a lower-triangular matrix in which the diagonal entries are 1 and the off-diagonal entries are nonpositive,*

[8]A submatrix whose row and column indices correspond to the same $k$ indices, $i_1 < i_2 < \cdots < i_k$, is called a *principal submatrix* of order $k$. If $i_1 = 1, i_2 = 2, \ldots, i_k = k$, it is called the *leading principal submatrix* of order $k$. The determinant of a principal submatrix is called a *principal minor*, and the determinant of a leading principal submatrix is called a *leading principal minor*.

[9]For details on the LU decomposition, see Remark 2.9.

- *$U$ is an upper-triangular matrix in which the diagonal entries are positive and the off-diagonal entries are nonpositive.*

**Proof.** We will show [(a) $\Rightarrow$ (b) $\Rightarrow$ (c) $\Rightarrow$ (d) $\Rightarrow$ (e) $\Rightarrow$ (a)] and [(d) $\Rightarrow$ (f) $\Rightarrow$ (g) $\Rightarrow$ (a)].

[(a) $\Rightarrow$ (b)] Since $B$ is nonsingular, we have $B\boldsymbol{x} = \boldsymbol{c}$ for $\boldsymbol{x} = B^{-1}\boldsymbol{c}$. From $B^{-1} \geqq O$ and $\boldsymbol{c} \geqq \boldsymbol{0}$, it follows that $\boldsymbol{x} \geqq \boldsymbol{0}$.

[(b) $\Rightarrow$ (c)] is obvious.

[(c) $\Rightarrow$ (d)] From (c), we can take $\boldsymbol{d} > \boldsymbol{0}$ and $\boldsymbol{y} \geqq \boldsymbol{0}$ that satisfy $B\boldsymbol{y} = \boldsymbol{d}$. For any positive number $\alpha$, we have $\boldsymbol{x} = \boldsymbol{y} + \alpha\boldsymbol{1} > \boldsymbol{0}$, and if $\alpha > 0$ is sufficiently small, we also have

$$B\boldsymbol{x} = B(\boldsymbol{y} + \alpha\boldsymbol{1}) = \boldsymbol{d} + \alpha(B\boldsymbol{1}) > \boldsymbol{0}.$$

Hence, (d) holds for $\boldsymbol{c} = \boldsymbol{d} + \alpha(B\boldsymbol{1})$.

[(d) $\Rightarrow$ (e)] Note first that $s > 0$ and $A = sI - B$ is a nonnegative matrix. By (d) there exist positive vectors $\boldsymbol{x}$ and $\boldsymbol{c}$ such that $(sI - A)\boldsymbol{x} = \boldsymbol{c}$. Putting $\overline{A} = s^{-1}A$ and $\boldsymbol{y} = s^{-1}\boldsymbol{c}$, we obtain $(I - \overline{A})\boldsymbol{x} = \boldsymbol{y}$ and $\boldsymbol{y} > \boldsymbol{0}$. Note the relation

$$\boldsymbol{y} + \overline{A}\boldsymbol{y} + \overline{A}^2\boldsymbol{y} + \cdots + \overline{A}^p\boldsymbol{y} = \boldsymbol{x} - \overline{A}^{p+1}\boldsymbol{x} \leqq \boldsymbol{x}.$$

The sum on the left-hand side above is bounded by $\boldsymbol{x}$, whereas each term in the sum is a nonnegative vector. Therefore, $\overline{A}^p\boldsymbol{y} \to \boldsymbol{0}$ as $p \to \infty$. This implies, in turn, that $\overline{A}^p \to O$, since $\boldsymbol{y}$ is a positive vector. Thus, $\rho(\overline{A}) < 1$, that is, $\rho(A) < s$.

[(e) $\Rightarrow$ (a)] By $\rho(A) < s$, the matrix $B = sI - A$ is nonsingular. Put $\overline{A} = s^{-1}A$. Then $\rho(\overline{A}) < 1$ and

$$B^{-1} = s^{-1}(I - \overline{A})^{-1} = s^{-1}(I + \overline{A} + \overline{A}^2 + \overline{A}^3 + \cdots)$$

is convergent. Since $\overline{A}$ is a nonnegative matrix, each term on the right-hand side is a nonnegative matrix, and thus, $B^{-1} \geqq O$.

[(d) $\Rightarrow$ (f)] We have $B\boldsymbol{x} = \boldsymbol{c}$ for some $\boldsymbol{x} > \boldsymbol{0}$ and $\boldsymbol{c} > \boldsymbol{0}$. Let $B_k$ denote the leading principal submatrix of order $k$ of $B$ and set $\boldsymbol{y} = (x_1, \ldots, x_k)^\top$. For $i = 1, \ldots, k$, we have:

$$(\text{the } i\text{th component of } B_k\boldsymbol{y}) = c_i - \sum_{j=k+1}^{n} b_{ij}x_j \geqq c_i > 0.$$

This shows that the matrix $B_k$ satisfies the condition (d). As we have already shown [(d) $\Rightarrow$ (e)], we can conclude that $\rho(s_kI - B_k) < s_k$ for $s_k = \max_{1 \leqq i \leqq k} b_{ii}$. This shows that, in the complex plane, all of the eigenvalues of $B_k$ are located inside the circle of radius $s_k$ with center at $s_k$. Thus, the real parts of the eigenvalues of $B_k$ are all positive. Since the determinant is equal to the product of the eigenvalues, we have $\det B_k > 0$.

[(f) ⇒ (g)] For the diagonal entries of $U = (u_{ij})$, we have

$$u_{11} = b_{11} > 0, \qquad u_{kk} = \det B_k / \det B_{k-1} > 0 \quad (k = 2, \dots, n),$$

where $B_k$ is the leading principal submatrix of order $k$ of $B$. This is a well-known fact about the LU decomposition.[10] The diagonal entries of $L = (\ell_{ij})$ are, by definition, $\ell_{kk} = 1 > 0$ $(k = 1, \dots, n)$. To show that the signs of the off-diagonal entries of $U$ and $L$ are nonpositive, we refer to the algorithm for the LU decomposition given in Remark 2.9 below, specifically, to the formula

$$b_{ij}^{(k+1)} = b_{ij}^{(k)} - \ell_{ik} u_{kj}.$$

This shows that $b_{ij}^{(k+1)} \leqq 0$ follows from $b_{ij}^{(k)} \leqq 0$, $\ell_{ik} \leqq 0$ and $u_{kj} \leqq 0$, whereas $b_{ij}^{(1)} = b_{ij} \leqq 0$.

[(g) ⇒ (a)] Both $L$ and $U$ are nonsingular, since their diagonal entries are all positive. From the sign patterns of $L$ and $U$, it then follows that $L^{-1} \geqq O$ and $U^{-1} \geqq O$. Thus, $B^{-1} = U^{-1} L^{-1} \geqq O$. □

**Remark 2.9.** The algorithm of the LU decomposition is shown here for a square matrix $B$ of order $n$. The given matrix $B$ is decomposed as $B = LU$, where $L = (\ell_{ij})$ is a lower-triangular matrix with diagonal entries equal to 1, *i.e.*, $\ell_{ii} = 1$, $\ell_{ij} = 0$ $(i < j)$, and $U = (u_{ij})$ is an upper-triangular matrix with $u_{ij} = 0$ $(i > j)$. Initially we set $B^{(1)} = B$, and compute $B^{(2)}, \dots, B^{(n)}$ as shown below, to finally obtain $\ell_{ik}$ $(i > k)$ and $u_{kj}$ $(k \leqq j)$.

---

**for** $k := 1$ **to** $n - 1$ **do**
  **begin**
    **for** $j := k$ **to** $n$ **do** $u_{kj} := b_{kj}^{(k)}$;
    **for** $i := k + 1$ **to** $n$ **do**
      **begin**
        $\ell_{ik} := b_{ik}^{(k)} / u_{kk}$;
        **for** $j := k + 1$ **to** $n$ **do** $b_{ij}^{(k+1)} := b_{ij}^{(k)} - \ell_{ik} u_{kj}$
      **end**
  **end**

---

See, *e.g.*, [3, 15, 30] for more details about the LU decomposition. ■

[10] See, *e.g.*, [15, Sec. 2.15.3], [3, Sec. 3.2], or [30, Theorem 2.3].

**Remark 2.10.** A symmetric matrix that is an M-matrix is necessarily positive-definite. This follows from (f) in Theorem 2.6. ■

## 2.5 Doubly Stochastic Matrices

### 2.5.1 *Definition*

A nonnegative matrix with each row sum and each column sum being equal to 1 is called a *doubly stochastic matrix*. A doubly stochastic matrix is a square matrix.[11] That is, an $n \times n$ matrix $A = (a_{ij})$ is said to be a doubly stochastic matrix if

$$\sum_{j=1}^{n} a_{ij} = 1 \qquad (i = 1, \ldots, n), \tag{2.52}$$

$$\sum_{i=1}^{n} a_{ij} = 1 \qquad (j = 1, \ldots, n), \tag{2.53}$$

$$a_{ij} \geqq 0 \qquad (i, j = 1, \ldots, n). \tag{2.54}$$

The first equation (2.52) can be written as $A\mathbf{1} = \mathbf{1}$, and the second equation (2.53) as $\mathbf{1}^\top A = \mathbf{1}^\top$.

For example, the matrix

$$A = \begin{bmatrix} 0.7 & 0.3 & 0 \\ 0 & 0.2 & 0.8 \\ 0.3 & 0.5 & 0.2 \end{bmatrix} \tag{2.55}$$

is a doubly stochastic matrix. A symmetric stochastic matrix is a doubly stochastic matrix. A *permutation matrix* (a square matrix with entries of 0 or 1, having exactly one 1 in each row and each column) is also a doubly stochastic matrix.

### 2.5.2 *Birkhoff's Theorem*

Consider a weighted average of permutation matrices $P_1, P_2, \ldots, P_k$, *i.e.*,

$$A = \sum_{i=1}^{k} w_i P_i, \tag{2.56}$$

---

[11]This follows from the identities: $\sum_i \left( \sum_j a_{ij} \right) = \sum_i 1 =$ (number of rows of $A$), and $\sum_j \left( \sum_i a_{ij} \right) = \sum_j 1 =$ (number of columns of $A$).

where the weighting should satisfy

$$w_i \geqq 0 \quad (i = 1, \ldots, k); \qquad \sum_{i=1}^{k} w_i = 1. \tag{2.57}$$

(A linear combination with weights $w_i$ satisfying (2.57) is called a *convex combination.*) A matrix $A$ represented in this way is a doubly stochastic matrix. Indeed, by (2.56) we have

$$A\mathbf{1} = \left( \sum_{i=1}^{k} w_i P_i \right) \mathbf{1} = \sum_{i=1}^{k} w_i (P_i \mathbf{1}) = \sum_{i=1}^{k} w_i \mathbf{1} = \mathbf{1},$$

and similarly, we have $\mathbf{1}^\top A = \mathbf{1}^\top$.

It is known that the converse of this statement is also true: every doubly stochastic matrix can be presented as a convex combination of permutation matrices. Before stating this fact as a theorem, let us look at an example.

**Example 2.7.** The doubly stochastic matrix $A$ in (2.55) is expressed as a convex combination of permutation matrices as

$$\begin{bmatrix} 0.7 & 0.3 & 0 \\ 0 & 0.2 & 0.8 \\ 0.3 & 0.5 & 0.2 \end{bmatrix} = 0.2 \begin{bmatrix} 1 & 0 & 0 \\ 0 & 1 & 0 \\ 0 & 0 & 1 \end{bmatrix} + 0.3 \begin{bmatrix} 0 & 1 & 0 \\ 0 & 0 & 1 \\ 1 & 0 & 0 \end{bmatrix} + 0.5 \begin{bmatrix} 1 & 0 & 0 \\ 0 & 0 & 1 \\ 0 & 1 & 0 \end{bmatrix}.$$

■

The following theorem is called *Birkhoff's theorem.*[12]

**Theorem 2.7 (Birkhoff's theorem).** *Every doubly stochastic matrix can be expressed as a convex combination of permutation matrices.*

**Proof.** Using Hall's theorem, stated in Remark 1.6 (Sec. 1.2.2), we will prove the claim by induction on the number of nonzero entries in a doubly stochastic matrix $A$. For a subset $X$ of the row set of matrix $A$, we define

$$\Gamma(X) = \{j \mid \exists i \in X : a_{ij} \neq 0\}.$$

For any $X$, the number $|\Gamma(X)|$ of elements of $\Gamma(X)$ is greater than or equal to the number $|X|$ of elements of $X$, since

$$\begin{aligned} |\Gamma(X)| &= \sum_{j \in \Gamma(X)} \sum_{i=1}^{n} a_{ij} \\ &\geqq \sum_{j \in \Gamma(X)} \sum_{i \in X} a_{ij} = \sum_{i \in X} \sum_{j \in \Gamma(X)} a_{ij} = \sum_{i \in X} \sum_{j=1}^{n} a_{ij} = |X|, \end{aligned}$$

[12]This is also called the *Birkhoff–von Neumann theorem.*

where the first equality follows from the fact that the column sum is 1, and the last equality follows from the fact that the row sum is 1. Then Hall's theorem guarantees the existence of a permutation $\sigma$ such that $a_{i\sigma(i)} \neq 0$ for $i = 1, \ldots, n$. Denote the corresponding permutation matrix by $P_1$, and let $w_1$ denote the minimum of $a_{i\sigma(i)}$ taken over $i = 1, \ldots, n$.

If $w_1 = 1$, then $A = P_1$, and this completes the proof. If $w_1 < 1$, then the matrix $A'$ defined by

$$A' = \frac{1}{1 - w_1}(A - w_1 P_1)$$

is a doubly stochastic matrix with strictly fewer nonzero entries than $A$. By the induction hypothesis, $A'$ can be expressed as a convex combination of permutation matrices, say, $P_2, \ldots, P_k$ as

$$A' = \sum_{i=2}^{k} w_i' P_i$$

with $w_i' \geqq 0$ $(i = 2, \ldots, k)$ and $\sum_{i=2}^{k} w_i' = 1$. By defining $w_i = (1 - w_1)w_i'$ $(i = 2, \ldots, k)$, we obtain

$$A = w_1 P_1 + (1 - w_1)A' = w_1 P_1 + (1 - w_1)\sum_{i=2}^{k} w_i' P_i = \sum_{i=1}^{k} w_i P_i,$$

$$\sum_{i=1}^{k} w_i = w_1 + (1 - w_1)\sum_{i=2}^{k} w_i' = w_1 + (1 - w_1) = 1.$$

This establishes (2.56) and (2.57). □

The set of all matrices $(a_{ij})$ that satisfy the conditions (2.52), (2.53), and (2.54) forms a polyhedron in the $n^2$-dimensional space $\mathbb{R}^{n^2}$. Denote this polyhedron by $\mathcal{D}$. Theorem 2.7 can be restated as follows,[13] which fact is used extensively in combinatorial optimization [27, 28].

**Theorem 2.8.** *Every extreme point of the polyhedron $\mathcal{D}$ formed by doubly stochastic matrices corresponds to a permutation matrix.*

**Proof.** We will give two proofs.

(i) Let $A$ be an extreme point of the polyhedron $\mathcal{D}$. By Theorem 2.7 we have the expression in (2.56). Since $A$ is an extreme point, there exists only one $i$ with $w_i > 0$. Then $A$ itself is a permutation matrix.

[13] An *extreme point* of a polyhedron $\mathcal{D}$ means a point $\boldsymbol{x}$ belonging to $\mathcal{D}$ that cannot be expressed as $\boldsymbol{x} = (\boldsymbol{y} + \boldsymbol{z})/2$ using two distinct points $\boldsymbol{y}, \boldsymbol{z} \in \mathcal{D}$ (see Sec. 3.4).

(ii) As the second proof, we will show a proof relying on fundamental facts about polyhedra (Chap. 3) and integer matrices (Chap. 4), and not on Theorem 2.7 (which we proved on the basis of Hall's theorem). With the use of the $n^2$-dimensional vector $\boldsymbol{x}$ consisting of the matrix entries $a_{ij}$ arranged in a suitable order, the conditions of (2.52) and (2.53) together can be written in the form of $B\boldsymbol{x} = \mathbf{1}$. The coefficient matrix $B$ here is a $2n \times n^2$ matrix having rank $2n - 1$. In the case of $n = 3$, for example, the matrix $B$ is given by

$$B = \left[\begin{array}{ccc|ccc|ccc} \underline{1} & 1 & 1 & 0 & 0 & 0 & 0 & 0 & 0 \\ 0 & 0 & 0 & \underline{1} & 1 & 1 & 0 & 0 & 0 \\ 0 & 0 & 0 & 0 & 0 & 0 & 1 & 1 & 1 \\ \hline 1 & 0 & 0 & 1 & 0 & 0 & \underline{1} & 0 & 0 \\ 0 & 1 & 0 & 0 & 1 & 0 & 0 & \underline{1} & 0 \\ 0 & 0 & 1 & 0 & 0 & 1 & 0 & 0 & \underline{1} \end{array}\right].$$

The submatrix of $B$ consisting of the rows and columns containing the underlined entries is a nonsingular matrix of order $2n - 1 = 5$.

Each extreme point $\boldsymbol{x}$ of the polyhedron $\mathcal{D}$ is constructed as follows. Choose a nonsingular submatrix $C$ of $B$ of order $2n-1$, determine a $(2n-1)$-dimensional vector $\boldsymbol{y}$ from $C\boldsymbol{y} = \mathbf{1}$, and then form an $n^2$-dimensional vector $\boldsymbol{x}$ by augmenting $\boldsymbol{y}$ with zero components. By Remark 4.7 in Sec. 4.6.2, we have $\det C \in \{1, -1\}$, from which it follows that $\boldsymbol{y}$ is an integer vector. Furthermore, since $0 \leqq a_{ij} \leqq 1$, the components of $\boldsymbol{y}$ are either 0 or 1. A doubly stochastic matrix with $a_{ij} \in \{0, 1\}$ for all $(i, j)$ is a permutation matrix. Thus we have shown that an extreme point of $\mathcal{D}$ corresponds to a permutation matrix. $\square$

Chapter 3

# Systems of Linear Inequalities

In this chapter we deal with systems of linear inequalities and the polyhedra described by them. The Fourier–Motzkin elimination method is employed to prove Farkas' lemma, which is used as the pivotal tool for deriving major results, including theorems of the alternative and the structure of solutions to systems of linear inequalities. These results form the mathematical foundation of linear programming for optimization.

## 3.1 Forms of Linear Inequalities

In engineering, inequalities are often employed to express various kinds of constraints. Optimal designs in engineering usually aim at maximizing some indices (objective functions to represent performance, profit, *etc.*) subject to those constraints. Formulation of the constraints and objective functions in terms of linear functions results in an *optimization problem* such as:

$$\text{Maximize} \quad \sum_{j=1}^{n} c_j x_j \quad \text{subject to} \quad \sum_{j=1}^{n} a_{ij} x_j \leqq b_i \quad (i = 1, \ldots, m).$$

In general, an optimization problem of maximizing or minimizing a linear objective function subject to constraints described by linear equations and/or inequalities is called a *linear program.* In this chapter we investigate the mathematical structures of systems of linear inequalities with a view to clarifying the mathematical structure of linear programs.

Three types can be distinguished among inequalities and equations expressed by linear functions:

$$\sum_j a_{ij}x_j \leqq b_i, \tag{3.1}$$

$$\sum_j a_{ij}x_j \geqq b_i, \tag{3.2}$$

$$\sum_j a_{ij}x_j = b_i. \tag{3.3}$$

Three types of sign constraints for variables can also be distinguished:

$$x_j \leqq 0, \tag{3.4}$$

$$x_j \geqq 0, \tag{3.5}$$

$$x_j \in \mathbb{R} \qquad \text{(no sign constraint)}. \tag{3.6}$$

(In the context of optimization, inequalities that allow equality are used almost always rather than strict inequalities.) The same constraint can be described in several different forms. For example, we can change the direction of an inequality by swapping the left- and right-hand sides, or by changing the signs of the variables.

For theoretical treatment, however, it is convenient to fix a convention for describing systems of inequalities. Therefore, in this chapter, we mainly use the following three forms:

$$A\boldsymbol{x} \leqq \boldsymbol{b}, \tag{3.7}$$

$$A\boldsymbol{x} \geqq \boldsymbol{b}, \tag{3.8}$$

$$A\boldsymbol{x} = \boldsymbol{b}, \qquad \boldsymbol{x} \geqq \boldsymbol{0}. \tag{3.9}$$

Here, inequalities between vectors mean inequalities between the corresponding components. For example, $A\boldsymbol{x} \leqq \boldsymbol{b}$ means inequalities $\sum_j a_{ij}x_j \leqq b_i$ for all $i$.

The three forms (3.7), (3.8), and (3.9) above are equivalent in terms of descriptive power. This is verified as follows. First, the equivalence

$$A\boldsymbol{x} \leqq \boldsymbol{b} \iff -A\boldsymbol{x} \geqq -\boldsymbol{b}$$

allows us to rewrite (3.7) in the form of (3.8). The converse is also obvious, *i.e.*, we can rewrite (3.8) in the form of (3.7). In order to put (3.9) in the form of (3.7), we can use the relation:

$$A\boldsymbol{x} = \boldsymbol{b},\ \boldsymbol{x} \geqq \boldsymbol{0} \iff \begin{bmatrix} A \\ -A \\ -I \end{bmatrix} \boldsymbol{x} \leqq \begin{bmatrix} \boldsymbol{b} \\ -\boldsymbol{b} \\ \boldsymbol{0} \end{bmatrix}.$$

The right-hand side of the above is in the form of (3.7). Conversely, to put (3.7) in the form of (3.9), we first introduce a new variable $\boldsymbol{s}$ to observe:

$$A\boldsymbol{x} \leqq \boldsymbol{b} \iff A\boldsymbol{x} + \boldsymbol{s} = \boldsymbol{b}, \quad \boldsymbol{s} \geqq \boldsymbol{0}.$$

We also express the variable $\boldsymbol{x}$, which is free from sign constraint, as $\boldsymbol{x} = \boldsymbol{y} - \boldsymbol{z}$ using new variables $\boldsymbol{y}$ and $\boldsymbol{z}$ with sign constraints $\boldsymbol{y} \geqq \boldsymbol{0}$ and $\boldsymbol{z} \geqq \boldsymbol{0}$. Then we obtain

$$A\boldsymbol{x} \leqq \boldsymbol{b} \iff \begin{bmatrix} A & -A & I \end{bmatrix} \begin{bmatrix} \boldsymbol{y} \\ \boldsymbol{z} \\ \boldsymbol{s} \end{bmatrix} = \boldsymbol{b}, \quad \begin{bmatrix} \boldsymbol{y} \\ \boldsymbol{z} \\ \boldsymbol{s} \end{bmatrix} \geqq \boldsymbol{0},$$

the right-hand side of which is in the form of (3.9). The precise meaning of "$\iff$" above is as follows. For every $(\boldsymbol{y}, \boldsymbol{z}, \boldsymbol{s})$ satisfying the condition on the right-hand side, $\boldsymbol{x} = \boldsymbol{y} - \boldsymbol{z}$ satisfies the condition $A\boldsymbol{x} \leqq \boldsymbol{b}$ on the left-hand side, and conversely, for every $\boldsymbol{x}$ satisfying the condition on the left-hand side, there exists a triple $(\boldsymbol{y}, \boldsymbol{z}, \boldsymbol{s})$ that satisfies the condition on the right-hand side as well as the relation $\boldsymbol{x} = \boldsymbol{y} - \boldsymbol{z}$.

## 3.2 Fourier–Motzkin Elimination Method

For systems of equations, the elimination of variables through a combination of two equations (*elimination method*) is extremely important in both theoretical analysis and numerical computation. Specifically, Gaussian elimination is an established and practiced method for systems of linear equations. Also for systems of linear inequalities, the idea of eliminating variables through combinations of inequalities is a natural idea, and a systematic method for this is the *Fourier–Motzkin elimination method.* It should be mentioned, however, that the significance of the Fourier–Motzkin elimination method is limited to being a theoretical tool for the analysis of systems of inequalities. As an algorithm for computations, it is very inefficient and hardly used in practice. For actual numerical computations, the simplex method for linear programming (Sec. 3.5) and its variants are employed instead.

Suppose that a system of inequalities is given in the form of

$$A\boldsymbol{x} \leqq \boldsymbol{b}.$$

We assume that $A$ is an $m \times n$ matrix and $\boldsymbol{b}$ is an $m$-dimensional vector. The componentwise expression of the above system of inequalities is given by

$$\sum_{j=1}^{n} a_{ij} x_j \leqq b_i \qquad (i = 1, \ldots, m). \tag{3.10}$$

To eliminate the first variable $x_1$, we classify the inequalities by the sign of the coefficient $a_{i1}$ of $x_1$. To avoid complication in notation, we assume that the first $m_1$ are positive ($a_{11} > 0, \ldots, a_{m_1 1} > 0$), the next $m_2$ are negative ($a_{m_1+1,1} < 0, \ldots, a_{m_1+m_2,1} < 0$), and the remaining $m_0 = m - m_1 - m_2$ are 0 ($a_{m_1+m_2+1,1} = 0, \ldots, a_{m1} = 0$). In addition, we may assume that the positive coefficients are 1 and the negative coefficients are $-1$, since inequalities can be normalized through division by positive numbers. Then, the system (3.10) of inequalities is as follows:

$$x_1 + a_{i2}x_2 + \cdots + a_{in}x_n \leqq b_i \quad (i = 1, \ldots, m_1), \tag{3.11}$$

$$-x_1 + a_{k2}x_2 + \cdots + a_{kn}x_n \leqq b_k \quad (k = m_1 + 1, \ldots, m_1 + m_2), \tag{3.12}$$

$$a_{l2}x_2 + \cdots + a_{ln}x_n \leqq b_l \quad (l = m_1 + m_2 + 1, \ldots, m). \tag{3.13}$$

The $m_1$ inequalities in (3.11) are equivalent to

$$x_1 \leqq \min_{1 \leqq i \leqq m_1} (b_i - (a_{i2}x_2 + \cdots + a_{in}x_n)).$$

The $m_2$ inequalities in (3.12) are equivalent to

$$\max_{m_1+1 \leqq k \leqq m_1+m_2} (a_{k2}x_2 + \cdots + a_{kn}x_n - b_k) \leqq x_1.$$

Combining both together we obtain

$$\max_{m_1+1 \leqq k \leqq m_1+m_2} \left( \sum_{j=2}^{n} a_{kj}x_j - b_k \right) \leqq x_1 \leqq \min_{1 \leqq i \leqq m_1} \left( b_i - \sum_{j=2}^{n} a_{ij}x_j \right), \tag{3.14}$$

from which we can derive

$$\max_{m_1+1 \leqq k \leqq m_1+m_2} \left( \sum_{j=2}^{n} a_{kj}x_j - b_k \right) \leqq \min_{1 \leqq i \leqq m_1} \left( b_i - \sum_{j=2}^{n} a_{ij}x_j \right).$$

This condition is satisfied if and only if

$$\sum_{j=2}^{n} a_{kj}x_j - b_k \leqq b_i - \sum_{j=2}^{n} a_{ij}x_j \tag{3.15}$$

holds for all $i = 1, \ldots, m_1$ and $k = m_1 + 1, \ldots, m_1 + m_2$. The condition (3.15) can be rewritten as

$$\sum_{j=2}^{n} (a_{ij} + a_{kj})x_j \leqq b_i + b_k \quad (i = 1, \ldots, m_1;\ k = m_1 + 1, \ldots, m_1 + m_2). \tag{3.16}$$

This system of inequalities coincides with the set of inequalities generated by adding the $i$th inequality of (3.11) and the $k$th inequality of (3.12) for all $(i, k)$.

Through the above transformation, the original system (3.10) of inequalities in $(x_1, x_2, \ldots, x_n)$ consisting of (3.11), (3.12), and (3.13) has been decomposed into two subsystems, one system of inequalities in $(x_2, \ldots, x_n)$ consisting of (3.16) and (3.13), and the other system (3.14) involving $x_1$. This decomposition provides a necessary and sufficient condition for $(x_1, x_2, \ldots, x_n)$ to satisfy (3.10). Indeed, if $(x_1, x_2, \ldots, x_n)$ satisfies (3.10), then it obviously satisfies the two subsystems, and conversely, if $x_1$ satisfies the inequality (3.14) for $(x_2, \ldots, x_n)$ satisfying the inequalities (3.16) and (3.13), then $(x_1, x_2, \ldots, x_n)$ satisfies all the given inequalities (3.11), (3.12), and (3.13), and hence (3.10).

Next we apply the same procedure to the system of inequalities of (3.16) and (3.13) in $(x_2, \ldots, x_n)$ to eliminate variable $x_2$ and derive a system of inequalities in $(x_3, \ldots, x_n)$. By repeating this process of eliminating variables, we eventually arrive at a system of inequalities in just one variable $x_n$. Then it is easy to find $x_n$ that satisfies this system of inequalities (or, confirm the nonexistence of such $x_n$). This is the idea of the Fourier–Motzkin elimination method.

**Remark 3.1.** Let us consider how the number of inequalities changes during the process of eliminating variables. Combinations of $m_1$ inequalities in (3.11) and $m_2$ inequalities in (3.12) generate $m_1 m_2$ inequalities in (3.16). Therefore, the number of inequalities changes from $m = m_1 + m_2 + m_0$ to $m_1 m_2 + m_0$. This is the change that occurs in the elimination of just one variable. Therefore, the number of inequalities rapidly increases as we repeat the process of eliminating variables. It is important to be aware of this possible explosion in the number of inequalities when using the Fourier–Motzkin elimination method. ∎

In geometric terms, the elimination of a variable corresponds to projection along a coordinate axis. Let $S$ ($\subseteqq \mathbb{R}^n$) denote the set (domain) consisting of points $(x_1, x_2, \ldots, x_n)$ satisfying the given inequality conditions in (3.11), (3.12), and (3.13). The system of inequalities (3.16) and (3.13) resulting from the elimination procedure for $x_1$ describes the *projection* of $S$ along the $x_1$-axis, which is defined by

$$\hat{S} = \{(x_2, \ldots, x_n) \mid (x_1, x_2, \ldots, x_n) \in S \text{ for some } x_1\}.$$

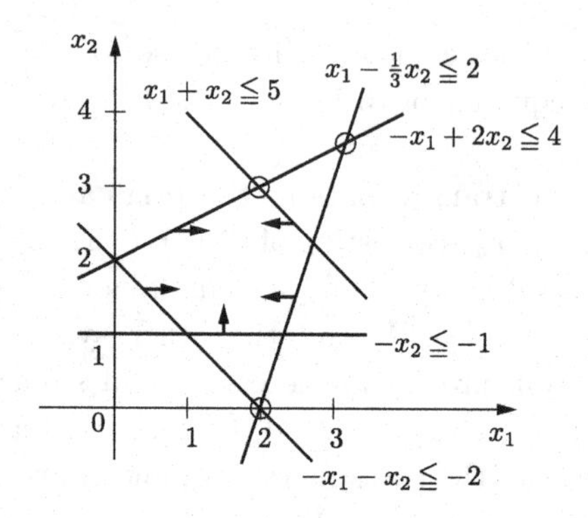


Fig. 3.1 An example for the Fourier–Motzkin elimination method.

**Example 3.1.** We apply the Fourier–Motzkin elimination method to the following system of inequalities in two variables $(x_1, x_2)$:

$$x_1 - \frac{1}{3}x_2 \leqq 2, \tag{3.17}$$

$$x_1 + x_2 \leqq 5, \tag{3.18}$$

$$-x_1 - x_2 \leqq -2, \tag{3.19}$$

$$-x_1 + 2x_2 \leqq 4, \tag{3.20}$$

$$-x_2 \leqq -1. \tag{3.21}$$

The set $S$ described by this system of inequalities is a polygonal domain shown in Fig. 3.1. As we can see from this figure, this system of inequalities admits a solution $(x_1, x_2)$. Let us derive this fact using the Fourier–Motzkin elimination method.

There are five inequalities in total; the coefficients of $x_1$ are positive for two of them, and negative for two of them ($m = 5$, $m_1 = m_2 = 2$). The inequality in (3.14) is given as

$$\max(2 - x_2, -4 + 2x_2) \leqq x_1 \leqq \min\left(2 + \frac{1}{3}x_2, 5 - x_2\right) \tag{3.22}$$

and the system (3.16) of inequalities resulting from the elimination of $x_1$ is

$$\left(-\frac{1}{3} - 1\right)x_2 \leqq (2 - 2), \qquad \left(-\frac{1}{3} + 2\right)x_2 \leqq (2 + 4),$$

$$(1 - 1)\,x_2 \leqq (5 - 2), \qquad (1 + 2)\,x_2 \leqq (5 + 4),$$

*i.e.*,

$$x_2 \geqq 0, \quad x_2 \leqq \frac{18}{5}, \quad \text{free } (-\infty < x_2 < +\infty), \quad x_2 \leqq 3. \tag{3.23}$$

Note the correspondence of these values to the $x_2$-coordinates of the points indicated by ○ in Fig. 3.1. Then, from (3.23) and (3.21) we obtain

$$1 = \max(1,0) \leqq x_2 \leqq \min\left(\frac{18}{5}, 3\right) = 3. \tag{3.24}$$

Therefore, the projection of $S$ is given by $\hat{S} = \{x_2 \mid 1 \leqq x_2 \leqq 3\}$. Since there exists $x_2$ that satisfies the reduced system of inequalities (3.24), a solution $(x_1, x_2)$ to the given inequalities also exists. For example, if $x_2 = 2$ is chosen, then (3.22) reads as follows:

$$0 = \max(0,0) \leqq x_1 \leqq \min\left(\frac{8}{3}, 3\right) = \frac{8}{3}.$$

So we can take $x_1 = 1$, for example. In this way, we can conclude that $(x_1, x_2) = (1, 2)$ satisfies the inequality conditions (3.17) to (3.21). ■

## 3.3 Existence of Solutions to Systems of Linear Inequalities

In this section we consider conditions for a system of linear inequalities $A\boldsymbol{x} \leqq \boldsymbol{b}$ to have a solution $\boldsymbol{x}$. It is assumed that $A$ is an $m \times n$ matrix and $\boldsymbol{b}$ is an $m$-dimensional vector.

For the sake of comparison, let us recall a theorem for a system of equations (with equalities, not inequalities) [6, Theorem 5.2].

**Theorem 3.1.** *The following two conditions* (a) *and* (b) *are equivalent for a matrix $A$ and a vector $\boldsymbol{b}$:*

(a) *There exists a vector $\boldsymbol{x}$ that satisfies $A\boldsymbol{x} = \boldsymbol{b}$.*
(b) *If $\boldsymbol{y}^\top A = \boldsymbol{0}^\top$, then $\boldsymbol{y}^\top \boldsymbol{b} = 0$.*

This pattern of Theorem 3.1 will be followed by the theorems on systems of inequalities to be presented in this section, although the specific conditions are naturally different for equations and inequalities. In this context, we mention that the condition (a) above can be rephrased in geometric terms as condition (c) below:

(c) $\boldsymbol{b}$ belongs to $\mathrm{Im}(A)$, where $\mathrm{Im}(A)$ denotes the subspace spanned by the column vectors of $A$.

### 3.3.1 *Farkas' Lemma*

In this section we discuss Farkas' lemma and related theorems, which provide the conditions for a system of linear inequalities to admit solutions.

Farkas' lemma is the most fundamental fact in the theory of linear inequalities, and will be applied to the structural analysis of solutions to a system of inequalities (Sec. 3.4) and to linear programming duality (Sec. 3.5.3).

First of all, we consider a system of inequalities of the form of $A\boldsymbol{x} \leqq \boldsymbol{b}$. Note the similarity of Theorem 3.2 below to Theorem 3.1 above.

**Theorem 3.2.** *The following two conditions* (a) *and* (b) *are equivalent for a matrix $A$ and a vector $\boldsymbol{b}$:*

(a) *There exists a vector $\boldsymbol{x}$ that satisfies $A\boldsymbol{x} \leqq \boldsymbol{b}$.*
(b) *If $\boldsymbol{y} \geqq \boldsymbol{0}$ and $\boldsymbol{y}^\top A = \boldsymbol{0}^\top$, then $\boldsymbol{y}^\top \boldsymbol{b} \geqq 0$.*

**Proof.** [(a) $\Rightarrow$ (b)]: If $A\boldsymbol{x} \leqq \boldsymbol{b}$, $\boldsymbol{y} \geqq \boldsymbol{0}$, $\boldsymbol{y}^\top A = \boldsymbol{0}^\top$, then $\boldsymbol{y}^\top \boldsymbol{b} \geqq \boldsymbol{y}^\top (A\boldsymbol{x}) = (\boldsymbol{y}^\top A)\boldsymbol{x} = \boldsymbol{0}^\top \boldsymbol{x} = 0$.

[(b) $\Rightarrow$ (a)]: This implication is equivalent to the following statement:

($*$) If no $\boldsymbol{x}$ exists that satisfies $A\boldsymbol{x} \leqq \boldsymbol{b}$, then there exists $\boldsymbol{y} \geqq \boldsymbol{0}$ that satisfies $\boldsymbol{y}^\top [A \mid \boldsymbol{b}] = [\boldsymbol{0}^\top \mid -1]$.

We prove the statement ($*$) by induction on the number $n$ of the columns of matrix $A$. We make use of the Fourier–Motzkin elimination method.

Assume that the given system $A\boldsymbol{x} \leqq \boldsymbol{b}$ is in the form of (3.11), (3.12), and (3.13). We continue to use the notation of Sec. 3.2.

When $n = 1$, if there is no $\boldsymbol{x}$ that satisfies $A\boldsymbol{x} \leqq \boldsymbol{b}$, then at least one of the following is true:

(i) there exist $i_0$ and $k_0$ such that $b_{i_0} + b_{k_0} < 0$ $(1 \leqq i_0 \leqq m_1, m_1 + 1 \leqq k_0 \leqq m_1 + m_2)$,
(ii) there exists $l_0$ such that $b_{l_0} < 0$ $(m_1 + m_2 + 1 \leqq l_0 \leqq m)$.

In case (i), define $\boldsymbol{y}$ by setting its $i_0$th and $k_0$th components to $1/|b_{i_0} + b_{k_0}|$ and the other components to 0. In case (ii), define $\boldsymbol{y}$ by setting its $l_0$th component to $1/|b_{l_0}|$ and the other components to 0. Then, in either case, we have $\boldsymbol{y} \geqq \boldsymbol{0}$ and $\boldsymbol{y}^\top [A \mid \boldsymbol{b}] = [\boldsymbol{0}^\top \mid -1]$.

When $n \geqq 2$, let

$$\tilde{A}\tilde{\boldsymbol{x}} \leqq \tilde{\boldsymbol{b}}$$

denote the system of inequalities, consisting of (3.16) and (3.13), that results from the elimination of $x_1$ by the Fourier–Motzkin elimination method. Here $\tilde{\boldsymbol{x}} = (x_2, \ldots, x_n)^\top$, $\tilde{A}$ is an $(m_1 m_2 + m_0) \times (n-1)$ matrix, and $\tilde{\boldsymbol{b}}$ is an $(m_1 m_2 + m_0)$-dimensional vector. The rows of $\tilde{A}$ are specified by $(i,k)$ or $l$ $(1 \leqq i \leqq m_1,\ m_1 + 1 \leqq k \leqq m_1 + m_2,\ m_1 + m_2 + 1 \leqq l \leqq m)$.

If there exists no $\boldsymbol{x}$ that satisfies $A\boldsymbol{x} \leqq \boldsymbol{b}$, then there exists no $\tilde{\boldsymbol{x}}$ that satisfies $\tilde{A}\tilde{\boldsymbol{x}} \leqq \tilde{\boldsymbol{b}}$. By the induction hypothesis, there exists an $(m_1m_2+m_0)$-dimensional vector $\tilde{\boldsymbol{y}} \geqq \mathbf{0}$ that satisfies

$$\tilde{\boldsymbol{y}}^\top [\tilde{A} \mid \tilde{\boldsymbol{b}}] = [\mathbf{0}^\top \mid -1].$$

By rewriting this using (3.16) we obtain

$$\sum_i \sum_k \tilde{y}_{ik}([a_{i2}, \ldots, a_{in} \mid b_i] + [a_{k2}, \ldots, a_{kn} \mid b_k]) \\ + \sum_l \tilde{y}_l [a_{l2}, \ldots, a_{ln} \mid b_l] = [0, \ldots, 0 \mid -1]. \tag{3.25}$$

Here, $\tilde{y}_{ik}$ and $\tilde{y}_l$ are the components of $\tilde{\boldsymbol{y}}$, and the indices $i$, $k$, and $l$ run in the ranges of $1 \leqq i \leqq m_1$, $m_1+1 \leqq k \leqq m_1+m_2$, and $m_1+m_2+1 \leqq l \leqq m$, respectively.

Next, from the vector $\tilde{\boldsymbol{y}}$ above, we define an $m$-dimensional vector $\boldsymbol{y}$ by

$$y_i = \sum_k \tilde{y}_{ik} \qquad (i = 1, \ldots, m_1),$$

$$y_k = \sum_i \tilde{y}_{ik} \qquad (k = m_1 + 1, \ldots, m_1 + m_2),$$

$$y_l = \tilde{y}_l \qquad (l = m_1 + m_2 + 1, \ldots, m).$$

Then $\boldsymbol{y} \geqq \mathbf{0}$ and

$$\sum_i y_i [1, a_{i2}, \ldots, a_{in} \mid b_i] + \sum_k y_k [-1, a_{k2}, \ldots, a_{kn} \mid b_k] \\ + \sum_l y_l [0, a_{l2}, \ldots, a_{ln} \mid b_l] = [0, 0, \ldots, 0 \mid -1]. \tag{3.26}$$

We note that the zero vector $[0, \ldots, 0]$ on the right-hand side of (3.25) is an $(n-1)$-dimensional vector, whereas the zero vector $[0, 0, \ldots, 0]$ on the right-hand side of (3.26) is an $n$-dimensional vector. It should be clear that the first component of (3.26) becomes 0 as a consequence of

$$\sum_i y_i = \sum_i \sum_k \tilde{y}_{ik} = \sum_k y_k.$$

The equation (3.26) shows

$$\boldsymbol{y}^\top [A \mid \boldsymbol{b}] = [\mathbf{0}^\top \mid -1],$$

which proves the statement $(*)$ for $n$. □

The following theorem is a famous fact, known as *Farkas' lemma.* It provides a condition for the existence of a nonnegative solution $\boldsymbol{x} \geqq \mathbf{0}$ to a system of equations $A\boldsymbol{x} = \boldsymbol{b}$. As can be understood from the proof, this theorem and Theorem 3.2 are almost equivalent. Here we derive Farkas' lemma from Theorem 3.2, but the converse is also straightforward.

**Theorem 3.3 (Farkas' lemma).** *The following two conditions* (a) *and* (b) *are equivalent for a matrix* $A$ *and a vector* $\boldsymbol{b}$*:*

(a) *There exists a nonnegative vector* $\boldsymbol{x}$ *that satisfies* $A\boldsymbol{x} = \boldsymbol{b}$.
(b) *If* $\boldsymbol{y}^\top A \geqq \mathbf{0}^\top$, *then* $\boldsymbol{y}^\top \boldsymbol{b} \geqq 0$.

**Proof.** We rewrite condition (a): $A\boldsymbol{x} = \boldsymbol{b}$, $\boldsymbol{x} \geqq \mathbf{0}$, to

$$\begin{bmatrix} A \\ -A \\ -I \end{bmatrix} \boldsymbol{x} \leqq \begin{bmatrix} \boldsymbol{b} \\ -\boldsymbol{b} \\ \mathbf{0} \end{bmatrix}$$

and apply Theorem 3.2, to see that (a) is equivalent to:
(b′) $\boldsymbol{u}, \boldsymbol{v}, \boldsymbol{w} \geqq \mathbf{0}, \ \boldsymbol{u}^\top A - \boldsymbol{v}^\top A - \boldsymbol{w}^\top = \mathbf{0}^\top \implies \boldsymbol{u}^\top \boldsymbol{b} - \boldsymbol{v}^\top \boldsymbol{b} \geqq 0$.
We will show that this condition (b′) is equivalent to (b).

To show [(b) ⇒ (b′)], take $\boldsymbol{u}$, $\boldsymbol{v}$, and $\boldsymbol{w}$ satisfying the assumption of (b′), *i.e.*, $\boldsymbol{u}, \boldsymbol{v}, \boldsymbol{w} \geqq \mathbf{0}$ and $\boldsymbol{u}^\top A - \boldsymbol{v}^\top A - \boldsymbol{w}^\top = \mathbf{0}^\top$. Put $\boldsymbol{y} = \boldsymbol{u} - \boldsymbol{v}$. Then $\boldsymbol{y}^\top A = \boldsymbol{w}^\top \geqq \mathbf{0}^\top$. By (b) it then follows that $0 \leqq \boldsymbol{y}^\top \boldsymbol{b} = \boldsymbol{u}^\top \boldsymbol{b} - \boldsymbol{v}^\top \boldsymbol{b}$. Thus, (b) implies (b′).

Conversely, to show [(b′) ⇒ (b)], take $\boldsymbol{y}$ satisfying the assumption of (b), *i.e.*, $\boldsymbol{y}^\top A \geqq \mathbf{0}^\top$. Decompose $\boldsymbol{y}$ as $\boldsymbol{y} = \boldsymbol{u} - \boldsymbol{v}$ with $\boldsymbol{u}, \boldsymbol{v} \geqq \mathbf{0}$ and put $\boldsymbol{w} = A^\top \boldsymbol{y}$. Then we have $\boldsymbol{u}^\top A - \boldsymbol{v}^\top A - \boldsymbol{w}^\top = \mathbf{0}^\top$. By (b′) it then follows that $0 \leqq \boldsymbol{u}^\top \boldsymbol{b} - \boldsymbol{v}^\top \boldsymbol{b} = \boldsymbol{y}^\top \boldsymbol{b}$. Thus, (b′) implies (b).

We mention here that the proof for [(a) ⇒ (b)] is quite easy (without relying on Theorem 3.2). Indeed it can be shown as $\boldsymbol{y}^\top \boldsymbol{b} = \boldsymbol{y}^\top (A\boldsymbol{x}) = (\boldsymbol{y}^\top A)\boldsymbol{x} \geqq 0$ by using $A\boldsymbol{x} = \boldsymbol{b}$, $\boldsymbol{x} \geqq \mathbf{0}$ and $\boldsymbol{y}^\top A \geqq \mathbf{0}^\top$. □

The geometric meaning of condition (a) of Theorem 3.3 (Farkas' lemma) is as follows. Consider the set of all nonnegative linear combinations of the column vectors $\boldsymbol{a}_j \in \mathbb{R}^m$ $(j = 1, \ldots, n)$ of matrix $A$, which is denoted as[1]

$$\mathrm{Cone}(A) = \mathrm{Cone}(\boldsymbol{a}_1, \ldots, \boldsymbol{a}_n) = \left\{ \sum_{j=1}^{n} x_j \boldsymbol{a}_j \;\middle|\; x_j \geqq 0 \ (j = 1, \ldots, n) \right\}. \tag{3.27}$$

[1] $\mathrm{Cone}(\boldsymbol{a}_1, \ldots, \boldsymbol{a}_n)$ is referred to as the *convex cone* generated by $\boldsymbol{a}_1, \ldots, \boldsymbol{a}_n$. See Sec. 3.4.2 for more details about convex cones.

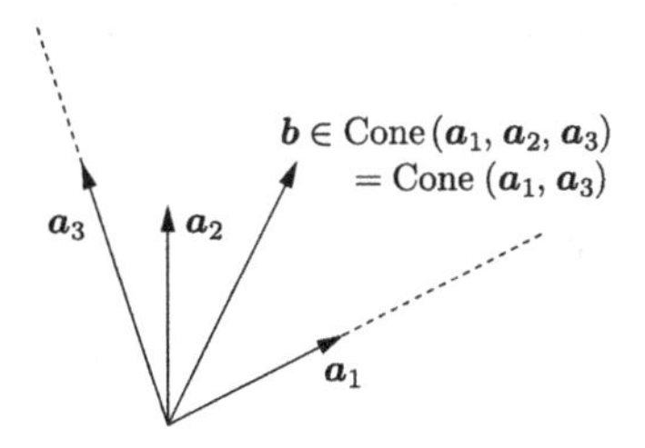


Fig. 3.2 Farkas' lemma: geometric meaning of condition (a).

Then the condition (a) can be written as

$$\boldsymbol{b} \in \mathrm{Cone}(A).$$

(Figure 3.2 shows an example for $m = 2$ and $n = 3$, where $\boldsymbol{b} \in \mathrm{Cone}(\boldsymbol{a}_1, \boldsymbol{a}_2, \boldsymbol{a}_3) = \mathrm{Cone}(\boldsymbol{a}_1, \boldsymbol{a}_3)$.) This alternative expression may be compared to a similar expression in Theorem 3.1 for a system of equations; the condition (a) of Theorem 3.1 can be rephrased to a geometric condition of $\boldsymbol{b} \in \mathrm{Im}(A)$ given as (c). In understanding Farkas' lemma, such geometric images are extremely useful. The geometric meaning of condition (b) of Farkas' lemma will also be explained later in Remark 3.4 in Sec. 3.3.2.

The following theorem gives a condition for the existence of a nonnegative solution $\boldsymbol{x} \geqq \boldsymbol{0}$ to a homogeneous system of equations $A\boldsymbol{x} = \boldsymbol{0}$. Note that the trivial solution $\boldsymbol{x} = \boldsymbol{0}$ is excluded, and that the inequality sign in (b) is not "$\geqq$" but a strict inequality "$>$" without equality.

**Theorem 3.4.** *The following two conditions* (a) *and* (b) *are equivalent for a matrix $A$:*

(a) *There exists a nonzero, nonnegative vector $\boldsymbol{x} \geqq \boldsymbol{0}$ (with $\boldsymbol{x} \neq \boldsymbol{0}$) that satisfies $A\boldsymbol{x} = \boldsymbol{0}$.*
(b) *There exists no vector $\boldsymbol{y}$ that satisfies $\boldsymbol{y}^\top A > \boldsymbol{0}^\top$.*

**Proof.** [(a) $\Rightarrow$ (b)]: Assume (a) and suppose that there exists $\boldsymbol{y}$ satisfying $\boldsymbol{y}^\top A > \boldsymbol{0}^\top$. Then $0 < (\boldsymbol{y}^\top A)\boldsymbol{x} = \boldsymbol{y}^\top(A\boldsymbol{x}) = \boldsymbol{y}^\top\boldsymbol{0} = 0$, which is a contradiction.

[(b) $\Rightarrow$ (a)]: Assume that $A$ is an $m \times n$ matrix, and, for $j = 1, \ldots, n$, denote the $j$th unit vector by $\boldsymbol{e}_j$, which is an $n$-dimensional vector. The condition (a) holds if and only if, for some $j$, there exists $\boldsymbol{x} \geqq \boldsymbol{0}$ satisfying

$$A\boldsymbol{x} = \boldsymbol{0}, \qquad \boldsymbol{e}_j^\top \boldsymbol{x} = 1. \tag{3.28}$$

Suppose that (a) does not hold. This means that, for each $j = 1, \ldots, n$, there exists no $\boldsymbol{x} \geqq \boldsymbol{0}$ that satisfies (3.28). It then follows from Farkas' lemma (Theorem 3.3) that there exist $\boldsymbol{y}_j \in \mathbb{R}^m$ and $z_j \in \mathbb{R}$ satisfying

$$\boldsymbol{y}_j^\top A + z_j \boldsymbol{e}_j^\top \geqq \boldsymbol{0}^\top, \qquad z_j < 0$$

for $j = 1, \ldots, n$. For the vector $\boldsymbol{y} = \boldsymbol{y}_1 + \cdots + \boldsymbol{y}_n$, we have

$$\boldsymbol{y}^\top A \geqq -(z_1 \boldsymbol{e}_1^\top + \cdots + z_n \boldsymbol{e}_n^\top) = (-z_1, \ldots, -z_n) > (0, \ldots, 0).$$

This shows that (b) does not hold. □

**Remark 3.2.** In the argument of stochastic matrices in Sec. 2.3, we have used the Perron–Frobenius theorem (Theorem 2.1) to derive the fact (Theorem 2.4(1)) that an arbitrary Markov chain possesses a stationary distribution $\boldsymbol{\pi}$. This fact can also be derived from Theorem 3.4, as follows.

Let $P = (p_{ij})$ be a stochastic matrix. We want to prove the existence of a nonnegative (nonzero) row vector $\boldsymbol{\pi}$ that satisfies $\boldsymbol{\pi} = \boldsymbol{\pi} P$, where we may ignore the condition of the components of $\boldsymbol{\pi}$ summing to 1. By defining $A = I - P^\top$ and $\boldsymbol{x} = \boldsymbol{\pi}^\top$, we can rewrite $\boldsymbol{\pi} = \boldsymbol{\pi} P$ to $A\boldsymbol{x} = \boldsymbol{0}$. We want to show the existence of $\boldsymbol{x} \geqq \boldsymbol{0}$ (with $\boldsymbol{x} \neq \boldsymbol{0}$) that satisfies $A\boldsymbol{x} = \boldsymbol{0}$, which is the condition (a) in Theorem 3.4.

We prove (b) in Theorem 3.4 by deriving a contradiction from the existence of $\boldsymbol{y}$ satisfying $\boldsymbol{y}^\top A > \boldsymbol{0}^\top$. The inequality $\boldsymbol{y}^\top A > \boldsymbol{0}^\top$ means that $y_i > \sum_j p_{ij} y_j$ holds for all $i$. Let $y_k$ be the minimum among the components of $\boldsymbol{y}$. Since $\sum_j p_{ij} = 1$ and $p_{ij} \geqq 0$, we obtain $\sum_j p_{ij} y_j \geqq y_k$, and hence

$$y_i > \sum_j p_{ij} y_j \geqq y_k$$

for all $i$. But this is a contradiction for $i = k$. Therefore, (b) holds. The equivalence of (a) and (b) in Theorem 3.4 establishes the existence of a stationary distribution $\boldsymbol{\pi}$.

In contrast to the Perron–Frobenius theorem, which relies on the continuity of real numbers,[2] Theorem 3.4 is a purely algebraic theorem. To be more specific, the statement of Theorem 3.4 is still true even when matrices and vectors are considered over the set $\mathbb{Q}$ of rational numbers (rather than over $\mathbb{R}$). ■

[2] Recall that the proof of Proposition 2.4 is based upon the propositions that a bounded closed set in $\mathbb{R}^n$ is a compact set and that a sequence in a compact set contains a convergent subsequence.

**Remark 3.3.** In this book, we have proved Farkas' lemma (or Theorem 3.2, which is a variant thereof) using the Fourier–Motzkin elimination method. There are other ways of proving Farkas' lemma, including an algorithmic proof using the simplex method in linear programming and a geometric proof using the separation theorem for convex sets. See Remark 3.4 (Sec. 3.3.2) for the latter. ■

### 3.3.2 *Theorems of the Alternative*

In general, a theorem stating that exactly one of two possibilities must occur is referred to as a *theorem of the alternative.* We can recast Theorems 3.2, 3.3, and 3.4 into theorems of the alternative as follows.

**Theorem 3.5.** *For any matrix $A$ and vector $\boldsymbol{b}$, exactly one of the following two conditions* (a) *and* ($\overline{\text{b}}$) *holds (and not both):*

(a) *There exists a vector $\boldsymbol{x}$ that satisfies $A\boldsymbol{x} \leqq \boldsymbol{b}$.*
($\overline{\text{b}}$) *There exists a vector $\boldsymbol{y}$ that satisfies $\boldsymbol{y} \geqq \boldsymbol{0}$, $\boldsymbol{y}^\top A = \boldsymbol{0}^\top$, and $\boldsymbol{y}^\top \boldsymbol{b} < 0$.*

**Theorem 3.6.** *For any matrix $A$ and vector $\boldsymbol{b}$, exactly one of the following two conditions* (a) *and* ($\overline{\text{b}}$) *holds (and not both):*

(a) *There exists a nonnegative vector $\boldsymbol{x} \geqq \boldsymbol{0}$ that satisfies $A\boldsymbol{x} = \boldsymbol{b}$.*
($\overline{\text{b}}$) *There exists a vector $\boldsymbol{y}$ that satisfies $\boldsymbol{y}^\top A \geqq \boldsymbol{0}^\top$ and $\boldsymbol{y}^\top \boldsymbol{b} < 0$.*

**Theorem 3.7.** *For any matrix $A$ and vector $\boldsymbol{b}$, exactly one of the following two conditions* (a) *and* ($\overline{\text{b}}$) *holds (and not both):*

(a) *There exists a nonzero, nonnegative vector $\boldsymbol{x} \geqq \boldsymbol{0}$ (with $\boldsymbol{x} \neq \boldsymbol{0}$) that satisfies $A\boldsymbol{x} = \boldsymbol{0}$.*
($\overline{\text{b}}$) *There exists a vector $\boldsymbol{y}$ that satisfies $\boldsymbol{y}^\top A > \boldsymbol{0}^\top$.*

The significance (usage) of the above theorems of the alternative is explained here. Suppose, for example, that a system of inequalities $A\boldsymbol{x} \leqq \boldsymbol{b}$ is given and we want to prove the existence or nonexistence of a solution $\boldsymbol{x}$ that satisfies the inequalities. To prove that there is a solution, we simply need to demonstrate a vector $\boldsymbol{x}$ that satisfies $A\boldsymbol{x} \leqq \boldsymbol{b}$. Putting aside the algorithmic problem of how to find $\boldsymbol{x}$, the principle of proving its existence is rather simple. That is, we only have to show $\boldsymbol{x}$ as an "evidence of the existence." Then, conversely, how can we prove that a solution does not exist (proof of nonexistence)? This is, actually, a very difficult task, but Theorem 3.5 offers a remedy for it. Since exactly one of (a) and ($\overline{\text{b}}$) holds, when no solution $\boldsymbol{x}$ exists, that is, when (a) fails, there exists a vector $\boldsymbol{y}$ satisfying the condition of ($\overline{\text{b}}$). If we can somehow find such a vector $\boldsymbol{y}$, we can use this $\boldsymbol{y}$ as an "evidence of the nonexistence" of $\boldsymbol{x}$. This demonstrates a typical use of a theorem of the alternative.

We note, however, that while Theorem 3.5 is useful for proving the existence or nonexistence of a solution, it does not provide us with a method of computing a solution $\boldsymbol{x}$. To actually find a solution to a system of linear inequalities, we can employ solution methods for linear programming such as the simplex method and the interior-point method [37, 38, 40–43].

**Remark 3.4.** A geometric interpretation of Theorem 3.6 (Farkas' lemma in the form of a theorem of the alternative) is explained here. As we have seen in Sec. 3.3.1, the condition (a) is equivalent to saying $\boldsymbol{b} \in \text{Cone}(A)$; see Fig. 3.3 (a). Here, $\text{Cone}(A) = \text{Cone}(\boldsymbol{a}_1, \ldots, \boldsymbol{a}_n)$ denotes the convex cone generated by the column vectors $\boldsymbol{a}_j$ $(j = 1, \ldots, n)$ of matrix $A$ (see (3.27)). When the condition (a) is not met, the vector $\boldsymbol{b}$ and $\text{Cone}(A)$ can be separated by a hyperplane; see Fig. 3.3 ($\overline{\text{b}}$). Denote by $\boldsymbol{y}$ the normal vector of this hyperplane, with its direction suitably chosen. Then $\boldsymbol{y}^\top \boldsymbol{a}_j \geqq 0$ $(j = 1, \ldots, n)$ and $\boldsymbol{y}^\top \boldsymbol{b} < 0$, satisfying the condition ($\overline{\text{b}}$). This is the geometric meaning of Theorem 3.6. ■

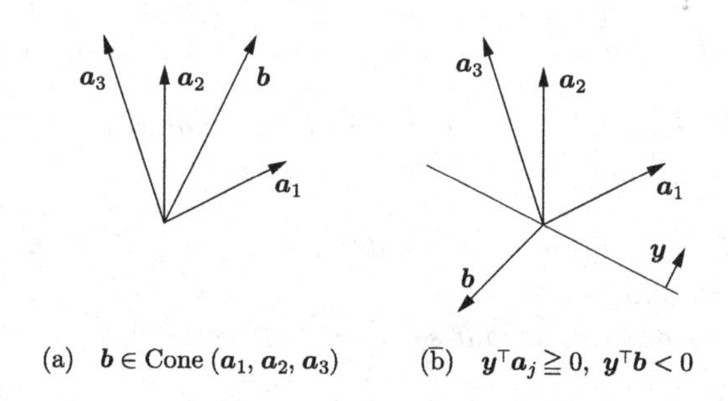


Fig. 3.3 Geometric meaning of Theorem 3.6 (Farkas' lemma in the form of a theorem of the alternative).

**Example 3.2.** For the matrix $A$ and vector $\boldsymbol{b}$ given by

$$A = \begin{bmatrix} 1 & -1 & 0 \\ 0 & 0 & 1 \end{bmatrix}, \qquad \boldsymbol{b} = \begin{bmatrix} 0 \\ -1 \end{bmatrix},$$

the convex cone, $\text{Cone}(A)$, generated by the vectors $\boldsymbol{a}_1 = (1, 0)^\top$, $\boldsymbol{a}_2 = (-1, 0)^\top$, and $\boldsymbol{a}_3 = (0, 1)^\top$ is the entire upper half-plane, and therefore,

$\boldsymbol{b} \notin \text{Cone}(A)$. The vector $\boldsymbol{y}$ in Theorem 3.6 ($\overline{\text{b}}$) is uniquely determined to $\boldsymbol{y} = c(0,1)^\top$ with $c > 0$, for which $\boldsymbol{y}^\top \boldsymbol{a}_1 = 0$, $\boldsymbol{y}^\top \boldsymbol{a}_2 = 0$, $\boldsymbol{y}^\top \boldsymbol{a}_3 = c > 0$, and $\boldsymbol{y}^\top \boldsymbol{b} = -c < 0$. As we can see from this example, we cannot strengthen the condition of Theorem 3.6 ($\overline{\text{b}}$) to "$\boldsymbol{y}^\top A > \boldsymbol{0}^\top$ and $\boldsymbol{y}^\top \boldsymbol{b} < 0$." ■

## 3.4 Structure of Solutions to Systems of Inequalities

### 3.4.1 *Inequalities and Polyhedra*

The set of all solutions to a system of linear inequalities forms a polyhedron. A (bounded) polyhedron can be represented not only as the set of solutions to a system of inequalities, but also as the set of all convex combinations of its extreme points.[3] This fact leads to a parametric representation of solutions to a system of linear inequalities. The objective of this section is to explain, for general unbounded polyhedra, these two representations and their equivalence and, in so doing, to reveal the structure of the set of solutions to a system of linear inequalities.

Before entering into the general argument, we first explain the key point using a simple example.

**Example 3.3.** Consider a system of inequalities[4] for variable $\boldsymbol{x} = (x_1, x_2)^\top$:

$$-x_1 + \frac{1}{3}x_2 \geqq -2, \quad -x_1 - x_2 \geqq -5, \quad x_1 + x_2 \geqq 2, \quad x_1 - 2x_2 \geqq -4, \quad x_2 \geqq 1. \tag{3.29}$$

This system of inequalities gives a description of a polygonal domain $S$, depicted in Fig. 3.4. There are five extreme points (vertices):

$$\boldsymbol{v}_1 = \begin{bmatrix} 0 \\ 2 \end{bmatrix}, \quad \boldsymbol{v}_2 = \begin{bmatrix} 1 \\ 1 \end{bmatrix}, \quad \boldsymbol{v}_3 = \begin{bmatrix} 7/3 \\ 1 \end{bmatrix}, \quad \boldsymbol{v}_4 = \begin{bmatrix} 11/4 \\ 9/4 \end{bmatrix}, \quad \boldsymbol{v}_5 = \begin{bmatrix} 2 \\ 3 \end{bmatrix},$$

as marked by • in the figure. In this case, as readily seen from the figure, a point $\boldsymbol{x}$ belongs to $S$ if and only if it can be represented as a *convex combination*[5] of $\boldsymbol{v}_1, \ldots, \boldsymbol{v}_5$, *i.e.*,

$$\boldsymbol{x} = \alpha_1 \boldsymbol{v}_1 + \alpha_2 \boldsymbol{v}_2 + \alpha_3 \boldsymbol{v}_3 + \alpha_4 \boldsymbol{v}_4 + \alpha_5 \boldsymbol{v}_5 \tag{3.30}$$

using nonnegative coefficients $\alpha_1, \ldots, \alpha_5$ ($\geqq 0$) with $\alpha_1 + \cdots + \alpha_5 = 1$.

---

[3] The meaning of an extreme point should be intuitively clear, but the precise definition is as follows: A point $\boldsymbol{x}$ of a polyhedron $S$ is called an *extreme point* of $S$, if it cannot be expressed as $\boldsymbol{x} = (\boldsymbol{y} + \boldsymbol{z})/2$ using two distinct points $\boldsymbol{y}$ and $\boldsymbol{z}$ in $S$.

[4] The inequality system (3.29) is equivalent to the system of inequalities in Example 3.1 in Sec. 3.2.

[5] See Sec. 3.4.2 for convex combination.

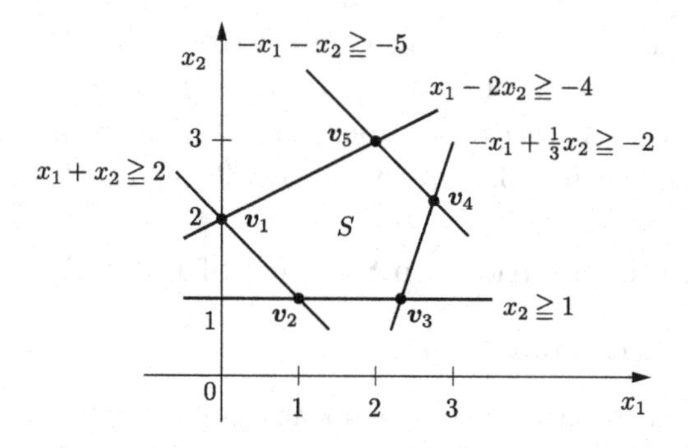


Fig. 3.4 Structure of solutions to a system of linear inequalities (bounded domain).

The equation (3.29) describes $S$ from the "outside", so to speak, by giving the conditions that should be satisfied by the points belonging to $S$. In contrast, the equation (3.30) describes $S$ from the "inside" by giving a parametric representation of the points in domain $S$. ■

The solution set of a system of inequalities sometimes forms an unbounded domain. In such cases, the extreme points alone will not be sufficient for the description of the domain. Let us examine such a case in the next example.

**Example 3.4.** The domain $S$ described by the system of inequalities

$$2x_1 - x_2 \geqq -2, \quad x_1 + x_2 \geqq 2, \quad x_2 \geqq 1, \quad -x_1 + 2x_2 \geqq -1 \tag{3.31}$$

for variable $\boldsymbol{x} = (x_1, x_2)^\top$ is an unbounded domain (extending to infinity toward the upper right) shown in Fig. 3.5. There are three extreme points

$$\boldsymbol{v}_1 = \begin{bmatrix} 0 \\ 2 \end{bmatrix}, \quad \boldsymbol{v}_2 = \begin{bmatrix} 1 \\ 1 \end{bmatrix}, \quad \boldsymbol{v}_3 = \begin{bmatrix} 3 \\ 1 \end{bmatrix},$$

as marked by • in the figure. It is not possible to make a parametric representation of the domain $S$ using only these extreme points $\boldsymbol{v}_1$, $\boldsymbol{v}_2$, and $\boldsymbol{v}_3$. To represent the directions extending to infinity, we introduce two additional vectors

$$\boldsymbol{d}_1 = \begin{bmatrix} 2 \\ 1 \end{bmatrix}, \quad \boldsymbol{d}_2 = \begin{bmatrix} 1 \\ 2 \end{bmatrix},$$

shown by arrows in the figure. Then, a point $\boldsymbol{x}$ belongs to $S$ if and only if it can be represented as the sum of a convex combination of $\boldsymbol{v}_1$, $\boldsymbol{v}_2$, and $\boldsymbol{v}_3$ and a nonnegative combination of $\boldsymbol{d}_1$ and $\boldsymbol{d}_2$, *i.e.*,

$$\boldsymbol{x} = (\alpha_1 \boldsymbol{v}_1 + \alpha_2 \boldsymbol{v}_2 + \alpha_3 \boldsymbol{v}_3) + (\beta_1 \boldsymbol{d}_1 + \beta_2 \boldsymbol{d}_2) \tag{3.32}$$

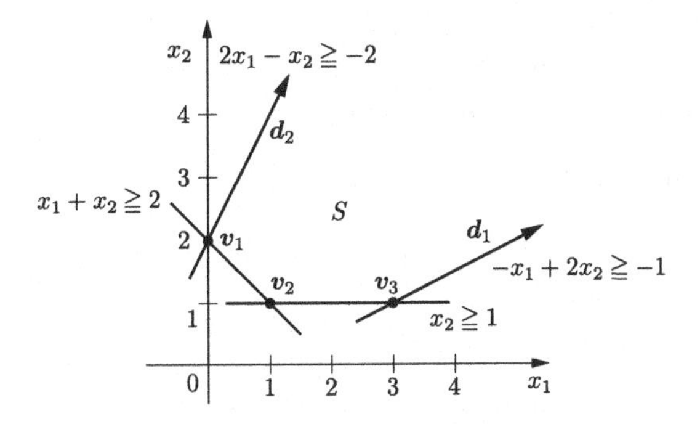


Fig. 3.5 Structure of solutions to a system of linear inequalities (unbounded domain).

using nonnegative coefficients $\alpha_1$, $\alpha_2$, and $\alpha_3$ summing to 1 and nonnegative coefficients $\beta_1$ and $\beta_2$ $(\alpha_1 + \alpha_2 + \alpha_3 = 1; \alpha_1, \alpha_2, \alpha_3 \geqq 0;\ \beta_1, \beta_2 \geqq 0)$.

The equation (3.31) describes $S$ from the "outside" by giving the conditions that should be satisfied by the points belonging to $S$. In contrast, the equation (3.32) describes $S$ from the "inside" by giving a parametric representation of the points in the domain $S$. The expression in (3.32) corresponds to the decomposition of the polygonal domain $S$ into the sum of a bounded polygon (triangle) and a convex cone, as shown in Fig. 3.6. This decomposition can be expressed as

$$S = \mathrm{Conv}(\boldsymbol{v}_1, \boldsymbol{v}_2, \boldsymbol{v}_3) + \mathrm{Cone}(\boldsymbol{d}_1, \boldsymbol{d}_2) \tag{3.33}$$

with the notations to be introduced in (3.36) and (3.38) in Sec. 3.4.2, where "+" on the right-hand side means forming all possible sums of the vectors belonging to the respective sets. ∎

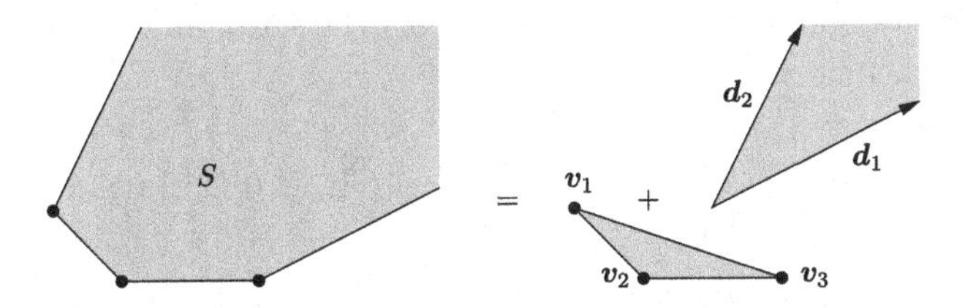


Fig. 3.6 Decomposition of an unbounded domain.

We have observed the following facts in the above two examples:

- If the solution set of a system of linear inequalities is a bounded polyhedron, it admits a parametric representation in terms of convex combinations of the extreme points of that polyhedron.

- If the solution set of a system of linear inequalities is an unbounded polyhedron, a parametric representation of the solution set can be obtained through the decomposition:

$$\text{unbounded polyhedron} = \text{bounded polyhedron} + \text{convex cone.}$$

We discuss these issues in full generality in the following sections.

### 3.4.2 *Convex Cones*

In this section we explain fundamental properties of convex cones, as preliminaries for investigating the structure of solution sets of systems of inequalities.

#### 3.4.2.1 *Definition of convex cones*

A set $S$ ($\subseteqq \mathbb{R}^n$) is called a *convex set* if the line segment connecting two arbitrary points in $S$ are included in $S$. That is, $S$ is a convex set if the following condition holds:

$$\boldsymbol{x}, \boldsymbol{y} \in S, \ \ 0 \leqq \alpha \leqq 1 \ \ \Longrightarrow \ \ \alpha \boldsymbol{x} + (1-\alpha)\boldsymbol{y} \in S. \tag{3.34}$$

For a finite number of points $\boldsymbol{x}_1, \ldots, \boldsymbol{x}_k$, the expression of the form of

$$\alpha_1 \boldsymbol{x}_1 + \cdots + \alpha_k \boldsymbol{x}_k \qquad \left(\text{where } \sum_{i=1}^{k} \alpha_i = 1, \ \alpha_i \geqq 0 \ (1 \leqq i \leqq k)\right) \tag{3.35}$$

is called a *convex combination* of $\boldsymbol{x}_1, \ldots, \boldsymbol{x}_k$. If $S$ is a convex set, the convex combination of an arbitrary finite number of points of $S$ belongs to $S$.

For a finite number of vectors $\boldsymbol{v}_1, \ldots, \boldsymbol{v}_k$, the set of all convex combinations thereof, *i.e.*,

$$\mathrm{Conv}(\boldsymbol{v}_1, \ldots, \boldsymbol{v}_k) = \left\{ \sum_{i=1}^{k} \alpha_i \boldsymbol{v}_i \ \middle| \ \sum_{i=1}^{k} \alpha_i = 1, \ \alpha_i \geqq 0 \ (1 \leqq i \leqq k) \right\}, \tag{3.36}$$

is the smallest convex set that contains $\boldsymbol{v}_1, \ldots, \boldsymbol{v}_k$. This is called the *convex hull* of $\boldsymbol{v}_1, \ldots, \boldsymbol{v}_k$.

A set $S$ is called a *cone*, if, for every point $\boldsymbol{x} \in S$ and every nonnegative real number $\alpha \geqq 0$, the point $\alpha \boldsymbol{x}$ belongs to $S$. A cone that is a convex set is called a *convex cone*. A necessary and sufficient condition for a set $S$ to be a convex cone is given by:

$$\boldsymbol{x}, \boldsymbol{y} \in S, \ \ \alpha, \beta \geqq 0 \ \ \Longrightarrow \ \ \alpha \boldsymbol{x} + \beta \boldsymbol{y} \in S. \tag{3.37}$$

The solution set of a homogeneous system of inequalities $\{\boldsymbol{x} \mid A\boldsymbol{x} \geqq \boldsymbol{0}\}$ is a convex cone. A convex cone that can be expressed in this form for some matrix $A$ is called a *polyhedral convex cone* or a *convex polyhedral cone.*

For a matrix $D = [\boldsymbol{d}_1, \ldots, \boldsymbol{d}_l]$, the set of all nonnegative linear combinations of its column vectors $\boldsymbol{d}_j$ $(j = 1, \ldots, l)$, *i.e.*,

$$\mathrm{Cone}(D) = \mathrm{Cone}(\boldsymbol{d}_1, \ldots, \boldsymbol{d}_l) = \left\{ \sum_{j=1}^{l} \beta_j \boldsymbol{d}_j \;\middle|\; \beta_j \geqq 0 \ (j = 1, \ldots, l) \right\}, \quad (3.38)$$

is a convex cone. This is called the convex cone *generated* by $\boldsymbol{d}_1, \ldots, \boldsymbol{d}_l$. Not every convex cone is generated by a finite number of vectors.[6] A convex cone is called a *finitely generated convex cone* if it is generated by a finite number of vectors. It is known that a convex cone is polyhedral if and only if it is finitely generated (to be stated in Theorem 3.9 below). The central theme of this section is to show this fact.

### 3.4.2.2 *Dual cones*

Next, we explain the concept of dual cones,[7] which serves as a prototype of various kinds of duality to be introduced later.

For a convex cone $K$ $(\subseteqq \mathbb{R}^n)$, the set defined by

$$K^* = \{\boldsymbol{y} \mid \boldsymbol{y}^\top \boldsymbol{x} \geqq 0 \quad \text{for all } \boldsymbol{x} \in K\}$$

is called the *dual cone* of $K$. The dual cone $K^*$ is a convex cone, satisfying the condition (3.37).

**Example 3.5.** The nonnegative orthant $K = \{\boldsymbol{x} \mid x_i \geqq 0 \ (i = 1, \ldots, n)\}$ is a convex cone. Its dual cone $K^*$ coincides with $K$ itself. ■

**Example 3.6.** In $\mathbb{R}^2$, an infinite sector $K = \mathrm{Cone}((1,0)^\top, (\cos\theta, \sin\theta)^\top)$ of angle $\theta$ $(0 < \theta < \pi)$ is a convex cone, and its dual cone is another infinite sector $K^* = \mathrm{Cone}((0,1)^\top, (\sin\theta, -\cos\theta)^\top)$ of angle $\pi - \theta$. ■

If a convex cone $K$ is given in the form of $K = \mathrm{Cone}(D)$ using a matrix $D$, its dual cone $K^*$ has a concrete representation as follows.

**Proposition 3.1.** *For any matrix $D$,*

$$\mathrm{Cone}(D)^* = \{\boldsymbol{y} \mid \boldsymbol{y}^\top D \geqq \boldsymbol{0}^\top\}.$$

---

[6]For example, $\{(x_1, x_2, x_3) \mid \sqrt{{x_1}^2 + {x_2}^2} \leqq x_3\}$ is a convex cone that cannot be generated by a finite number of vectors. This is not a polyhedral convex cone.

[7]Dual cones are also used in the proof of Theorem 3.9.

**Proof.** By the definition (3.38) of Cone($D$), $\boldsymbol{y} \in \text{Cone}(D)^*$ is equivalent to $\boldsymbol{y}^\top \boldsymbol{d}_j \geqq 0$ $(j = 1, \ldots, l)$, *i.e.*, $\boldsymbol{y}^\top D \geqq \mathbf{0}^\top$. □

Since the dual cone $K^*$ is again a convex cone, we can conceive of its dual cone $(K^*)^*$, which we denote as $K^{**}$. Then, $K^{**}$ is equal to $K$ under the (reasonably natural) assumption that $K$ is finitely generated.[8]

**Theorem 3.8 (Duality).** *If a convex cone $K$ is finitely generated, then*

$$K = K^{**}.$$

**Proof.** $[K \subseteqq K^{**}]$:[9] Assume $\boldsymbol{x} \in K$. Then $\boldsymbol{x}^\top \boldsymbol{y} = \boldsymbol{y}^\top \boldsymbol{x} \geqq 0$ for all $\boldsymbol{y} \in K^*$. This shows that $\boldsymbol{x} \in (K^*)^* = K^{**}$.

$[K \supseteqq K^{**}]$: Since $K$ is finitely generated, it can be represented as $K = \text{Cone}(D)$. Then

$$K^* = \text{Cone}(D)^* = \{\boldsymbol{y} \mid \boldsymbol{y}^\top D \geqq \mathbf{0}^\top\}$$

by Proposition 3.1 above. Take any $\boldsymbol{x} \in K^{**}$. By definition, we have the implication $[\boldsymbol{y} \in K^* \Longrightarrow \boldsymbol{y}^\top \boldsymbol{x} \geqq 0]$, or equivalently, $[\boldsymbol{y}^\top D \geqq \mathbf{0}^\top \Longrightarrow \boldsymbol{y}^\top \boldsymbol{x} \geqq 0]$. The latter is in the form of (b) in Theorem 3.3 (Farkas' lemma), which is equivalent to the condition (a) that there exists a nonnegative vector $\boldsymbol{\beta} \geqq \mathbf{0}$ such that $\boldsymbol{x} = D\boldsymbol{\beta}$. This shows that $\boldsymbol{x} \in \text{Cone}(D) = K$. Therefore, $K^{**} \subseteqq K$. □

### 3.4.3 *Solutions to Homogeneous Systems of Inequalities*

In this section, we consider the structure of the solution set of a *homogeneous system of inequalities $A\boldsymbol{x} \geqq \mathbf{0}$*. It is assumed that $A$ is an $m \times n$ matrix and $\boldsymbol{x}$ is an $n$-dimensional vector.

#### 3.4.3.1 *Theorem description*

Polyhedral convex cones are finitely generated, and conversely, finitely generated convex cones are polyhedral. That is,

$$\text{polyhedral convex cone} \iff \text{finitely generated convex cone.}$$

[8]We may think of the dual cone as the convex cone seen from an "opposite side." We may further say (symbolically) that Theorem 3.8 shows that the rear of the rear is the front.

[9]The proof of this direction does not depend on the assumption that $K$ is a finitely generated convex cone. Therefore, $K \subseteqq K^{**}$ is true for every $K$.

The following theorem establishes this important fact, stating the equivalence of the two kinds of descriptions of convex cones.

**Theorem 3.9.**

(1) *For any matrix $A$, there exists a matrix $D$ such that*

$$\{\boldsymbol{x} \mid A\boldsymbol{x} \geqq \boldsymbol{0}\} = \mathrm{Cone}(D). \tag{3.39}$$

*That is, every polyhedral convex cone is a finitely generated convex cone.*

(2) *For any matrix $D$, there exists a matrix $A$ such that*

$$\mathrm{Cone}(D) = \{\boldsymbol{x} \mid A\boldsymbol{x} \geqq \boldsymbol{0}\}. \tag{3.40}$$

*That is, every finitely generated convex cone is a polyhedral convex cone.*

By Theorem 3.9(1), solutions $\boldsymbol{x}$ of the system of inequalities $A\boldsymbol{x} \geqq \boldsymbol{0}$ admits a parametric representation

$$\boldsymbol{x} = \sum_{j=1}^{l} \beta_j \boldsymbol{d}_j \qquad (\beta_j \geqq 0 \ (j = 1, \ldots, l)) \tag{3.41}$$

using a finite number of suitably chosen vectors $\boldsymbol{d}_1, \ldots, \boldsymbol{d}_l$.

The proof of Theorem 3.9 is given below, but the reader may skip the proof and go on to Sec. 3.4.4 ("Solutions to Inhomogeneous Systems of Inequalities"). However, reading the details of the proof will bring about a deeper understanding of the mathematical arguments based upon Farkas' lemma, duality of convex cones, *etc.*

#### 3.4.3.2 *Proof of Theorem 3.9(1)*

The matrix $D$ for the representation in (3.39) of Theorem 3.9(1) will be constructed as $D = [D_0 \mid D_1]$ from the matrix $D_0$ in Proposition 3.2 below and the matrix $D_1$ in Proposition 3.3 below. We recall notations:

$$\mathrm{Ker}(A) = \{\boldsymbol{x} \mid A\boldsymbol{x} = \boldsymbol{0}\},$$
$$\mathrm{Im}(A) = \{\boldsymbol{y} \mid \boldsymbol{y} = A\boldsymbol{x} \text{ for some } \boldsymbol{x}\}$$

for a matrix $A$.

**Proposition 3.2.** *For any matrix $A$, there exists a matrix $D_0$ such that*

$$\mathrm{Ker}(A) = \mathrm{Cone}(D_0). \tag{3.42}$$

**Proof.** $\mathrm{Ker}(A)$ forms a linear subspace. Take a basis $\boldsymbol{d}_1, \ldots, \boldsymbol{d}_l$ of this subspace and set $D_0 = [\boldsymbol{d}_1, \ldots, \boldsymbol{d}_l, -\boldsymbol{d}_1, \ldots, -\boldsymbol{d}_l]$. □

**Proposition 3.3.** *For any matrix $A$, there exists a matrix $D_1$ such that*

$$\{\boldsymbol{y} \mid \boldsymbol{y} \in \mathrm{Im}(A),\ \boldsymbol{y} \geqq \boldsymbol{0}\} = A \cdot \mathrm{Cone}(D_1), \tag{3.43}$$

*where $A \cdot \mathrm{Cone}(D_1)$ means the set of all $\boldsymbol{y}$ that can be written as $\boldsymbol{y} = A\boldsymbol{x}$ for some $\boldsymbol{x} \in \mathrm{Cone}(D_1)$, i.e.,*

$$A \cdot \mathrm{Cone}(D_1) = \{\boldsymbol{y} \mid \boldsymbol{y} = A\boldsymbol{x},\ \boldsymbol{x} \in \mathrm{Cone}(D_1)\}.$$

**Proof.** Since $\mathrm{Im}(A)$ is a linear subspace,

$$\{\boldsymbol{y} \mid \boldsymbol{y} \in \mathrm{Im}(A),\ \boldsymbol{y} \geqq \boldsymbol{0}\} = \mathrm{Cone}(F)$$

holds for a suitably chosen matrix $F = [\boldsymbol{f}_1, \ldots, \boldsymbol{f}_l]$, which is proved in Proposition 3.4 below at the end of this section. Then, for each $j$, we have $\boldsymbol{f}_j \in \mathrm{Im}(A)$, and therefore, there exists a vector $\boldsymbol{d}_j$ such that $\boldsymbol{f}_j = A\boldsymbol{d}_j$. The matrix $D_1 = [\boldsymbol{d}_1, \ldots, \boldsymbol{d}_l]$ satisfies (3.43). □

Consider the matrix $D = [D_0 \mid D_1]$ consisting of the matrix $D_0$ in Proposition 3.2 and the matrix $D_1$ in Proposition 3.3. We will show that this matrix $D$ satisfies (3.39) in Theorem 3.9(1).

First we prove

$$\{\boldsymbol{x} \mid A\boldsymbol{x} \geqq \boldsymbol{0}\} \supseteqq \mathrm{Cone}(D).$$

Take any $\boldsymbol{x} \in \mathrm{Cone}(D)$. Then we can express $\boldsymbol{x}$ as $\boldsymbol{x} = D_0\boldsymbol{\beta}_0 + D_1\boldsymbol{\beta}_1$ using some nonnegative vectors $\boldsymbol{\beta}_0, \boldsymbol{\beta}_1 \geqq \boldsymbol{0}$, and therefore,

$$A\boldsymbol{x} = AD_0\boldsymbol{\beta}_0 + AD_1\boldsymbol{\beta}_1.$$

For the first term on the right-hand side, we have $AD_0\boldsymbol{\beta}_0 = \boldsymbol{0}$ by (3.42). For the second term, we have $AD_1\boldsymbol{\beta}_1 \in A \cdot \mathrm{Cone}(D_1)$, and hence $AD_1\boldsymbol{\beta}_1 \geqq \boldsymbol{0}$ by (3.43). Therefore, $A\boldsymbol{x} \geqq \boldsymbol{0}$.

Next, we prove the reverse inclusion:

$$\{\boldsymbol{x} \mid A\boldsymbol{x} \geqq \boldsymbol{0}\} \subseteqq \mathrm{Cone}(D).$$

Assume $A\boldsymbol{x} \geqq \boldsymbol{0}$ and set $\boldsymbol{y} = A\boldsymbol{x}$. By (3.43) we have $\boldsymbol{y} = AD_1\boldsymbol{\beta}_1$ for some nonnegative vector $\boldsymbol{\beta}_1 \geqq \boldsymbol{0}$. Then it follows from $A\boldsymbol{x} = \boldsymbol{y} = AD_1\boldsymbol{\beta}_1$ that $A(\boldsymbol{x} - D_1\boldsymbol{\beta}_1) = \boldsymbol{0}$. By (3.42) we have $\boldsymbol{x} - D_1\boldsymbol{\beta}_1 = D_0\boldsymbol{\beta}_0$ with some nonnegative vector $\boldsymbol{\beta}_0 \geqq \boldsymbol{0}$. Therefore, $\boldsymbol{x} = D_0\boldsymbol{\beta}_0 + D_1\boldsymbol{\beta}_1 \in \mathrm{Cone}(D)$.

We have thus shown that $D = [D_0 \mid D_1]$ is valid for (3.39). Finally, it remains to prove the following fact used in the proof of Proposition 3.3.

**Proposition 3.4.** *For any linear subspace $L$, there exists a matrix $F$ such that*

$$\{\boldsymbol{y} \mid \boldsymbol{y} \in L,\ \boldsymbol{y} \geqq \boldsymbol{0}\} = \mathrm{Cone}(F). \tag{3.44}$$

**Proof.** We prove this by induction on the dimension $\dim L$ of the linear subspace $L$. If $\dim L = 0$, we can take $F = O$ (zero matrix). When $\dim L \geqq 1$, we denote the left-hand side of (3.44) as

$$L^+ = \{\boldsymbol{y} \mid \boldsymbol{y} \in L,\ \boldsymbol{y} \geqq \boldsymbol{0}\}$$

and define a set $J \subseteqq \{1, \ldots, n\}$ of indices by

$$J = \{j \mid y_j > 0 \text{ for some } \boldsymbol{y} \in L^+\}.$$

If $J$ is empty, then $L^+ = \{\boldsymbol{0}\}$ and we can take $F = O$.

If $J$ is nonempty, then for each $j \in J$, we (arbitrarily) take a vector $\hat{\boldsymbol{y}}^{(j)} \in L^+$ with the $j$th component positive, and form their sum

$$\hat{\boldsymbol{y}} = \sum_{j \in J} \hat{\boldsymbol{y}}^{(j)}.$$

Then we have $\hat{\boldsymbol{y}} \in L^+$, and moreover, $\hat{y}_j > 0$ for all $j \in J$, where $\hat{y}_j$ means the $j$th component of $\hat{\boldsymbol{y}}$. For each $j \in J$, we define

$$L_j = \{\boldsymbol{y} \mid \boldsymbol{y} \in L,\ y_j = 0\}, \qquad L_j^+ = \{\boldsymbol{y} \mid \boldsymbol{y} \in L_j,\ \boldsymbol{y} \geqq \boldsymbol{0}\}.$$

We have $\dim L_j < \dim L$, since $\hat{\boldsymbol{y}} \in L$, $\hat{\boldsymbol{y}} \notin L_j$, and both $L$ and $L_j$ are linear subspaces. Therefore, by the induction hypothesis, there exists a matrix $F_j$ satisfying

$$L_j^+ = \mathrm{Cone}(F_j).$$

Using the vector $\hat{\boldsymbol{y}}$ and the matrices $F_j$ $(j \in J)$, define a matrix $F$ as

$$F = [\hat{\boldsymbol{y}} \mid F_{j_1} \mid F_{j_2} \mid \cdots \mid F_{j_q}],$$

where $J = \{j_1, j_2, \ldots, j_q\}$. We will show that this $F$ satisfies $L^+ = \mathrm{Cone}(F)$ in (3.44).

First, we show $\mathrm{Cone}(F) \subseteqq L^+$. Take any $\boldsymbol{y} \in \mathrm{Cone}(F)$, and then

$$\boldsymbol{y} = \beta_0 \hat{\boldsymbol{y}} + \sum_{j \in J} F_j \boldsymbol{\beta}_j$$

for some nonnegative real number $\beta_0$ and nonnegative vectors $\boldsymbol{\beta}_j$ $(j \in J)$. This shows $\boldsymbol{y} \in L^+$, since $\hat{\boldsymbol{y}} \in L^+$, $F_j \boldsymbol{\beta}_j \in \mathrm{Cone}(F_j) = L_j^+ \subseteqq L^+$, and $L^+$ is a convex cone.

Next, we show $L^+ \subseteqq \mathrm{Cone}(F)$. Take any $\boldsymbol{y} \in L^+$. Define a real number $\theta \geqq 0$ by

$$\theta = \min_{j \in J} \frac{y_j}{\hat{y}_j} \tag{3.45}$$

and consider a vector

$$\boldsymbol{z} = \boldsymbol{y} - \theta \hat{\boldsymbol{y}}.$$

By the definition of $\theta$ we have $\boldsymbol{z} \geqq \mathbf{0}$, whereas $\boldsymbol{z} \in L$ by $\boldsymbol{y} \in L$ and $\hat{\boldsymbol{y}} \in L$. Let $j^*$ be an index $j \in J$ for which the minimum is attained on the right-hand side of (3.45). Then the $j^*$th component of $\boldsymbol{z}$ is 0, and therefore,

$$\boldsymbol{z} \in L_{j^*}^{+} = \mathrm{Cone}(F_{j^*}) \subseteqq \mathrm{Cone}(F).$$

Since $\hat{\boldsymbol{y}} \in \mathrm{Cone}(F)$ also holds, we can conclude that

$$\boldsymbol{y} = \boldsymbol{z} + \theta \hat{\boldsymbol{y}} \in \mathrm{Cone}(F).$$

Thus, we have shown $L^{+} \subseteqq \mathrm{Cone}(F)$, thereby establishing the relation (3.44). □

This completes the proof of Theorem 3.9(1).

#### 3.4.3.3 *Proof of Theorem 3.9(2)*

By Proposition 3.1 we have

$$\mathrm{Cone}(D)^* = \{\boldsymbol{y} \mid \boldsymbol{y}^{\top} D \geqq \mathbf{0}^{\top}\}.$$

This shows that $\mathrm{Cone}(D)^*$ is a polyhedral convex cone, and therefore, by Theorem 3.9(1), there exists a matrix $B$ such that $\mathrm{Cone}(D)^* = \mathrm{Cone}(B)$. By considering the dual cones of these cones, we obtain

$$\mathrm{Cone}(D) = \mathrm{Cone}(D)^{**} = \mathrm{Cone}(B)^* = \{\boldsymbol{x} \mid \boldsymbol{x}^{\top} B \geqq \mathbf{0}^{\top}\}$$

using Theorem 3.8 and Proposition 3.1. Therefore, (3.40) is satisfied for $A = B^{\top}$.

### 3.4.4 *Solutions to Inhomogeneous Systems of Inequalities*

In this section, we consider the structure of the solution set of an *inhomogeneous system of inequalities* $A\boldsymbol{x} \geqq \boldsymbol{b}$. It is assumed that $A$ is an $m \times n$ matrix, $\boldsymbol{b}$ is an $m$-dimensional vector, and $\boldsymbol{x}$ is an $n$-dimensional vector.

#### 3.4.4.1 *Theorem description*

The solution set of an inhomogeneous system of inequalities $A\boldsymbol{x} \geqq \boldsymbol{b}$ is described in the following theorem, where the notations $\mathrm{Conv}(\cdots)$ and $\mathrm{Cone}(\cdots)$ used in (3.46) are defined in (3.36) and (3.38), respectively. The operation "+" on the right-hand side of (3.46) means forming all possible sums of the vectors belonging to the respective sets. A concrete example of (3.46) is given in (3.33) of Example 3.4 (Sec. 3.4.1) and its geometric meaning is the decomposition shown in Fig. 3.6.

**Theorem 3.10.** *For any matrix $A$ and vector $\boldsymbol{b}$, there exist a finite number of vectors $\boldsymbol{v}_1, \ldots, \boldsymbol{v}_k$ and $\boldsymbol{d}_1, \ldots, \boldsymbol{d}_l$ such that*

$$\{\boldsymbol{x} \mid A\boldsymbol{x} \geqq \boldsymbol{b}\} = \mathrm{Conv}(\boldsymbol{v}_1, \ldots, \boldsymbol{v}_k) + \mathrm{Cone}(\boldsymbol{d}_1, \ldots, \boldsymbol{d}_l). \tag{3.46}$$

Theorem 3.10 implies that the solutions $\boldsymbol{x}$ to a system of inequalities $A\boldsymbol{x} \geqq \boldsymbol{b}$ can be parametrized as

$$\boldsymbol{x} = \sum_{i=1}^{k} \alpha_i \boldsymbol{v}_i + \sum_{j=1}^{l} \beta_j \boldsymbol{d}_j \tag{3.47}$$

using a finite number of suitably chosen vectors $\boldsymbol{v}_1, \ldots, \boldsymbol{v}_k$ and $\boldsymbol{d}_1, \ldots, \boldsymbol{d}_l$, where

$$\sum_{i=1}^{k} \alpha_i = 1, \quad \alpha_i \geqq 0 \ \ (i = 1, \ldots, k), \quad \beta_j \geqq 0 \ \ (j = 1, \ldots, l).$$

In particular, if the solution set is bounded, the parametric representation (3.47) reduces to

$$\boldsymbol{x} = \sum_{i=1}^{k} \alpha_i \boldsymbol{v}_i, \tag{3.48}$$

where

$$\sum_{i=1}^{k} \alpha_i = 1, \qquad \alpha_i \geqq 0 \ \ (i = 1, \ldots, k).$$

The expression (3.32) in Example 3.4 (Sec. 3.4.1) is an example of the parametric representation (3.47) and the expression (3.30) in Example 3.3 is an example of (3.48).

#### 3.4.4.2 *Proof of Theorem 3.10*

Let $S$ denote the solution set of an inhomogeneous system of inequalities $A\boldsymbol{x} \geqq \boldsymbol{b}$, *i.e.*, $S = \{\boldsymbol{x} \mid A\boldsymbol{x} \geqq \boldsymbol{b}\}$. Define an $(m+1) \times (n+1)$ matrix $\tilde{A}$ and an $(n+1)$-dimensional vector $\tilde{\boldsymbol{x}}$ as

$$\tilde{A} = \begin{bmatrix} A & -\boldsymbol{b} \\ \boldsymbol{0}^\top & 1 \end{bmatrix}, \qquad \tilde{\boldsymbol{x}} = \begin{bmatrix} \boldsymbol{x} \\ t \end{bmatrix}$$

and let $\tilde{S} = \{\tilde{\boldsymbol{x}} \mid \tilde{A}\tilde{\boldsymbol{x}} \geqq \boldsymbol{0}\}$. Then we have:

$$\boldsymbol{x} \in S \iff \begin{bmatrix} \boldsymbol{x} \\ 1 \end{bmatrix} \in \tilde{S}. \tag{3.49}$$

Theorem 3.9(1) applied to the homogeneous system of inequalities $\tilde{A}\tilde{\boldsymbol{x}} \geqq \mathbf{0}$ shows that there exist a finite number of vectors $\tilde{\boldsymbol{x}}_1, \ldots, \tilde{\boldsymbol{x}}_p$ such that

$$\tilde{S} = \text{Cone}(\tilde{\boldsymbol{x}}_1, \ldots, \tilde{\boldsymbol{x}}_p).$$

Each $\tilde{\boldsymbol{x}}_i$ belongs to $\tilde{S}$, and hence its $(n+1)$st component is nonnegative. With a suitable renumbering of the vectors, we may assume that the $(n+1)$st component is positive for $\tilde{\boldsymbol{x}}_1, \ldots, \tilde{\boldsymbol{x}}_k$ and 0 for $\tilde{\boldsymbol{x}}_{k+1}, \ldots, \tilde{\boldsymbol{x}}_{k+l}$, where $k + l = p$. In addition, we can normalize the $(n+1)$st components of $\tilde{\boldsymbol{x}}_1, \ldots, \tilde{\boldsymbol{x}}_k$ to 1. Thus, we may assume the following:

$$\tilde{\boldsymbol{x}}_i = \begin{bmatrix} \boldsymbol{v}_i \\ 1 \end{bmatrix} \quad (i = 1, \ldots, k), \qquad \tilde{\boldsymbol{x}}_{k+j} = \begin{bmatrix} \boldsymbol{d}_j \\ 0 \end{bmatrix} \quad (j = 1, \ldots, l).$$

Then we have:

$$\tilde{\boldsymbol{x}} = \begin{bmatrix} \boldsymbol{x} \\ 1 \end{bmatrix} \in \tilde{S} \iff \tilde{\boldsymbol{x}} = \sum_{i=1}^{k} \alpha_i \begin{bmatrix} \boldsymbol{v}_i \\ 1 \end{bmatrix} + \sum_{j=1}^{l} \beta_j \begin{bmatrix} \boldsymbol{d}_j \\ 0 \end{bmatrix} \tag{3.50}$$

$(\alpha_i \geqq 0\ (i = 1, \ldots, k);\ \beta_j \geqq 0\ (j = 1, \ldots, l))$. The combination of (3.49) and (3.50) shows (3.47). In addition, the normalization condition $\sum_{i=1}^{k} \alpha_i = 1$ is obtained from the $(n+1)$st component of (3.50).

## 3.5 Linear Programming

### 3.5.1 *Problem Descriptions*

Optimization problems in which a linear function is maximized or minimized under conditions described by linear equations and/or inequalities are called *linear programming problems* or simply *linear programs*. The conditions that should be satisfied by the variables are called *constraints* and the functions to be maximized or minimized are called *objective functions*.

#### 3.5.1.1 *Example*

Let us start with a simple problem:

---

Find the maximum value of $f(x_1, x_2) = x_1 + 2x_2$ when the variables $(x_1, x_2)$ lie in the region described by

$$x_1 + 4x_2 \leqq 12, \qquad x_1 + x_2 \leqq 4, \qquad x_1 \leqq 3. \tag{3.51}$$

---

In this problem, (3.51) gives the constraints and $f(x_1, x_2) = x_1 + 2x_2$ is the objective function. According to the convention of the theory of optimization, this problem is usually written in the form of

$$\begin{array}{lrcl} \text{Maximize} & x_1 + 2x_2 & & \\ \text{subject to} & x_1 + 4x_2 & \leqq & 12 \\ & x_1 + x_2 & \leqq & 4 \\ & x_1 & \leqq & 3 \end{array} \tag{3.52}$$

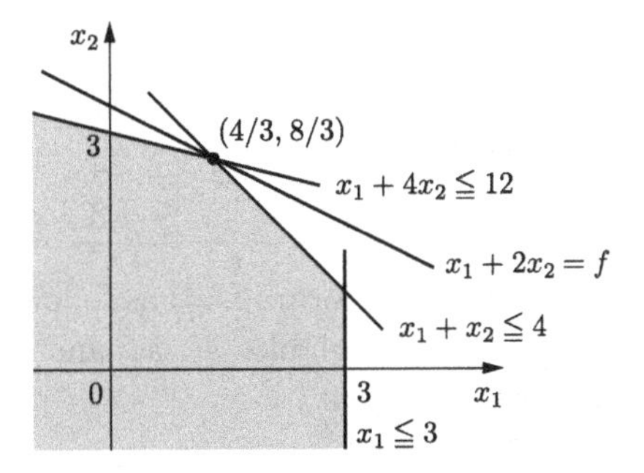


Fig. 3.7 A linear program in two variables.

This problem, involving only two variables, can be solved by drawing a figure as in Fig. 3.7. The shaded domain is the set of points that satisfy the conditions in (3.51). From this figure, we can see that the *optimal solution* is $(x_1, x_2) = (4/3, 8/3)$ and the maximum value of function $f$ is

$$f_{\max} = \frac{4}{3} + 2 \cdot \frac{8}{3} = \frac{20}{3}. \tag{3.53}$$

### 3.5.1.2 *Standard form*

A linear programming problem can be formulated equivalently in several different ways. For example, we can flip the sign of the objective function of a maximization problem to write it as a minimization problem. For theoretical treatment and software development, it is convenient to fix a format to describe problems. The following form

$$\begin{array}{ll} \text{Minimize} & \boldsymbol{c}^\top \boldsymbol{x} \\ \text{subject to} & A\boldsymbol{x} = \boldsymbol{b} \\ & \boldsymbol{x} \geqq \boldsymbol{0} \end{array} \tag{3.54}$$

is often used and is called the *standard form* of linear programs.[10] The characteristic property of this form is that the inequality conditions are

[10]The terminology for this form varies in the literature.

limited to the sign (nonnegativity) constraints of variables. We assume below that $A$ is an $m \times n$ matrix, $\boldsymbol{b}$ is an $m$-dimensional vector, and $\boldsymbol{c}$ is an $n$-dimensional vector.

Every linear programming problem can be rewritten in the standard form by introducing suitable variables and so forth. We note, however, that the transformation to the standard form is not uniquely determined.

**Example 3.7.** The problem in (3.52):

$$\begin{array}{ll} \text{Maximize} & x_1 + 2x_2 \\ \text{subject to} & x_1 + 4x_2 \leqq 12 \\ & x_1 + \phantom{4}x_2 \leqq 4 \\ & x_1 \phantom{+ 4x_2} \leqq 3 \end{array} \tag{3.55}$$

can be transformed to the standard form (3.54) as follows. For the inequality constraints, we introduce new variables $s_1$, $s_2$, and $s_3$, to rewrite them in the form of equality constraints as

$$x_1 + 4x_2 + s_1 = 12, \qquad x_1 + x_2 + s_2 = 4, \qquad x_1 + s_3 = 3$$

with sign constraints $s_1, s_2, s_3 \geqq 0$. For variables $x_1$ and $x_2$, which are not subject to sign constraints, we introduce new variables $y_1$, $y_2$, $z_1$, and $z_2$ to express them as

$$x_1 = y_1 - z_1, \qquad x_2 = y_2 - z_2$$

and impose sign constraints $y_1, y_2, z_1, z_2 \geqq 0$. In addition, we can change the sign of the objective function to convert the maximization problem to a minimization problem. In doing so, the problem (3.55) is rewritten to the following linear programming problem

$$\begin{array}{llll} \text{Minimize} & -(y_1 - z_1) - 2(y_2 - z_2) & & \\ \text{subject to} & (y_1 - z_1) + 4(y_2 - z_2) + s_1 & = & 12 \\ & (y_1 - z_1) + \phantom{4}(y_2 - z_2) + s_2 & = & 4 \\ & (y_1 - z_1) \phantom{+ 4(y_2 - z_2)} + s_3 & = & 3 \\ & y_1,\ y_2,\ z_1,\ z_2,\ s_1,\ s_2,\ s_3 & \geqq & 0 \end{array} \tag{3.56}$$

with $(y_1, y_2, z_1, z_2, s_1, s_2, s_3)$ being the decision variables for optimization. This problem is in the standard form and, as an optimization problem, it is equivalent to the problem (3.55). The variables introduced to transform inequality constraints to equality constraints, such as $s_i$ in the above, are called *slack variables*. ■

**Example 3.8.** The method of transformation illustrated in Example 3.7 above can be generalized as follows. A linear programming problem in the form of

$$\text{Maximize } \ \boldsymbol{c}^\top \boldsymbol{x} \quad \text{subject to} \quad A\boldsymbol{x} \leqq \boldsymbol{b} \tag{3.57}$$

is equivalent to a problem in the standard form (3.54) in which $A$, $\boldsymbol{b}$, and $\boldsymbol{c}$ are replaced as follows:

$$\tilde{A} = \begin{bmatrix} A & -A & I \end{bmatrix}, \qquad \tilde{\boldsymbol{b}} = \boldsymbol{b}, \qquad \tilde{\boldsymbol{c}} = \begin{bmatrix} -\boldsymbol{c}^\top & \boldsymbol{c}^\top & \boldsymbol{0}^\top \end{bmatrix}^\top. \tag{3.58}$$

This can be derived as follows. Rewrite $A\boldsymbol{x} \leqq \boldsymbol{b}$ as $A\boldsymbol{x} + \boldsymbol{s} = \boldsymbol{b}$, $\boldsymbol{s} \geqq \boldsymbol{0}$, and put $\boldsymbol{x} = \boldsymbol{y} - \boldsymbol{z}$ with sign constraints $\boldsymbol{y} \geqq \boldsymbol{0}$ and $\boldsymbol{z} \geqq \boldsymbol{0}$. Then flip the sign of the objective function to obtain the following minimization problem:

| | |
|---|---|
| Minimize | $-\boldsymbol{c}^\top \boldsymbol{y} + \boldsymbol{c}^\top \boldsymbol{z}$ |
| subject to | $A\boldsymbol{y} - A\boldsymbol{z} + \boldsymbol{s} = \boldsymbol{b}$ |
| | $\boldsymbol{y}, \boldsymbol{z}, \boldsymbol{s} \geqq \boldsymbol{0}$ |

Here, $\boldsymbol{s}$ is the vector of slack variables. ■

A point (vector) $\boldsymbol{x}$ that satisfies the constraints is called a *feasible solution* or *admissible solution.*[11] A problem that has a feasible solution is called *feasible* and a problem that does not have one is called *infeasible.* The set $S$ of points $\boldsymbol{x}$ that satisfy the constraints is call the *feasible region.* For example, for a problem in the standard form (3.54),

$$S = \{\boldsymbol{x} \mid A\boldsymbol{x} = \boldsymbol{b}, \boldsymbol{x} \geqq \boldsymbol{0}\}$$

is the feasible region.

**Remark 3.5.** In this section we have explained *linear programs.* There is another similar term *linear programming.* This term was proposed by George B. Dantzig around 1947 and has since been used to denote the methodology including the following aspects:

- Modeling for formulating actual problems into the form of linear programs,
- Theory for the mathematical structures of linear programs,
- Algorithms for finding numerical solutions to linear programs.

[11]Since we are considering a problem to minimize the objective function, it would probably be natural to let the word "solution" mean the point at which the objective function is at its minimum. In the field of optimization, however, "solution" is used to simply mean a point (a vector consisting of variables for optimization), and the former is called an "optimal solution".

The solution algorithms for linear programs can be classified, roughly, into three categories: simplex methods, interior-point methods, and ellipsoid methods. Algorithmic issues lead to a variety of topics of mathematical interest, and modeling techniques are extremely important to bridge reality and mathematics. However, in this book, we do not address modeling and algorithms, but instead restrict our discussion to the structures of linear inequalities and related topics. For various aspects of linear programming the reader is referred to textbooks [37,38,40,41,43]. ■

### 3.5.2 *Existence of Optimal Solutions*

Consider a feasible linear programming problem in the standard form (3.54). There is a possibility of the objective function becoming indefinitely small, and therefore, an optimal solution not existing. For example, this is the case with

$$\text{Minimize} \quad -x_1 \quad \text{subject to} \quad x_1 - x_2 = 1, \quad x_1, x_2 \geqq 0.$$

Such a linear programming problem is said to be *unbounded*, whereas a problem with a finite minimum value is called *bounded*.

In Theorem 3.10 we have investigated the structure of the solution set of an inhomogeneous system of inequalities, thereby obtaining a parametric representation of the solution set. This general result can be applied to the system of linear inequalities arising from a given linear programming problem. The constraints $A\boldsymbol{x} = \boldsymbol{b}$ and $\boldsymbol{x} \geqq \boldsymbol{0}$ of the given linear program in the standard form (3.54) can be rewritten into a form to which Theorem 3.10 can be applied. Then, by Theorem 3.10, the feasible region

$$S = \{\boldsymbol{x} \mid A\boldsymbol{x} = \boldsymbol{b}, \boldsymbol{x} \geqq \boldsymbol{0}\}$$

can be represented as

$$S = \text{Conv}(\boldsymbol{v}_1, \ldots, \boldsymbol{v}_k) + \text{Cone}(\boldsymbol{d}_1, \ldots, \boldsymbol{d}_l) \tag{3.59}$$

using a finite number of vectors $\boldsymbol{v}_1, \ldots, \boldsymbol{v}_k$ and $\boldsymbol{d}_1, \ldots, \boldsymbol{d}_l$. In accordance with this, the feasible solutions $\boldsymbol{x}$ are parametrized as

$$\boldsymbol{x} = \sum_{i=1}^{k} \alpha_i \boldsymbol{v}_i + \sum_{j=1}^{l} \beta_j \boldsymbol{d}_j, \tag{3.60}$$

where

$$\sum_{i=1}^{k} \alpha_i = 1, \quad \alpha_i \geqq 0 \ \ (i = 1, \ldots, k), \quad \beta_j \geqq 0 \ \ (j = 1, \ldots, l).$$

From the parametric representation (3.60), we can express the objective function as

$$c^\top x = \sum_{i=1}^{k} \alpha_i (c^\top v_i) + \sum_{j=1}^{l} \beta_j (c^\top d_j).$$

Therefore, the linear programming problem (3.54) is bounded (has a finite minimum value) if and only if

$$c^\top d_j \geqq 0 \qquad (j = 1, \ldots, l). \tag{3.61}$$

**Remark 3.6.** Suppose that the feasible region $S$ of a linear programming problem (3.54) is nonempty, and denote $\inf\{c^\top x \mid x \in S\}$ by $\mu$. Then $\mu \in \mathbb{R}$ (finite value) or $\mu = -\infty$. The problem (3.54) is defined to be *bounded* if $\mu \in \mathbb{R}$, and *unbounded* if $\mu = -\infty$. With this definition, we can prove a proposition that "there exists an optimal solution (*i.e.*, a vector $x \in S$ that satisfies $\mu = c^\top x$) in a bounded problem." Therefore, a bounded linear programming problem has an optimal solution. However, this property is a special feature of linear programs and does not hold true for general nonlinear optimization problems. For example, in the problem of minimizing $f(x) = \exp(-x)$ under the constraint $x \geqq 0$, we have $\mu = \inf\{\exp(-x) \mid x \geqq 0\} = 0$, which shows that the problem is bounded, but there exists no (optimal solution) $x$ that satisfies $\exp(-x) = 0$. ■

### 3.5.3 *Duality*

#### 3.5.3.1 *Dual problems*

When given a problem in the standard form (3.54), we consider another linear programming problem

$$\begin{array}{ll} \text{Maximize} & b^\top y \\ \text{subject to} & A^\top y \leqq c \end{array} \tag{3.62}$$

determined by the same data $A$, $b$, and $c$ for the given problem. We call this problem (3.62) the *dual problem* of the problem (3.54). The original problem (3.54) is then referred to as the *primal problem*. For later reference, we display the two problems side-by-side:

$$\begin{array}{ll} \text{[Primal Problem P]} & \\ \hline \text{Minimize} & c^\top x \\ \text{subject to} & Ax = b \\ & x \geqq 0 \\ \hline \end{array} \qquad \begin{array}{ll} \text{[Dual Problem D]} & \\ \hline \text{Maximize} & b^\top y \\ \text{subject to} & A^\top y \leqq c \\ \hline \end{array} \tag{3.63}$$

The primal problem P is a minimization problem and the dual problem D is a maximization problem. Since $A$ is an $m \times n$ matrix, $\boldsymbol{b}$ is an $m$-dimensional vector, and $\boldsymbol{c}$ is an $n$-dimensional vector, the variable $\boldsymbol{x}$ of the primal problem P is an $n$-dimensional vector and the variable $\boldsymbol{y}$ of the dual problem D is an $m$-dimensional vector.

We have defined the dual problem for a problem in the standard form, whereas every linear programming problem can be transformed to the standard form. This suggests that we could consider the dual of the dual problem. Put more precisely, we will construct a problem in the standard form that is equivalent to the problem D and then consider its dual problem according to the above definition applied to that problem. We carry out this calculation below.

Problem D is equivalent to a problem in the standard form (3.54) with $A$, $\boldsymbol{b}$, and $\boldsymbol{c}$ replaced with

$$\tilde{A} = \left[ A^\top \; -A^\top \; I \right], \qquad \tilde{\boldsymbol{b}} = \boldsymbol{c}, \qquad \tilde{\boldsymbol{c}} = \left[ -\boldsymbol{b}^\top \; \boldsymbol{b}^\top \; \boldsymbol{0}^\top \right]^\top . \tag{3.64}$$

(To see this, note that Problem D is identical with (3.57) in Example 3.8 when $(A, \boldsymbol{b}, \boldsymbol{c})$ is changed to $(A^\top, \boldsymbol{c}, \boldsymbol{b})$; then (3.64) is obtained from (3.58).) The corresponding dual problem is:

$$\text{Maximize } \; \boldsymbol{c}^\top \tilde{\boldsymbol{y}} \quad \text{subject to} \quad \begin{bmatrix} A \\ -A \\ I \end{bmatrix} \tilde{\boldsymbol{y}} \leqq \begin{bmatrix} -\boldsymbol{b} \\ \boldsymbol{b} \\ \boldsymbol{0} \end{bmatrix}.$$

With the use of variable $\tilde{\boldsymbol{x}} = -\tilde{\boldsymbol{y}}$ we can rewrite the above problem as

$$\text{Minimize } \; \boldsymbol{c}^\top \tilde{\boldsymbol{x}} \quad \text{subject to} \quad A\tilde{\boldsymbol{x}} = \boldsymbol{b}, \;\; \tilde{\boldsymbol{x}} \geqq \boldsymbol{0}.$$

This problem is nothing but the original problem P.

This is a very remarkable fact, which is stated below as a theorem; the precise meaning of the statement should be understood as explained above.

**Theorem 3.11.** *In linear programming, the dual of the dual problem is equal to the primal problem.*

**Remark 3.7.** Theorem 3.11 for linear programs is similar in form to Theorem 3.8 for convex cones, and both assert that "the dual of the dual is equal to the original." This similarity, however, is not a coincidence, but it can be understood in a unified manner as the duality concerning the *Legendre transformation*. For more details, see [39] and other references that deal with convex analysis. ■

#### 3.5.3.2 *Duality theorem*

For the primal problem P and its dual problem D in (3.63), the respective feasible regions are denoted as follows:

$$P = \{\boldsymbol{x} \in \mathbb{R}^n \mid A\boldsymbol{x} = \boldsymbol{b}, \boldsymbol{x} \geqq \boldsymbol{0}\}, \qquad D = \{\boldsymbol{y} \in \mathbb{R}^m \mid A^\top \boldsymbol{y} \leqq \boldsymbol{c}\}. \tag{3.65}$$

The following *duality theorem* holds.

**Theorem 3.12 (Duality Theorem).** *For any (primal) problem* P *in the standard form and its dual problem* D *with* $P \neq \emptyset$ *and* $D \neq \emptyset$*, the following hold:*

(1) (Weak Duality) $\boldsymbol{c}^\top \boldsymbol{x} \geqq \boldsymbol{b}^\top \boldsymbol{y}$ *for any* $\boldsymbol{x} \in P$ *and* $\boldsymbol{y} \in D$.
(2) (Strong Duality)

$$\inf\{\boldsymbol{c}^\top \boldsymbol{x} \mid \boldsymbol{x} \in P\} = \sup\{\boldsymbol{b}^\top \boldsymbol{y} \mid \boldsymbol{y} \in D\}. \tag{3.66}$$

*Moreover, there exist* $\boldsymbol{x}$ *and* $\boldsymbol{y}$ *that attain* inf *and* sup*, respectively; in particular, the values of both sides in* (3.66) *are finite.*

**Proof.** (1) Assume $\boldsymbol{x} \in P$ and $\boldsymbol{y} \in D$. The equation $\boldsymbol{y}^\top \boldsymbol{b} = \boldsymbol{y}^\top (A\boldsymbol{x}) = (\boldsymbol{y}^\top A)\boldsymbol{x}$ follows from $\boldsymbol{b} = A\boldsymbol{x}$, whereas $(\boldsymbol{y}^\top A)\boldsymbol{x} \leqq \boldsymbol{c}^\top \boldsymbol{x}$ results from $\boldsymbol{y}^\top A \leqq \boldsymbol{c}^\top$ and $\boldsymbol{x} \geqq \boldsymbol{0}$. Therefore, $\boldsymbol{b}^\top \boldsymbol{y} \leqq \boldsymbol{c}^\top \boldsymbol{x}$.

(2) Since the weak duality is proved in (1), it suffices to show the existence of $\boldsymbol{x} \in P$ and $\boldsymbol{y} \in D$ that satisfy $\boldsymbol{c}^\top \boldsymbol{x} \leqq \boldsymbol{b}^\top \boldsymbol{y}$. Define auxiliary variables $w$, $\boldsymbol{y}'$, $\boldsymbol{y}''$, and $\boldsymbol{z}$ as

$$w = \boldsymbol{b}^\top \boldsymbol{y} - \boldsymbol{c}^\top \boldsymbol{x} \geqq 0, \quad \boldsymbol{y} = \boldsymbol{y}' - \boldsymbol{y}'' \quad (\boldsymbol{y}', \boldsymbol{y}'' \geqq \boldsymbol{0}), \quad \boldsymbol{z} = \boldsymbol{c} - A^\top \boldsymbol{y} \geqq \boldsymbol{0}.$$

Then the existence of $\boldsymbol{x} \in P$ and $\boldsymbol{y} \in D$ satisfying $\boldsymbol{c}^\top \boldsymbol{x} \leqq \boldsymbol{b}^\top \boldsymbol{y}$ is equivalent to the system of equations

$$\begin{bmatrix} 1 & \boldsymbol{c}^\top & -\boldsymbol{b}^\top & \boldsymbol{b}^\top & \boldsymbol{0}^\top \\ \boldsymbol{0} & A & O & O & O \\ \boldsymbol{0} & O & A^\top & -A^\top & I \end{bmatrix} \begin{bmatrix} w \\ \boldsymbol{x} \\ \boldsymbol{y}' \\ \boldsymbol{y}'' \\ \boldsymbol{z} \end{bmatrix} = \begin{bmatrix} 0 \\ \boldsymbol{b} \\ \boldsymbol{c} \end{bmatrix}$$

having a nonnegative solution. By Farkas' lemma (Theorem 3.3), this condition is equivalent further to the condition that the implication

$$\alpha \geqq 0, \ \ \alpha \boldsymbol{c} + A^\top \boldsymbol{\beta} \geqq \boldsymbol{0}, \ \ \alpha \boldsymbol{b} = A\boldsymbol{\gamma}, \ \ \boldsymbol{\gamma} \geqq \boldsymbol{0} \ \Longrightarrow \ \boldsymbol{\beta}^\top \boldsymbol{b} + \boldsymbol{\gamma}^\top \boldsymbol{c} \geqq 0 \tag{3.67}$$

holds for all $\alpha$, $\boldsymbol{\beta}$ and $\boldsymbol{\gamma}$. We prove this implication by dividing into two cases: $\alpha > 0$ or $\alpha = 0$.

(i) [Case of $\alpha > 0$] We have

$$\alpha(\boldsymbol{\beta}^\top \boldsymbol{b} + \boldsymbol{\gamma}^\top \boldsymbol{c}) = \boldsymbol{\beta}^\top (\alpha \boldsymbol{b}) + \boldsymbol{\gamma}^\top (\alpha \boldsymbol{c}) \geqq \boldsymbol{\beta}^\top (A\boldsymbol{\gamma}) + \boldsymbol{\gamma}^\top (-A^\top \boldsymbol{\beta}) = 0.$$

If $\alpha > 0$, this implies that $\boldsymbol{\beta}^\top \boldsymbol{b} + \boldsymbol{\gamma}^\top \boldsymbol{c} \geqq 0$.

(ii) [Case of $\alpha = 0$] Since $P \neq \emptyset$ and $A^\top \boldsymbol{\beta} \geqq \boldsymbol{0}$, it follows from Farkas' lemma (Theorem 3.3) that $\boldsymbol{\beta}^\top \boldsymbol{b} \geqq 0$. Since $D \neq \emptyset$, $A\boldsymbol{\gamma} = \boldsymbol{0}$ and $\boldsymbol{\gamma} \geqq \boldsymbol{0}$, we can take (any) $\boldsymbol{y}_0 \in D$ to obtain

$$\boldsymbol{\gamma}^\top \boldsymbol{c} \geqq \boldsymbol{\gamma}^\top (A^\top \boldsymbol{y}_0) = (A\boldsymbol{\gamma})^\top \boldsymbol{y}_0 = 0.$$

Therefore, $\boldsymbol{\beta}^\top \boldsymbol{b} + \boldsymbol{\gamma}^\top \boldsymbol{c} \geqq 0$. □

Theorem 3.12 shows that solving the given problem P is equivalent to finding $(\boldsymbol{x}, \boldsymbol{y})$ that satisfies the following linear equality and inequality conditions:

$$A\boldsymbol{x} = \boldsymbol{b}, \qquad \boldsymbol{x} \geqq \boldsymbol{0}, \qquad \boldsymbol{c}^\top \boldsymbol{x} \leqq \boldsymbol{b}^\top \boldsymbol{y}, \qquad A^\top \boldsymbol{y} \leqq \boldsymbol{c}. \tag{3.68}$$

In this way, the duality theorem enables us to transform an optimization problem to a problem of finding a solution to a system of inequalities.

**Remark 3.8.** Theorem 3.12 provides us with a method for verifying optimality. Consider an optimization problem, in general, of minimizing an objective function $f(\boldsymbol{x})$. To prove that a given feasible solution $\boldsymbol{x}$ is *not* the optimal solution, we only have to show another feasible solution $\boldsymbol{x}'$ that satisfies $f(\boldsymbol{x}) > f(\boldsymbol{x}')$. However, what can we do to prove that $\boldsymbol{x}$ *is* actually the optimal solution? It is out of the question to check inequalities $f(\boldsymbol{x}) \leqq f(\boldsymbol{x}')$ for all (infinitely many) $\boldsymbol{x}'$. To prove that $\boldsymbol{x}$ is optimal, we need a "proof of nonexistence" that there exists no better $\boldsymbol{x}'$. Of course, this could be a very difficult problem in general, but in linear programming, Theorem 3.12 offers a remedy. We can prove the optimality of $\boldsymbol{x}$ by demonstrating a feasible solution $\boldsymbol{y} \in D$ of the dual problem D that satisfies $\boldsymbol{c}^\top \boldsymbol{x} = \boldsymbol{b}^\top \boldsymbol{y}$, which is always possible by Theorem 3.12. In general terms such information (vectors, sets, *etc.*) as $\boldsymbol{y}$ here is referred to as a *certificate of optimality*, which, once available, can prove the optimality. Note the similarity of this argument for the duality theorem (Theorem 3.12) to that for theorems of the alternative explained in Sec. 3.3.2. ■

#### 3.5.3.3 *Example*

Let us verify the duality theorem through concrete calculations for a simple example. We consider the following linear programming problem in the standard form:

$$
\begin{array}{llll}
\text{Minimize} & 12\,x_1 + 4\,x_2 + 3\,x_3 & & \cdots \text{(i)} \\
\text{subject to} & x_1 + \ x_2 + \ x_3 & = 1 & \cdots \text{(ii)} \\
 & 4\,x_1 + \ x_2 & = 2 & \cdots \text{(iii)} \\
 & x_1,\ x_2,\ x_3 & \geqq 0 & \cdots \text{(iv)}
\end{array}
\tag{3.69}
$$

Here, $x_1$, $x_2$, and $x_3$ are the decision variables for optimization. Problems with three variables cannot be solved so easily by visualizing with a figure. But this particular problem can be solved fairly easily by eliminating the variables in an appropriate manner. First,

$$x_2 = 2 - 4x_1 \quad \geqq 0$$

follows from (iii) and (iv). Then, from this with (ii) and (iv) follows

$$x_3 = 1 - x_1 - x_2 = -1 + 3x_1 \quad \geqq 0.$$

Therefore, the objective function $g$ can be expressed as

$$g = 12x_1 + 4x_2 + 3x_3 = 12x_1 + 4(2 - 4x_1) + 3(-1 + 3x_1) = 5 + 5x_1.$$

The range of variable $x_1$ is described by the inequalities shown above as well as $x_1 \geqq 0$ in (iv), which amounts to $1/3 \leqq x_1 \leqq 1/2$. Therefore, $g$ takes the minimum at $x_1 = 1/3$. Accordingly, the optimal solution is $(x_1, x_2, x_3) = (1/3, 2/3, 0)$ and the minimum value of the objective function $g$ is given as

$$g_{\min} = 5 + 5 \cdot \frac{1}{3} = \frac{20}{3}. \tag{3.70}$$

Next, we consider the problem dual to (3.69). According to (3.62), the dual problem is

$$
\begin{array}{lll}
\text{Maximize} & y_1 + 2y_2 & \\
\text{subject to} & y_1 + 4y_2 & \leqq 12 \\
 & y_1 + y_2 & \leqq 4 \\
 & y_1 & \leqq 3
\end{array}
\tag{3.71}
$$

This problem is essentially the same as the problem in (3.52), with the only difference being that the variables are changed to $(y_1, y_2)$. Therefore, the optimal solution of the problem (3.71) is $(y_1, y_2) = (4/3, 8/3)$ and the maximum value of the objective function $f_{\max}$ is

$$f_{\max} = \frac{4}{3} + 2 \cdot \frac{8}{3} = \frac{20}{3}, \tag{3.72}$$

as is shown in (3.53).

With (3.70) and (3.72), we obtain the strong duality

$$\begin{aligned}&\text{minimum value } g_{\text{min}} \text{ of the primal problem (3.69)}\\&= \text{maximum value } f_{\text{max}} \text{ of the dual problem (3.71)}\end{aligned}$$

as a particular instance of (3.66). We have thus verified the duality theorem for a simple example by solving the primal and dual problems with direct calculations.

The duality theorem is also used as a theoretical basis for practical algorithms such as the simplex method and the interior-point method.

# Chapter 4

# Integer Matrices

In this chapter we focus on matrices with integer entries, with particular emphasis on integer solutions to systems of linear equations and inequalities. Since division is not always possible between integers, it is necessary to consider divisibility conditions in addition to the rank condition. Fundamental facts about integer solutions to systems of linear equations and inequalities are discussed by using integer elementary transformations, the Hermite normal form, and the Smith normal form.

## 4.1 Unimodular Matrices

### 4.1.1 *Inverse of Integer Matrices*

A matrix is called an *integer matrix* if its entries are all integers. Even when a square integer matrix has a nonzero determinant, and is thus nonsingular, it is not necessarily true that its inverse is an integer matrix. This stems from the fact that the multiplicative inverse (the reciprocal) of an integer is not necessarily an integer.

**Example 4.1.** Matrix $A = \begin{bmatrix} 1 & -1 \\ 1 & 1 \end{bmatrix}$ is nonsingular, but its inverse $A^{-1} = \begin{bmatrix} 1/2 & 1/2 \\ -1/2 & 1/2 \end{bmatrix}$ is not an integer matrix. ■

First, let us derive a necessary condition for an integer matrix to have an integer inverse matrix. Suppose that an integer matrix $A$ is nonsingular and that its inverse $B$ is an integer matrix. Since the determinant of an integer matrix is an integer, both $\det A$ and $\det B$ are integers, whereas

$$\det A \cdot \det B = \det(AB) = 1$$

by $AB = I$ (unit matrix). It then follows that $\det A = 1$ or $-1$, which is a necessary condition for $A$ to have an integer inverse matrix.

As is stated in Theorem 4.1 below, this necessary condition is, in fact, a sufficient condition. A (square) integer matrix is called a *unimodular matrix* if its determinant is equal to 1 or $-1$.

**Theorem 4.1.** *For an integer matrix $A$, the inverse of $A$ exists and is an integer matrix if and only if $A$ is a unimodular matrix.*

**Proof.** The necessity has already been shown. To show the sufficiency, assume $A$ to be a unimodular matrix. Then $A$ is nonsingular and the inverse $A^{-1}$ exists. Each entry of the matrix $A^{-1}$ is expressed in the form of "cofactor of $A$ / $\det A$" where a cofactor of an integer matrix $A$ is an integer and $\det A = \pm 1$. Therefore, each entry of $A^{-1}$ is an integer. □

**Proposition 4.1.** *If an integer matrix $A$ is unimodular, then its inverse $A^{-1}$ is also unimodular.*

**Proof.** By Theorem 4.1, $A^{-1}$ is an integer matrix. We also have $\det(A^{-1}) = 1/\det A = \pm 1$. Therefore, $A^{-1}$ is a unimodular matrix. □

**Example 4.2.** Matrix $A = \begin{bmatrix} 1 & 2 \\ 1 & 1 \end{bmatrix}$ is a unimodular matrix. Its inverse $A^{-1} = \begin{bmatrix} -1 & 2 \\ 1 & -1 \end{bmatrix}$ is an integer matrix and, moreover, a unimodular matrix. Indeed we have $\det A = \det(A^{-1}) = -1$. ■

### 4.1.2 *Integer Lattice Points*

A vector is called an *integer vector* if all of its components are integers. The set of all $n$-dimensional integer vectors will be denoted as $\mathbb{Z}^n$, which can be regarded as a subset of the $n$-dimensional real vector space $\mathbb{R}^n$. Geometrically, $\mathbb{Z}^n$ is an *integer lattice* lying in the space $\mathbb{R}^n$. We can also write this as

$$\mathbb{Z}^n = \{x_1 \boldsymbol{e}_1 + x_2 \boldsymbol{e}_2 + \cdots + x_n \boldsymbol{e}_n \mid x_j \in \mathbb{Z} \ (j = 1, 2, \ldots, n)\}, \tag{4.1}$$

where $\boldsymbol{e}_j = (0, \ldots, 0, 1, 0, \ldots, 0)^\top$ is the $j$th unit vector for $j = 1, \ldots, n$.

Suppose that $n$ integer vectors $\boldsymbol{a}_1, \boldsymbol{a}_2, \ldots, \boldsymbol{a}_n$, are given, where each $\boldsymbol{a}_j$ is an $n$-dimensional vector. The set of all linear combinations with integer coefficients thereof, *i.e.*,

$$\{x_1 \boldsymbol{a}_1 + x_2 \boldsymbol{a}_2 + \cdots + x_n \boldsymbol{a}_n \mid x_j \in \mathbb{Z} \ (j = 1, 2, \ldots, n)\},$$

is called the *lattice* generated by $\boldsymbol{a}_1, \boldsymbol{a}_2, \ldots, \boldsymbol{a}_n$. As shown in (4.1), $\mathbb{Z}^n$ is generated by the unit vectors $\boldsymbol{e}_1, \boldsymbol{e}_2, \ldots, \boldsymbol{e}_n$, but it can also be generated by other sets of vectors. The next theorem gives a necessary and sufficient condition for vectors $\boldsymbol{a}_1, \boldsymbol{a}_2, \ldots, \boldsymbol{a}_n$ to generate $\mathbb{Z}^n$. It also implies that the transformation between two different generating sets of $\mathbb{Z}^n$ is expressed by a unimodular matrix.

**Theorem 4.2.** *A set of $n$-dimensional integer vectors $\boldsymbol{a}_1, \boldsymbol{a}_2, \ldots, \boldsymbol{a}_n$ generates $\mathbb{Z}^n$ if and only if $A = [\boldsymbol{a}_1, \boldsymbol{a}_2, \ldots, \boldsymbol{a}_n]$ is a unimodular matrix.*

**Proof.** Suppose that $A$ is a unimodular matrix. For every $\boldsymbol{b} \in \mathbb{Z}^n$, the vector $\boldsymbol{x} = A^{-1}\boldsymbol{b}$ satisfies $A\boldsymbol{x} = \boldsymbol{b}$ and $\boldsymbol{x} \in \mathbb{Z}^n$. Therefore, $\boldsymbol{a}_1, \boldsymbol{a}_2, \ldots, \boldsymbol{a}_n$ generate $\mathbb{Z}^n$.

Conversely, suppose that integer vectors $\boldsymbol{a}_1, \boldsymbol{a}_2, \ldots, \boldsymbol{a}_n$ generate $\mathbb{Z}^n$. For the $j$th unit vector $\boldsymbol{e}_j$, there exists an integer vector $\boldsymbol{x}_j$ that satisfies $A\boldsymbol{x}_j = \boldsymbol{e}_j$ $(j = 1, \ldots, n)$. The matrix $X = [\boldsymbol{x}_1, \boldsymbol{x}_2, \ldots, \boldsymbol{x}_n]$ formed by the vectors $\boldsymbol{x}_1, \boldsymbol{x}_2, \ldots, \boldsymbol{x}_n$ is an integer matrix and satisfies $AX = I$. Therefore, $A$ is a unimodular matrix by Theorem 4.1. □

**Example 4.3.** The column vectors $\boldsymbol{a}_1 = \begin{bmatrix} 1 \\ 1 \end{bmatrix}$ and $\boldsymbol{a}_2 = \begin{bmatrix} -1 \\ 1 \end{bmatrix}$ of the matrix $A$ of Example 4.1, which is not unimodular, do not generate $\mathbb{Z}^2$. For example, the vector $\boldsymbol{b} = \begin{bmatrix} 0 \\ 1 \end{bmatrix}$ cannot be expressed as $\boldsymbol{b} = x_1\boldsymbol{a}_1 + x_2\boldsymbol{a}_2$ with $x_1, x_2 \in \mathbb{Z}$. ■

In Theorem 4.2 above, we have explained the significance of a unimodular matrix as a transformation of generating sets. The same mathematical content can also be expressed as a property of the mapping defined by a unimodular matrix as follows.

Let $A$ be a square integer matrix of order $n$. Since $A\boldsymbol{x}$ is an integer vector for every integer vector $\boldsymbol{x}$, the correspondence $\boldsymbol{x} \mapsto A\boldsymbol{x}$ defines a mapping from $\mathbb{Z}^n$ to $\mathbb{Z}^n$. The unimodularity of the matrix $A$ is a necessary and sufficient condition for the mapping $A : \mathbb{Z}^n \to \mathbb{Z}^n$ defined by $\boldsymbol{x} \mapsto A\boldsymbol{x}$ to be a bijection (one-to-one correspondence).

**Theorem 4.3.** *For a square integer matrix $A$ of order $n$, the mapping $A : \mathbb{Z}^n \to \mathbb{Z}^n$ is a bijection if and only if $A$ is a unimodular matrix.*

**Proof.** Suppose that $A$ is a unimodular matrix. For every $\boldsymbol{y} \in \mathbb{Z}^n$, the vector $\boldsymbol{x} = A^{-1}\boldsymbol{y}$ satisfies $A\boldsymbol{x} = \boldsymbol{y}$ and $\boldsymbol{x} \in \mathbb{Z}^n$; therefore, the mapping defined by $A$ is a surjection. If $A\boldsymbol{x} = A\boldsymbol{y}$, then $A(\boldsymbol{x} - \boldsymbol{y}) = \boldsymbol{0}$, and $\boldsymbol{x} = \boldsymbol{y}$

follows from the nonsingularity of $A$. Therefore, the mapping defined by $A$ is an injection.

Conversely, suppose that the mapping defined by $A$ is a bijection. For the $j$th unit vector $\boldsymbol{e}_j$, there exists an integer vector $\boldsymbol{x}_j$ that satisfies $A\boldsymbol{x}_j = \boldsymbol{e}_j$ $(j = 1, \ldots, n)$. The matrix $X = [\boldsymbol{x}_1, \boldsymbol{x}_2, \ldots, \boldsymbol{x}_n]$ formed by the vectors $\boldsymbol{x}_1, \boldsymbol{x}_2, \ldots, \boldsymbol{x}_n$ is an integer matrix and satisfies $AX = I$. Therefore, $A$ is a unimodular matrix by Theorem 4.1. □

## 4.2 Integer Elementary Transformations

We consider elementary transformations for integer matrices. Elementary transformations are basic operations in dealing with integer solutions of systems of linear equations and/or inequalities.

### 4.2.1 *Definition*

Let us first recall elementary transformations for real matrices (matrices whose entries are real numbers). The elementary column transformations consist of the following three kinds of operations [6, 15]:

(1) exchanging two columns,
(2) multiplying a certain column by a nonzero number, and
(3) adding a scalar multiple of a certain column to another column.

For integer matrices, however, we need to preserve integrality in elementary transformations and their inverse transformations. With this in mind we introduce the following three kinds of operations:

(1) exchanging two columns,
(2) multiplying a certain column by $-1$, and
(3) adding an integer multiple of a certain column to another column

as *elementary column transformations* for integer matrices, which are also referred to as *integer elementary column transformations.*

The above three types of integer elementary column transformations are represented, respectively, by matrices of the following forms:

$$E_1 = \left[\begin{array}{cc|ccc} & 1 & & & \\ 1 & & & & \\ \hline & & 1 & & \\ & & & \ddots & \\ & & & & 1 \end{array}\right], \quad E_2 = \left[\begin{array}{c|cccc} -1 & & & & \\ \hline & 1 & & & \\ & & 1 & & \\ & & & \ddots & \\ & & & & 1 \end{array}\right], \quad E_3 = \left[\begin{array}{cc|ccc} 1 & c & & & \\ & 1 & & & \\ \hline & & 1 & & \\ & & & \ddots & \\ & & & & 1 \end{array}\right], \tag{4.2}$$

where $c$ stands for an integer. Such matrices are called *integer elementary matrices*. More precisely, the matrices $E_1 = E_1(p, q)$, $E_2 = E_2(p)$, and $E_3 = E_3(p, q; c)$ are defined as follows ($p \neq q$):

(1) $E_1(p, q)$: The submatrix corresponding to row and column indices $\{p, q\}$ is $\begin{bmatrix} 0 & 1 \\ 1 & 0 \end{bmatrix}$, and the rest is a unit matrix,
(2) $E_2(p)$: The submatrix corresponding to row and column index $\{p\}$ is $[\,-1\,]$, and the rest is a unit matrix,
(3) $E_3(p, q; c)$: The $(p, q)$ entry is $c$, the other off-diagonal entries are all 0, and the diagonal entries are all 1. The submatrix corresponding to row and column indices $\{p, q\}$ is $\begin{bmatrix} 1 & c \\ 0 & 1 \end{bmatrix}$ if $p < q$, and $\begin{bmatrix} 1 & 0 \\ c & 1 \end{bmatrix}$ if $q < p$.

With the above definitions we have the following:

(1) If we multiply $A$ by $E_1(p, q)$ from the right, the $p$th column and the $q$th column are exchanged.
(2) If we multiply $A$ by $E_2(p)$ from the right, the $p$th column is multiplied by $-1$.
(3) If we multiply $A$ by $E_3(p, q; c)$ from the right, then $c$ times the $p$th column is added to the $q$th column.

Integer elementary matrices are unimodular matrices.[1] Indeed,

$$\det E_1(p, q) = -1, \qquad \det E_2(p) = -1, \qquad \det E_3(p, q; c) = 1,$$

and moreover,

$$E_1(p, q)^{-1} = E_1(p, q), \quad E_2(p)^{-1} = E_2(p), \quad E_3(p, q; c)^{-1} = E_3(p, q; -c).$$

Similarly, the following three kinds of operations are defined as *elementary row transformations* for integer matrices:

(1) exchanging two rows,

[1]In Theorem 4.5 in Sec. 4.3, it will be shown that every unimodular matrix can be expressed as a product of integer elementary matrices.

(2) multiplying a certain row by $-1$, and
(3) adding an integer multiple of a certain row to another row.

These operations are also referred to as *integer elementary row transformations.* Integer elementary row transformations are equivalent to multiplying $A$ by $E_1(p,q)$, $E_2(p)$, and $E_3(p,q;c)$ from the left of the matrix as follows:

(1) If we multiply $A$ by $E_1(p,q)$ from the left, the $p$th row and $q$th row are exchanged.
(2) If we multiply $A$ by $E_2(p)$ from the left, the $p$th row is multiplied by $-1$.
(3) If we multiply $A$ by $E_3(p,q;c)$ from the left, then $c$ times the $q$th row is added to the $p$th row.

Integer elementary row and column transformations are together referred to as *integer elementary transformations.*

The significance of elementary transformations is summarized in the following proposition.[2] For integers $a_1, a_2, \ldots, a_n$ ($n \geqq 1$), in general, we denote by $\gcd(a_1, a_2, \ldots, a_n)$ their *greatest common divisor* ($\geqq 0$). In the exceptional case where all the $a_j$ are 0, we set their greatest common divisor to be 0. Note that $\gcd(a_1, a_2, \ldots, a_n) = \gcd(|a_1|, |a_2|, \ldots, |a_n|)$.

**Proposition 4.2.**

(1) *A row vector* $(a_1, a_2, \ldots, a_n) \neq \mathbf{0}^\top$ *with integer components can be transformed to* $(b, 0, \ldots, 0)$ *through repeated integer elementary column transformations, where* $b = \gcd(a_1, a_2, \ldots, a_n)$.
(2) *A column vector* $(a_1, a_2, \ldots, a_n)^\top \neq \mathbf{0}$ *with integer components can be transformed to* $(b, 0, \ldots, 0)^\top$ *through repeated integer elementary row transformations, where* $b = \gcd(a_1, a_2, \ldots, a_n)$.

**Proof.** (1) First, we can make all the components nonnegative (*i.e.*, $a_j \geqq 0$ for all $j$) by multiplying negative components, if any, by $-1$. Next, we permute the components (or columns) to bring the minimum of the positive components to the first component $a_1$. For $j = 2, \ldots, n$, divide $a_j$ by $a_1$ to obtain quotient $q_j$ and remainder $r_j$, *i.e.*,

$$a_j = a_1 q_j + r_j, \qquad a_1 > r_j \geqq 0. \tag{4.3}$$

By adding $-q_j$ times the first component to the $j$th component for $j = 2, \ldots, n$, the row vector $(a_1, a_2, \ldots, a_n)$ is transformed to $(a_1, r_2, \ldots, r_n)$. If

[2] Proposition 4.2 will be used in the proofs for the Hermite normal form (Theorem 4.4) and the Smith normal form (Theorem 4.8).

$r_j = 0$ for $j = 2, \ldots, n$, then the assertion of (1) holds with $b = a_1$. If this is not the case, we repeat the same transformation. In the course of such transformations, the minimum value of the positive components decreases strictly since $a_1 > r_j \geqq 0$ ($j = 2, \ldots, n$), and so the process ends after a finite number of iterations. Note an important fact that

$$\gcd(a_1, r_2, \ldots, r_n) = \gcd(a_1, a_2, \ldots, a_n),$$

which follows from (4.3).

The proof for (2) is similar to that for (1). □

**Example 4.4.** Row vector $(6, 4, 10)$ is transformed through integer elementary column transformations as follows:

$$\begin{aligned}(6,4,10) \;&\Rightarrow\; (4,6,10) \;\rightarrow\; (4,2,10) \;\rightarrow\; (4,2,2)\\ &\Rightarrow\; (2,4,2) \;\rightarrow\; (2,0,2) \;\rightarrow\; (2,0,0) = (b,0,0).\end{aligned}$$

Here, "$\Rightarrow$" stands for a swap of components and "$\rightarrow$" for a transformation via division in (4.3), and $b = 2 = \gcd(6, 4, 10)$. ■

**Remark 4.1.** The key fact in the proof of Proposition 4.2 is the reduction through "division" shown in (4.3). Such division and divisibility relation can also be defined for univariate polynomials, which fact enables us to develop a similar theory for matrices whose entries are univariate polynomials. In particular, we can formulate a proposition for polynomial matrices (Proposition 5.2) that corresponds to Proposition 4.2 above. On the basis of this proposition, theorems corresponding to the Smith normal form and the Hermite normal form can be established also for polynomial matrices. See Chap. 5 for more details. ■

### 4.2.2 *Determinantal Divisors*

The proof of Proposition 4.2 shows that the greatest common divisor of the components of a vector remains invariant under integer elementary transformations. Such a relationship between integer elementary transformations and the greatest common divisor is extended to matrices as follows.

For an $m \times n$ integer matrix $A$ and a natural number $k$, the greatest common divisor of all minors (subdeterminants) of order $k$ is called the $k$th *determinantal divisor* of $A$, which we denote by $d_k(A)$. That is,

$$d_k(A) = \gcd\{\det A[I, J] \mid |I| = |J| = k\}, \tag{4.4}$$

where the notation $A[I, J]$ is explained in Remark 4.2 below. On the right-hand side above, $I$ runs over all subsets of size $k$ of the row set of $A$, and $J$

runs over all subsets of size $k$ of the column set of $A$. The number of such $I$ is equal to $\binom{m}{k}$ (binomial coefficient) and the number of such $J$ is to $\binom{n}{k}$, and therefore, the number of submatrices $A[I,J]$ in (4.4) is $\binom{m}{k} \times \binom{n}{k}$. The index $k$ stays in the range of $1 \leqq k \leqq \min(m,n)$, but it is often convenient to define $d_0(A) = 1$ by allowing $k = 0$.

**Example 4.5.** For matrix $A = \begin{bmatrix} 6 & 4 & 10 \\ -1 & 1 & -5 \end{bmatrix}$ we have

$$d_1(A) = \gcd(6, 4, 10, -1, 1, -5) = 1,$$

$$d_2(A) = \gcd\left(\det\begin{bmatrix} 6 & 4 \\ -1 & 1 \end{bmatrix}, \det\begin{bmatrix} 6 & 10 \\ -1 & -5 \end{bmatrix}, \det\begin{bmatrix} 4 & 10 \\ 1 & -5 \end{bmatrix}\right) = 10.$$ ■

**Remark 4.2.** The notations for *submatrices* and *minors* (or *subdeterminants*) used in this book are explained here. For a row subset $I$ and a column subset $J$, $A[I,J]$ means the submatrix of $A$ obtained by deleting the rows not in $I$ and the columns not in $J$. If $I = \{i_1, i_2, \ldots, i_k\}$ $(i_1 < i_2 < \cdots < i_k)$ and $J = \{j_1, j_2, \ldots, j_k\}$ $(j_1 < j_2 < \cdots < j_k)$, $A[I,J]$ is a $k \times k$ matrix such that the $(p,q)$ entry is equal to $a_{i_p j_q}$, the $(i_p, j_q)$ entry of $A$. Incidentally, the ordering of rows and columns affects the sign of a subdeterminant, but it does not matter when the greatest common divisors are considered. ■

The determinantal divisors remain invariant under integer elementary transformations.

**Proposition 4.3.** *Let $A$ be an $m \times n$ integer matrix. For any integer elementary matrices $E$ and $F$ of orders $m$ and $n$, respectively, we have*

$$d_k(EAF) = d_k(A) \qquad (1 \leqq k \leqq \min(m,n)). \tag{4.5}$$

**Proof.** It suffices to prove the invariance for column transformations only; so we assume $E$ to be the identity matrix. The equation (4.5) holds obviously for $F = E_1(p,q)$ or $E_2(p)$. To consider the remaining case of $F = E_3(p,q;c)$, we put $B = A\,E_3(p,q;c)$ and compare $B[I,J]$ and $A[I,J]$, where $|I| = |J| = k$. There are four cases depending on the inclusion relation of $p$ and $q$ in the subset $J$: (i) $p \in J$, $q \in J$, (ii) $p \in J$, $q \notin J$, (iii) $p \notin J$, $q \notin J$, and (iv) $p \notin J$, $q \in J$. In the cases of (i), (ii), and (iii), we have $\det B[I,J] = \det A[I,J]$, which is a multiple of $d_k(A)$. In the case of (iv), we have[3]

$$\det B[I,J] = \det A[I,J] \pm c \cdot \det A[I, (J \setminus \{q\}) \cup \{p\}],$$

[3]The notation $(J \setminus \{q\}) \cup \{p\}$ means the set (of column indices) obtained from $J$ by deleting $q$ and adding $p$. The sign of the coefficient $\pm c$ should be chosen appropriately.

which shows that $\det B[I, J]$ is a multiple of $d_k(A)$. Therefore, $d_k(B)$ is a multiple of $d_k(A)$. Since integer elementary transformations are invertible, we also have that $d_k(A)$ is a multiple of $d_k(B)$. Therefore, $d_k(B) = d_k(A)$. □

## 4.3 Hermite Normal Form

We consider transforming a given integer matrix to a simplest possible form by means of integer elementary column transformations. Let $A$ be an $m \times n$ integer matrix of full row rank ($\operatorname{rank} A = m$). Since $m \leqq n$, $A$ is either a square matrix or a rectangular matrix with more columns than rows.

The targeted "simplest form" is an $m \times n$ integer matrix $B = (b_{ij})$ that satisfies the following conditions:

$$b_{ii} > 0 \qquad (1 \leqq i \leqq m), \tag{4.6}$$

$$0 \leqq b_{ij} < b_{ii} \qquad (1 \leqq j < i \leqq m), \tag{4.7}$$

$$b_{ij} = 0 \qquad (1 \leqq i \leqq m;\ i < j \leqq n). \tag{4.8}$$

That is, the targeted form is a lower (left) triangular matrix with non-negative integer entries such that the diagonal entries are greater than the off-diagonal entries in the same row. A matrix that satisfies these conditions is said to be in *Hermite normal form*.

The following important theorem, Theorem 4.4, states that every integer matrix can be brought to a matrix in Hermite normal form via integer elementary column transformations. Recall Proposition 4.2(1), which states that every row vector can be brought to the form of $(b, 0, \ldots, 0)$ via integer elementary column transformations. This proposition is certainly a special case ($m = 1$) of Theorem 4.4, but at the same time, it plays the essential role in the proof of Theorem 4.4.

**Theorem 4.4.** *Every integer matrix of full row rank can be transformed to Hermite normal form through repeated integer elementary column transformations.*

**Proof.** Let $A = (a_{ij})$ be an integer matrix of full row rank. By Proposition 4.2(1), we can transform the first row vector $(a_{11}, a_{12}, \ldots, a_{1n})$ of $A$, which is nonzero by the rank assumption, to a vector of the form of $(b_{11}, 0, \ldots, 0)$ with $b_{11} > 0$ through repeated integer elementary column transformations. Denote by $A^{(1)} = (a_{ij}^{(1)})$ the matrix resulting from these column transformations applied to the matrix $A$. Then

$$a_{11}^{(1)} = b_{11}, \qquad a_{1j}^{(1)} = 0 \quad (j \geqq 2).$$

Next, consider the row vector $(a_{22}^{(1)}, a_{23}^{(1)}, \ldots, a_{2n}^{(1)})$ consisting of the entries of $A^{(1)}$ in the second row with column indices greater than or equal to 2. This row vector is nonzero, again by the rank assumption, and we can transform it to the form of $(b_{22}, 0, \ldots, 0)$ with $b_{22} > 0$. Moreover, by division of $a_{21}^{(1)}$ by $b_{22}$, we can find integers $q$ and $r$ such that

$$a_{21}^{(1)} = b_{22}q + r, \qquad b_{22} > r \geqq 0.$$

Let $b_{21} = r$. By adding $-q$ times the second column to the first column, we can transform the second row vector to $(b_{21}, b_{22}, 0, \ldots, 0)$ where $0 \leqq b_{21} < b_{22}$. Denote by $A^{(2)} = (a_{ij}^{(2)})$ the resulting matrix. Then

$$a_{11}^{(2)} = b_{11}, \quad a_{1j}^{(2)} = 0 \ \ (j \geqq 2), \quad a_{21}^{(2)} = b_{21}, \quad a_{22}^{(2)} = b_{22}, \quad a_{2j}^{(2)} = 0 \ \ (j \geqq 3).$$

By repeating such a process, we arrive at $A^{(m)}$, which is a matrix in Hermite normal form. □

For an integer matrix $A$, the matrix in Hermite normal form given in Theorem 4.4 is referred to as the Hermite normal form of $A$. (The uniqueness of this normal form will be shown later in Theorem 4.7.)

**Example 4.6.** Matrix $A = \begin{bmatrix} 6 & 4 & 10 \\ -1 & 1 & -5 \end{bmatrix}$ is transformed to a matrix in Hermite normal form via elementary column transformations as follows:

$$\begin{bmatrix} 6 & 4 & 10 \\ -1 & 1 & -5 \end{bmatrix} \xrightarrow[\times E_1(1,2)]{} \begin{bmatrix} 4 & 6 & 10 \\ 1 & -1 & -5 \end{bmatrix} \xrightarrow[\times E_3(1,2;-1)]{} \begin{bmatrix} 4 & 2 & 10 \\ 1 & -2 & -5 \end{bmatrix}$$

$$\xrightarrow[\times E_3(1,3;-2)]{} \begin{bmatrix} 4 & 2 & 2 \\ 1 & -2 & -7 \end{bmatrix} \xrightarrow[\times E_1(1,2)]{} \begin{bmatrix} 2 & 4 & 2 \\ -2 & 1 & -7 \end{bmatrix}$$

$$\xrightarrow[\times E_3(1,2;-2)]{} \begin{bmatrix} 2 & 0 & 2 \\ -2 & 5 & -7 \end{bmatrix} \xrightarrow[\times E_3(1,3;-1)]{} \begin{bmatrix} 2 & 0 & 0 \\ -2 & 5 & -5 \end{bmatrix} = A^{(1)}$$

$$\xrightarrow[\times E_3(2,3;1)]{} \begin{bmatrix} 2 & 0 & 0 \\ -2 & 5 & 0 \end{bmatrix} \xrightarrow[\times E_3(2,1;1)]{} \begin{bmatrix} 2 & 0 & 0 \\ 3 & 5 & 0 \end{bmatrix} = A^{(2)} = B$$

(Hermite normal form). ■

As a corollary to Theorem 4.4 (existence of Hermite normal form) we obtain a characterization of unimodular matrices.

**Theorem 4.5.** *The following three conditions* (a) *to* (c) *are equivalent for a square integer matrix $A$:*

(a) *$A$ is a unimodular matrix.*

(b) *A is nonsingular and its Hermite normal form is the unit matrix I.*
(c) *A is expressed as a product of integer elementary matrices.*

**Proof.** Let $A$ be an $n \times n$ matrix. In each of cases (a), (b), and (c), $A$ is a nonsingular matrix, for which the Hermite normal form $B = (b_{ij})$ exists by Theorem 4.4. According to the construction of the Hermite normal form, there exist integer elementary matrices $E_1, E_2, \ldots, E_k$ such that $B = AE_1E_2 \cdots E_k$.

[(a) ⇒ (b)] Since the matrix $B$ is a lower-triangular matrix and $|\det E_i| = 1$ for $i = 1, \ldots, k$, we have

$$|b_{11}b_{22} \cdots b_{nn}| = |\det B| = |\det A| = 1.$$

This shows $b_{11} = b_{22} = \cdots = b_{nn} = 1$, since the diagonal entries $b_{11}, b_{22}, \ldots, b_{nn}$ are positive integers. Furthermore, we have $|b_{ij}| < b_{ii}$ $(j < i)$, which shows $b_{ij} = 0$ $(j < i)$. Hence $B = I$ (unit matrix).

[(b) ⇒ (c)] It follows from $I = B = AE_1E_2 \cdots E_k$ that $A = E_k^{-1} \cdots E_2^{-1}E_1^{-1}$, where each $E_j^{-1}$ is an integer elementary matrix.

[(c) ⇒ (a)] This is obvious, since an integer elementary matrix is unimodular. □

Theorem 4.5 enables us to rephrase Theorem 4.4 as follows.

**Theorem 4.6.** *For any integer matrix A of full row rank, there exists a unimodular matrix V such that the matrix AV is in Hermite normal form.*

**Proof.** An integer elementary column transformation for a matrix is equivalent to multiplying the matrix with the corresponding integer elementary matrix from the right. By Theorem 4.5, on the other hand, the product of integer elementary matrices is a unimodular matrix. □

**Example 4.7.** In Example 4.6, the product of the integer elementary matrices used to bring $A$ to its Hermite normal form $B$ is given by

$$\begin{aligned} V &= E_1(1,2) \cdot E_3(1,2;-1) \cdot E_3(1,3;-2) \cdot E_1(1,2) \\ &\quad \cdot E_3(1,2;-2) \cdot E_3(1,3;-1) \cdot E_3(2,3;1) \cdot E_3(2,1;1) \\ &= \begin{bmatrix} -1 & -2 & -3 \\ 2 & 3 & 2 \\ 0 & 0 & 1 \end{bmatrix}. \end{aligned}$$

This is a unimodular matrix and the transformation $AV = B$ is as follows:

$$\begin{bmatrix} 6 & 4 & 10 \\ -1 & 1 & -5 \end{bmatrix} \begin{bmatrix} -1 & -2 & -3 \\ 2 & 3 & 2 \\ 0 & 0 & 1 \end{bmatrix} = \begin{bmatrix} 2 & 0 & 0 \\ 3 & 5 & 0 \end{bmatrix}.$$

■

The existence of the Hermite normal form has been established in Theorem 4.4 (or Theorem 4.6). The Hermite normal form is, in fact, uniquely determined.

**Theorem 4.7.** *The Hermite normal form of an integer matrix of full row rank is uniquely determined.*

**Proof.** We prove the uniqueness by contradiction. Suppose that $B = (b_{ij})$ and $B' = (b'_{ij})$ are two different matrices in Hermite normal form that are obtained from a given $m \times n$ integer matrix $A$ of rank $m$ by unimodular column transformations. Define

$$\Lambda(A) = \{A\boldsymbol{x} \mid \boldsymbol{x} \in \mathbb{Z}^n\}.$$

For any unimodular matrix $V$ of order $n$ we have

$$\begin{aligned}\Lambda(AV) &= \{AV\boldsymbol{x} \mid \boldsymbol{x} \in \mathbb{Z}^n\} = \{A\boldsymbol{y} \mid \boldsymbol{y} = V\boldsymbol{x}, \boldsymbol{x} \in \mathbb{Z}^n\} \\ &= \{A\boldsymbol{y} \mid \boldsymbol{y} \in \mathbb{Z}^n\} = \Lambda(A),\end{aligned}$$

from which it follows that $\Lambda(A) = \Lambda(B) = \Lambda(B')$.

Since $B$ and $B'$ are distinct, there exists $(i, j)$ with $b_{ij} \neq b'_{ij}$, where $i \geqq j$. Take such $(i, j)$ with $i$ minimum. We may assume $b_{ii} \geqq b'_{ii}$; otherwise exchange $B$ and $B'$. If $i \neq j$, then by the property of Hermite normal form we have $0 \leqq b_{ij} < b_{ii}$ and $0 \leqq b'_{ij} < b'_{ii} \leqq b_{ii}$, from which follows $0 < |b'_{ij} - b_{ij}| < b_{ii}$. This inequality is also valid when $i = j$.

We denote the column vectors of the matrix $B$ as $\boldsymbol{b}_1, \boldsymbol{b}_2, \ldots, \boldsymbol{b}_n$ and those of $B'$ as $\boldsymbol{b}'_1, \boldsymbol{b}'_2, \ldots, \boldsymbol{b}'_n$, where $\boldsymbol{b}_k = \boldsymbol{b}'_k = \boldsymbol{0}$ $(k > m)$. We now examine the $j$th column vectors for the column index $j$ chosen above. Since $\boldsymbol{b}'_j \in \Lambda(B') = \Lambda(B)$ and $\boldsymbol{b}_j \in \Lambda(B)$, we have $\boldsymbol{b}'_j - \boldsymbol{b}_j \in \Lambda(B)$, and therefore, there exists $\boldsymbol{x} \in \mathbb{Z}^n$ such that

$$\boldsymbol{b}'_j - \boldsymbol{b}_j = B\boldsymbol{x} = x_1\boldsymbol{b}_1 + x_2\boldsymbol{b}_2 + \cdots + x_m\boldsymbol{b}_m.$$

By the choice of the row index $i$, the first $(i-1)$ components of the vector $\boldsymbol{b}'_j - \boldsymbol{b}_j$ are all 0. Since $B$ is a lower (left) triangular matrix, this implies that $x_1 = x_2 = \cdots = x_{i-1} = 0$. Then, the $i$th component in the above equation shows that $b'_{ij} - b_{ij} = x_i b_{ii}$. Since $x_i$ is an integer, this contradicts the inequality $0 < |b'_{ij} - b_{ij}| < b_{ii}$ shown above.

Thus a contradiction has been derived from the assumed existence of two different Hermite normal forms. This proves the uniqueness of the Hermite normal form. □

In this section, we have discussed Hermite normal forms for integer matrices of full row rank. For the general case without the rank assumption, see [15, Sec. 2.15.6].

## 4.4 Smith Normal Form

Hermite normal form is a triangularization realized by column transformations. By using both column and row elementary transformations, we can obtain a diagonal form. Since repeated integer elementary transformations are equivalent to multiplying with a unimodular matrix (Theorem 4.5), we can alternatively say that a given integer matrix can be transformed to a diagonal form by multiplying unimodular matrices from the left and the right.

**Example 4.8.** Matrix $A = \begin{bmatrix} 5 & 0 \\ 5 & 7 \end{bmatrix}$ is already in Hermite normal form and cannot be transformed to a simpler form with column transformations only. However, if we employ a row transformation of subtracting the first row from the second row, we can transform it to a diagonal matrix as follows:

$$\begin{bmatrix} 1 & 0 \\ -1 & 1 \end{bmatrix} \begin{bmatrix} 5 & 0 \\ 5 & 7 \end{bmatrix} = \begin{bmatrix} 5 & 0 \\ 0 & 7 \end{bmatrix}.$$

■

**Proposition 4.4.** *For any $m \times n$ integer matrix $A$, there exist a unimodular matrix $U$ of order $m$ and a unimodular matrix $V$ of order $n$ such that*

$$UAV = \left[\begin{array}{ccc|c} \alpha_1 & & 0 & \\ & \ddots & & O_{r,n-r} \\ 0 & & \alpha_r & \\ \hline & O_{m-r,r} & & O_{m-r,n-r} \end{array}\right]. \tag{4.9}$$

*Here $r = \operatorname{rank} A$, and $\alpha_1 \leqq \alpha_2 \leqq \cdots \leqq \alpha_r$ are positive integers.*

**Proof.** Set $A^{(0)} = A$. If $A^{(0)} = O$, we already have the form of (4.9) with $r = 0$. If $A^{(0)} \neq O$, we choose an entry with the smallest nonzero absolute value and move it to the $(1, 1)$ entry by permuting the rows and columns. If necessary, we change the sign of the first column to obtain $a_{11}^{(0)} > 0$.

We focus on the first row. For each $j = 2, \ldots, n$ we take integers $q_j$ and $r_j$ such that

$$a_{1j}^{(0)} = a_{11}^{(0)} q_j + r_j, \qquad a_{11}^{(0)} > r_j \geqq 0,$$

and add $-q_j$ times the first column to the $j$th column to change the $(1, j)$ entry to $r_j$. Next, we focus on the first column. For each $i = 2, \ldots, m$ we take integers $q_i'$ and $r_i'$ such that

$$a_{i1}^{(0)} = a_{11}^{(0)} q_i' + r_i', \qquad a_{11}^{(0)} > r_i' \geqq 0,$$

and add $-q'_i$ times the first row to the $i$th row to change the $(i, 1)$ entry to $r'_i$.

If, as a result of the above operations, the first row is $(a_{11}^{(0)}, 0, \ldots, 0)$ and the first column is $(a_{11}^{(0)}, 0, \ldots, 0)^\top$, then we define

$$A^{(1)} = (a_{ij}^{(1)} \mid 2 \leqq i \leqq m, 2 \leqq j \leqq n)$$

to be the $(m-1) \times (n-1)$ matrix consisting of the second row to the $m$th row and the second column to the $n$th column, thereby completing the first stage with

$$\left[\begin{array}{c|ccc} a_{11}^{(0)} & 0 & \cdots & 0 \\ \hline 0 & & & \\ \vdots & & A^{(1)} & \\ 0 & & & \end{array}\right].$$

Otherwise, there is a positive integer, other than $a_{11}^{(0)}$, in the first row or in the first column, and it is strictly smaller than $a_{11}^{(0)}$. We move this to the $(1, 1)$ entry and repeat the above process. In so doing, after a finite number of iterations, the first stage is completed with an integer $a_{11}^{(0)}$ and an $(m-1) \times (n-1)$ matrix $A^{(1)}$.

Then we apply the same procedure as above to this matrix $A^{(1)}$ and obtain the first diagonal entry $a_{22}^{(1)}$ and an $(m-2) \times (n-2)$ matrix

$$A^{(2)} = (a_{ij}^{(2)} \mid 3 \leqq i \leqq m, 3 \leqq j \leqq n),$$

to complete the second stage. Continuing in this way, we end up with diagonal entries $a_{11}^{(0)}, a_{22}^{(1)}, \ldots, a_{rr}^{(r-1)}$ and $A^{(r)} = O$. Finally, we permute the rows and columns to rearrange the diagonal entries in ascending order (smallest to largest). □

In the diagonal form (4.9), we can impose a further condition, a divisibility condition among the diagonal entries $\alpha_1, \alpha_2, \ldots, \alpha_r$:

$$\alpha_1 \mid \alpha_2 \mid \cdots \mid \alpha_r, \tag{4.10}$$

where the notation $a \mid b$ means that $a$ is a factor of $b$. The condition (4.10) says that $\alpha_i$ divides $\alpha_{i+1}$ for $i = 1, \ldots, r-1$. A diagonal matrix (4.9) that satisfies the divisibility condition (4.10) is said to be in *Smith normal form.*

As shown in Theorem 4.8 below, we can transform every integer matrix $A$ to Smith normal form by multiplication with unimodular matrices from the left and the right. Furthermore, the Smith normal form is uniquely determined by $A$. We note that the diagonal matrix in Proposition 4.4

(without the divisibility condition) is not uniquely determined, as demonstrated in Example 4.9 below.

**Example 4.9.** The diagonal matrix $\begin{bmatrix} 5 & 0 \\ 0 & 7 \end{bmatrix}$ is not in Smith normal form, since it does not satisfy the divisibility condition (4.10). The Smith normal form of this matrix can be obtained by integer elementary (row and column) transformations as follows:

$$\begin{bmatrix} 5 & 0 \\ 0 & 7 \end{bmatrix} \xrightarrow[E_3(1,2;3)]{\text{column}} \begin{bmatrix} 5 & 15 \\ 0 & 7 \end{bmatrix} \xrightarrow[E_3(1,2;-2)]{\text{row}} \begin{bmatrix} 5 & 1 \\ 0 & 7 \end{bmatrix}$$

$$\xrightarrow[E_1(1,2)]{\text{column}} \begin{bmatrix} 1 & 5 \\ 7 & 0 \end{bmatrix} \xrightarrow[E_3(1,2;-5)]{\text{column}} \begin{bmatrix} 1 & 0 \\ 7 & -35 \end{bmatrix}$$

$$\xrightarrow[E_3(2,1;-7)]{\text{row}} \begin{bmatrix} 1 & 0 \\ 0 & -35 \end{bmatrix} \xrightarrow[E_2(2)]{\text{column}} \begin{bmatrix} 1 & 0 \\ 0 & 35 \end{bmatrix} \quad \text{(Smith normal form).}$$

A key fact underlying the above transformation is that, for two integers $\alpha$ and $\beta$ in general, there exist integers $v$ and $u$ that satisfy

$$\alpha v + \beta u = \gcd(\alpha, \beta).$$

In the above example, $(\alpha, \beta) = (5, 7)$ and $\gcd(\alpha, \beta) = \gcd(5, 7) = 1$, and $5v + 7u = 1$ for $(v, u) = (3, -2)$. In the transformation shown above, we have employed a column transformation using $E_3(1, 2; 3) = E_3(1, 2; v)$ first and then a row transformation using $E_3(1, 2; -2) = E_3(1, 2; u)$ to create $1 = \gcd(\alpha, \beta)$ as the $(1, 2)$ entry. ■

**Theorem 4.8.** *Every integer matrix can be transformed to Smith normal form through unimodular row and column transformations. That is, for any $m \times n$ integer matrix $A$, there exist a unimodular matrix $U$ of order $m$ and a unimodular matrix $V$ of order $n$ such that*

$$UAV = \left[\begin{array}{ccc|c} \alpha_1 & & 0 & \\ & \ddots & & O_{r,n-r} \\ 0 & & \alpha_r & \\ \hline & O_{m-r,r} & & O_{m-r,n-r} \end{array}\right]. \tag{4.11}$$

*Here $r = \operatorname{rank} A$, and $\alpha_1 \leqq \alpha_2 \leqq \cdots \leqq \alpha_r$ are positive integers satisfying*

$$\textit{Divisibility condition } (4.10): \quad \alpha_1 \mid \alpha_2 \mid \cdots \mid \alpha_r$$

*and are uniquely determined by $A$.*

**Proof.** By Proposition 4.4 we can assume that $A$ is a diagonal matrix. Let $\alpha_1 \leqq \alpha_2 \leqq \cdots \leqq \alpha_r$ be the positive diagonal entries. If the divisibility condition (4.10) is satisfied, we are done. Otherwise, let $i$ be the minimum index for which $\alpha_i \mid \alpha_{i+1}$ fails.[4] Consider the submatrix diag $(\alpha_i, \alpha_{i+1})$ with rows and columns in $\{i, i+1\}$. Put $\alpha = \alpha_i$, $\beta = \alpha_{i+1}$, and $g = \gcd(\alpha, \beta)$, and take integers $v$ and $u$ that satisfy

$$\alpha v + \beta u = g$$

to transform the matrix as

$$\begin{bmatrix} 1 & u \\ 0 & 1 \end{bmatrix} \begin{bmatrix} \alpha & 0 \\ 0 & \beta \end{bmatrix} \begin{bmatrix} 1 & v \\ 0 & 1 \end{bmatrix} = \begin{bmatrix} \alpha & \alpha v + \beta u \\ 0 & \beta \end{bmatrix} = \begin{bmatrix} \alpha & g \\ 0 & \beta \end{bmatrix}.$$

Next, we exchange the columns and transform the resulting matrix as

$$\begin{bmatrix} 1 & 0 \\ -\beta/g & 1 \end{bmatrix} \begin{bmatrix} g & \alpha \\ \beta & 0 \end{bmatrix} \begin{bmatrix} 1 & -\alpha/g \\ 0 & 1 \end{bmatrix} = \begin{bmatrix} g & 0 \\ 0 & -\alpha\beta/g \end{bmatrix}.$$

Finally, we change the sign of the second column to obtain diag $(g, \alpha\beta/g)$.

The diagonal entries are thus changed from $(\alpha_1, \alpha_2, \ldots, \alpha_r)$ to

$$(\alpha_1, \ldots, \alpha_{i-1};\ g, \alpha_i \alpha_{i+1}/g;\ \alpha_{i+2}, \ldots, \alpha_r).$$

Let $(\alpha'_1, \alpha'_2, \ldots, \alpha'_r)$ be the above sequence rearranged in ascending order. Since $g < \alpha_i$, the sequence $(\alpha'_1, \alpha'_2, \ldots, \alpha'_r)$ is strictly smaller in lexicographical order than $(\alpha_1, \alpha_2, \ldots, \alpha_r)$. After a finite number of such transformations, the divisibility condition (4.10) is eventually satisfied, and then the Smith normal form is obtained.

The uniqueness can be shown as follows. In the Smith normal form $UAV$, the $k$th determinantal divisor $d_k(UAV)$ is given as

$$d_k(UAV) = \alpha_1 \alpha_2 \cdots \alpha_k$$

as a consequence of the divisibility condition (4.10). On the other hand, $d_k(UAV)$ is equal to $d_k(A)$, since the determinantal divisors remain invariant under integer elementary transformations (Proposition 4.3). Therefore,

$$d_k(A) = \alpha_1 \alpha_2 \cdots \alpha_k \qquad (k = 1, \ldots, r),$$

which shows

$$\alpha_k = d_k(A) \ /\ d_{k-1}(A) \qquad (k = 1, \ldots, r).$$

The right-hand side is determined by $A$, independently of $U$ and $V$. □

[4]What follows is a generalization of the argument in Example 4.9.

For an integer matrix $A$, the matrix in (4.11) in Theorem 4.8 is called the Smith normal form of $A$. The diagonal entries $\alpha_1, \alpha_2, \ldots, \alpha_r$ in the Smith normal form (4.11) are referred to as the *elementary divisors* of $A$, and are denoted as $e_1(A), e_2(A), \ldots, e_r(A)$. The following relations hold between elementary divisors and determinantal divisors:

$$d_k(A) = e_1(A)e_2(A)\cdots e_k(A), \quad e_k(A) = \frac{d_k(A)}{d_{k-1}(A)} \quad (k = 1, \ldots, r). \quad (4.12)$$

Smith normal form is also called (in Japan) "elementary divisor normal form."

**Example 4.10.** The Smith normal form of matrix $A = \begin{bmatrix} 6 & 4 & 10 \\ -1 & 1 & -5 \end{bmatrix}$ is given by $\begin{bmatrix} 1 & 0 & 0 \\ 0 & 10 & 0 \end{bmatrix}$. The elementary divisors are $e_1(A) = 1$ and $e_2(A) = 10$, and the determinantal divisors are $d_1(A) = 1$ and $d_2(A) = 10$ (Example 4.5). Also recall that the Hermite normal form of $A$ is $\begin{bmatrix} 2 & 0 & 0 \\ 3 & 5 & 0 \end{bmatrix}$ (Example 4.6). ■

As a corollary to Theorem 4.8, we obtain a characterization of unimodular matrices in terms of the Smith normal form.

**Theorem 4.9.** *The following two conditions* (a) *and* (b) *are equivalent for a square integer matrix $A$:*

(a) *$A$ is a unimodular matrix.*
(b) *The Smith normal form of $A$ is the unit matrix $I$.*

**Proof.** [(a) $\Rightarrow$ (b)] If $A$ is unimodular, then $UAV = I$ for $U = I$ and $V = A^{-1}$, where $V = A^{-1}$ is unimodular by Proposition 4.1.

[(b) $\Rightarrow$ (a)] We have $UAV = I$ for some unimodular matrices $U$ and $V$. Since $\det U = \pm 1$, $\det V = \pm 1$, and $\det U \cdot \det A \cdot \det V = 1$, we have $\det A = \pm 1$. □

## 4.5 Integer Solutions to Systems of Linear Equations

In this section, we consider a system of linear equations $A\boldsymbol{x} = \boldsymbol{b}$ with particular interest in the existence of an integer vector as the solution $\boldsymbol{x}$. It is assumed that $A$ is an $m \times n$ integer matrix and $\boldsymbol{b}$ is an $m$-dimensional integer vector. Divisibility conditions play an important role in addition to the rank condition, the latter of which is sufficient for the existence of a real vector solution.

First of all, let us recall, for the sake of comparison, the theorem for real vector solutions [6, Theorem 5.2].

**Theorem 4.10.** *The following four conditions* (a) *to* (d) *are equivalent for an $m \times n$ real matrix $A$ and an $m$-dimensional real vector $\boldsymbol{b}$:*

(a) $A\boldsymbol{x} = \boldsymbol{b}$ *has a real vector solution* $\boldsymbol{x}$.
(b) *If* $\boldsymbol{y}^\top A = \boldsymbol{0}^\top$, *then* $\boldsymbol{y}^\top \boldsymbol{b} = 0$, *where* $\boldsymbol{y}$ *is taken from real vectors.*
(c) $A$ *and* $[A \mid \boldsymbol{b}]$ *have the same rank.*
(d) $\boldsymbol{b} \in \mathrm{Im}(A)$, *where* $\mathrm{Im}(A)$ ($\subseteq \mathbb{R}^m$) *means the subspace of all linear combinations of the column vectors of $A$ with real coefficients.*

The following fact is fundamental to the integrality of vectors.

**Proposition 4.5.** *Let $V$ be a unimodular matrix of order $n$ and $\boldsymbol{x}$ be an $n$-dimensional real vector. Then $V\boldsymbol{x}$ is an integer vector if and only if $\boldsymbol{x}$ is an integer vector.*

**Proof.** Set $\boldsymbol{y} = V\boldsymbol{x}$. If $\boldsymbol{x}$ is an integer vector, then the vector $\boldsymbol{y}$, which is the product of an integer matrix and an integer vector, is an integer vector. Conversely, if $\boldsymbol{y}$ is an integer vector, then $\boldsymbol{x} = V^{-1}\boldsymbol{y}$ is an integer vector, since the inverse of a unimodular matrix $V$ is an integer matrix (Theorem 4.1). □

In considering an integer solution $\boldsymbol{x}$ for a system of equations $A\boldsymbol{x} = \boldsymbol{b}$, we can rewrite the equation $A\boldsymbol{x} = \boldsymbol{b}$ to

$$(UAV)(V^{-1}\boldsymbol{x}) = U\boldsymbol{b}$$

using unimodular matrices $U$ and $V$. By Proposition 4.5, $\boldsymbol{x}$ is an integer vector if and only if $\tilde{\boldsymbol{x}} = V^{-1}\boldsymbol{x}$ is an integer vector. With suitable unimodular matrices $U$ and $V$, we can transform the coefficient matrix $A$ to the Smith normal form (4.11) (or the diagonal form in (4.9)) $\tilde{A} = UAV$, thereby making the analysis easier. The transformation of the variables is summarized here for easy reference:

$$\tilde{A} = UAV, \qquad \tilde{\boldsymbol{x}} = V^{-1}\boldsymbol{x}, \qquad \tilde{\boldsymbol{b}} = U\boldsymbol{b}. \tag{4.13}$$

Then $A\boldsymbol{x} = \boldsymbol{b}$ is equivalent to $\tilde{A}\tilde{\boldsymbol{x}} = \tilde{\boldsymbol{b}}$.

The above discussion leads naturally to the following theorem.

**Theorem 4.11.** *The following four conditions* (a) *to* (d) *are equivalent for an $m \times n$ integer matrix $A$ and an $m$-dimensional integer vector $\boldsymbol{b}$:*

(a) $A\boldsymbol{x} = \boldsymbol{b}$ *has an integer vector solution* $\boldsymbol{x}$.

(b) *If* $\boldsymbol{y}^\top A$ *is an integer vector, then* $\boldsymbol{y}^\top \boldsymbol{b}$ *is an integer, where* $\boldsymbol{y}$ *is taken from real vectors.*

(c) *Matrices* $A$ *and* $[A \mid \boldsymbol{b}]$ *have the same elementary divisors.*[5]

(d) $\boldsymbol{b} \in \Lambda(A)$, *where* $\Lambda(A)$ $(\subseteqq \mathbb{Z}^m)$ *is the set of all integer vectors (lattice points) that can be represented as a linear combination of the column vectors of* $A$ *with integer coefficients.*

**Proof.** [(a) $\Leftrightarrow$ (b)] We make use of the diagonal form in (4.9). Assume that $\tilde{A} = UAV$ is in the diagonal form (4.9) with suitable unimodular matrices $U$ and $V$. Let $r = \operatorname{rank} A$ and let $\alpha_1, \alpha_2, \ldots, \alpha_r$ be the positive diagonal entries of $\tilde{A}$. By the transformation (4.13), $A\boldsymbol{x} = \boldsymbol{b}$ becomes

$$\alpha_i \tilde{x}_i = \tilde{b}_i \quad (i = 1, \ldots, r), \qquad 0 = \tilde{b}_i \quad (i = r+1, \ldots, m)$$

using notations in (4.13). Then, the condition (a) is obviously equivalent to

$$\alpha_i \mid \tilde{b}_i \quad (i = 1, \ldots, r), \qquad \tilde{b}_i = 0 \quad (i = r+1, \ldots, m). \tag{4.14}$$

Next, with the notation $\tilde{\boldsymbol{y}} = (U^{-1})^\top \boldsymbol{y} = (\tilde{y}_1, \tilde{y}_2, \ldots, \tilde{y}_m)^\top$, the condition (b) can be rewritten as

$$\tilde{y}_i \alpha_i \in \mathbb{Z} \quad (i = 1, \ldots, r) \implies \sum_{i=1}^{r} \tilde{y}_i \tilde{b}_i + \sum_{i=r+1}^{m} \tilde{y}_i \tilde{b}_i \in \mathbb{Z},$$

which is equivalent to the condition in (4.14).

[(a) $\Leftrightarrow$ (c)] For this proof it is convenient to choose $\tilde{A}$ to be the Smith normal form of $A$ in (4.11). Since the elementary divisors remain invariant under unimodular transformations, we have

$$e_k(A) = e_k(\tilde{A}), \quad e_k([A \mid \boldsymbol{b}]) = e_k([\tilde{A} \mid \tilde{\boldsymbol{b}}]) \qquad (k = 1, \ldots, r).$$

If the condition (4.14) holds, then we can transform $[\tilde{A} \mid \tilde{\boldsymbol{b}}]$ to $[\tilde{A} \mid \boldsymbol{0}]$ by the elementary column transformations using integer elementary matrices $E_3(i, n+1; -\tilde{b}_i/\alpha_i)$ $(i = 1, \ldots, r)$. This shows

$$e_k(\tilde{A}) = e_k([\tilde{A} \mid \tilde{\boldsymbol{b}}]) \qquad (k = 1, \ldots, r).$$

We have thus shown [(a) $\Rightarrow$ (c)]. Conversely, suppose that (c) holds. This implies in particular that $\operatorname{rank} \tilde{A} = \operatorname{rank} [\tilde{A} \mid \tilde{\boldsymbol{b}}]$, from which follows $\tilde{b}_i = 0$ $(i = r+1, \ldots, m)$. As shown in (4.12), the elementary divisors uniquely

[5] Since the number of elementary divisors is equal to the rank, condition (c) contains the condition that $A$ and $[A \mid \boldsymbol{b}]$ have the same rank.

determine the determinantal divisors. Hence it follows from the condition (c) that

$$d_k([\tilde{A} \mid \tilde{\boldsymbol{b}}]) = d_k(\tilde{A}) = \alpha_1 \cdots \alpha_{k-1}\alpha_k \qquad (k = 1, \ldots, r).$$

The subdeterminant of $[\tilde{A} \mid \tilde{\boldsymbol{b}}]$ with row indices $\{1, \ldots, k\}$ and column indices $\{1, \ldots, k-1; n+1\}$ is equal to $\alpha_1 \cdots \alpha_{k-1}\tilde{b}_k$. This must be a multiple of the $k$th determinantal divisor $\alpha_1 \cdots \alpha_{k-1}\alpha_k$. Hence follows that $\alpha_k \mid \tilde{b}_k$ $(k = 1, \ldots, r)$ in (4.14).

[(a) $\Leftrightarrow$ (d)] This is obviously true from the definition of $\Lambda(A)$. □

The solvability conditions for varying right-hand side vector $\boldsymbol{b}$ are as follows.

**Theorem 4.12.** *The following four conditions* (a) *to* (d) *are equivalent for an $m \times n$ integer matrix $A$ with* $\operatorname{rank} A = m$*:*

(a) *For each $m$-dimensional integer vector $\boldsymbol{b}$, $A\boldsymbol{x} = \boldsymbol{b}$ has an integer vector solution $\boldsymbol{x}$.*
(b) *If $\boldsymbol{y}^\top A$ is an integer vector, then $\boldsymbol{y}$ is an integer vector, where $\boldsymbol{y}$ is taken from real vectors.*
(c) *The elementary divisors $e_1(A), e_2(A), \ldots, e_m(A)$ are all equal to one.*[6]
(d) $\Lambda(A) = \mathbb{Z}^m$.

**Proof.** The condition (a) above says (literally) that the condition (a) in Theorem 4.11 holds for all $m$-dimensional integer vectors $\boldsymbol{b}$. The conditions (b), (c), and (d) above are equivalent to the corresponding conditions in Theorem 4.11 holding for all $m$-dimensional integer vectors $\boldsymbol{b}$ (in particular, for all the unit vectors $\boldsymbol{e}_1, \ldots, \boldsymbol{e}_m$). □

The equivalence of (a) and (b) in Theorem 4.11 can be reformulated in a theorem of the alternative.[7]

**Theorem 4.13.** *For an integer matrix $A$ and an integer vector $\boldsymbol{b}$, exactly one of the following two conditions* (a) *and* ($\overline{\text{b}}$) *holds (but not both at the same time).*

(a) *$A\boldsymbol{x} = \boldsymbol{b}$ has an integer vector solution $\boldsymbol{x}$.*
($\overline{\text{b}}$) *There exists a real vector $\boldsymbol{y}$ such that $\boldsymbol{y}^\top A$ is an integer vector and $\boldsymbol{y}^\top \boldsymbol{b}$ is not an integer.*

---

[6]This condition is equivalent to $e_m(A) = 1$ and, also to $d_m(A) = 1$.

[7]In general, a theorem stating that exactly one of two possibilities (not both or none) must occur is referred to as a *theorem of the alternative* (see Sec. 3.3.2).

**Remark 4.3.** The significance of theorems of the alternative was explained in general terms in Sec. 3.3.2. The significance of Theorem 4.13, in particular, may be described as follows. To prove the existence of an integer solution $\boldsymbol{x}$ to the equation $A\boldsymbol{x} = \boldsymbol{b}$, we can demonstrate $\boldsymbol{x}$ itself as an "evidence of existence." To prove the nonexistence, on the other hand, we can demonstrate a real vector $\boldsymbol{y}$ that satisfies condition ($\overline{\text{b}}$) as an "evidence of nonexistence" of $\boldsymbol{x}$. ■

**Example 4.11.** Consider a system of equations $A\boldsymbol{x} = \boldsymbol{b}$ given as follows:

$$\begin{bmatrix} 6 & 4 & 10 \\ -1 & 1 & -5 \end{bmatrix} \begin{bmatrix} x_1 \\ x_2 \\ x_3 \end{bmatrix} = \begin{bmatrix} 3 \\ 1 \end{bmatrix}. \tag{4.15}$$

For $\boldsymbol{y} = (1/2, 0)^\top$, $\boldsymbol{y}^\top A = (3, 2, 5)$ is an integer vector and $\boldsymbol{y}^\top \boldsymbol{b} = 3/2$ is not an integer. This $\boldsymbol{y}$ serves as an evidence to show the nonexistence of an integer solution $(x_1, x_2, x_3)$. We may also calculate the elementary divisors to find $e_1(A) = 1$, $e_2(A) = 10$, and $e_1([A \mid \boldsymbol{b}]) = e_2([A \mid \boldsymbol{b}]) = 1$, which shows that the condition (c) in Theorem 4.11 is not met. ■

Theorem 4.13 is certainly useful in proving the existence or nonexistence of an integer solution, but it does not provide us with a method for computing a solution $\boldsymbol{x}$. For computing such solutions, the Hermite normal form is relevant and useful. Transforming the coefficient matrix $A$ to the Hermite normal form $\tilde{A}$ is equivalent to rewriting the equation $A\boldsymbol{x} = \boldsymbol{b}$ to $(AV)(V^{-1}\boldsymbol{x}) = \boldsymbol{b}$ using a unimodular matrix $V$. Since the transformed coefficient matrix $\tilde{A} = AV$ is lower-triangular, we can determine the components of the solution $\tilde{\boldsymbol{x}}$ for $\tilde{A}\tilde{\boldsymbol{x}} = \boldsymbol{b}$ in the order of $\tilde{x}_1, \tilde{x}_2, \ldots$, and then the original unknown vector $\boldsymbol{x}$ can be computed by $\boldsymbol{x} = V\tilde{\boldsymbol{x}}$. Since $\boldsymbol{x}$ is an integer vector if and only if $\tilde{\boldsymbol{x}}$ is an integer vector, once we obtain $\tilde{\boldsymbol{x}}$, we can tell whether an integer solution $\boldsymbol{x}$ exists or not.

**Example 4.12.** The Hermite normal form of the coefficient matrix $A$ in (4.15) is given by $\tilde{A} = \begin{bmatrix} 2 & 0 & 0 \\ 3 & 5 & 0 \end{bmatrix}$ (Example 4.6), and accordingly the equation in (4.15) can be rewritten as

$$\begin{bmatrix} 2 & 0 & 0 \\ 3 & 5 & 0 \end{bmatrix} \begin{bmatrix} \tilde{x}_1 \\ \tilde{x}_2 \\ \tilde{x}_3 \end{bmatrix} = \begin{bmatrix} 3 \\ 1 \end{bmatrix}. \tag{4.16}$$

From this we obtain

$$\tilde{x}_1 = 3/2, \qquad \tilde{x}_2 = (1 - 3\tilde{x}_1)/5 = -7/10, \qquad \tilde{x}_3 = t \quad (t \in \mathbb{R}).$$

Therefore, no integer solution exists.

If the right-hand side vector $\boldsymbol{b}$ in (4.15) is changed to $[4, 11]^\top$, the equation (4.16) changes to

$$\begin{bmatrix} 2 & 0 & 0 \\ 3 & 5 & 0 \end{bmatrix} \begin{bmatrix} \tilde{x}_1 \\ \tilde{x}_2 \\ \tilde{x}_3 \end{bmatrix} = \begin{bmatrix} 4 \\ 11 \end{bmatrix}. \tag{4.17}$$

The integer solutions of this equation are given by

$$\tilde{x}_1 = 4/2 = 2, \qquad \tilde{x}_2 = (11 - 3\tilde{x}_1)/5 = 1, \qquad \tilde{x}_3 = t \quad (t \in \mathbb{Z}).$$

With the transformation matrix

$$V = \begin{bmatrix} -1 & -2 & -3 \\ 2 & 3 & 2 \\ 0 & 0 & 1 \end{bmatrix}$$

given in Example 4.7, the original vector $\boldsymbol{x} = V\tilde{\boldsymbol{x}}$ is calculated as

$$\begin{bmatrix} x_1 \\ x_2 \\ x_3 \end{bmatrix} = \begin{bmatrix} -1 & -2 & -3 \\ 2 & 3 & 2 \\ 0 & 0 & 1 \end{bmatrix} \begin{bmatrix} 2 \\ 1 \\ t \end{bmatrix} = \begin{bmatrix} -4 - 3t \\ 7 + 2t \\ t \end{bmatrix}$$

with an integer-valued parameter $t \in \mathbb{Z}$. ■

**Remark 4.4.** The implication [(a) ⇒ (b)] in Theorems 4.11 and 4.12 can be shown easily without resorting to diagonalization (transformation to a normal form). If $\boldsymbol{y}^\top A$ and $\boldsymbol{x}$ are integer vectors, then $\boldsymbol{y}^\top \boldsymbol{b} = \boldsymbol{y}^\top (A\boldsymbol{x}) = (\boldsymbol{y}^\top A)\boldsymbol{x}$ is an integer. Thus [(a) ⇒ (b)] in Theorem 4.11 is proved. In Theorem 4.12, we only have to note that, if $\boldsymbol{y}^\top \boldsymbol{b}$ is an integer for all integer vectors $\boldsymbol{b}$, then $\boldsymbol{y}$ must be an integer vector.

The proofs of [(a) ⇔ (b)] in Theorems 4.11 and 4.12 employ an argument consisting of the following steps:

- Rewrite the condition (a) to a condition for the normal form.[8]
- Rewrite the condition (b) to a condition for the normal form.
- Show the equivalence [(a) ⇔ (b)] for the normal form.

In this book we consistently employ this type of argument to demonstrate an effective use of normal forms. ■

[8] Here, we use the term "normal form" in a loose sense to refer to a particular normal form or anything similar.

## 4.6 Integrality of Systems of Linear Inequalities

In Chap. 3, we discussed some fundamental issues about systems of linear inequalities and the application thereof to linear programming. In this section, we consider systems of linear inequalities with an additional feature of integrality.

### 4.6.1 *Integer Programs and Linear Programs*

Optimization problems in the form of linear programs in which the variables are restricted to take integer values are called *integer programming problems* or *integer programs.* For example, the following problem

$$\begin{array}{ll} \text{Maximize} & 3x_1 + 4x_2 + 5x_3 \\ \text{subject to} & x_1 + x_2 \qquad\;\; \leqq 1 \\ & \qquad\; x_2 + x_3 \leqq 1 \\ & x_1 \qquad\; + x_3 \leqq 1 \\ & x_1, x_2, x_3 \in \mathbb{Z} \end{array} \tag{4.18}$$

is an instance of integer programs. The optimal solution for this problem is $(x_1, x_2, x_3) = (0, 0, 1)$ and the maximum value of the objective function is 5 (see Remark 4.5). If we remove integrality condition $x_1, x_2, x_3 \in \mathbb{Z}$ from the problem (4.18), we obtain

$$\begin{array}{ll} \text{Maximize} & 3x_1 + 4x_2 + 5x_3 \\ \text{subject to} & x_1 + x_2 \qquad\;\; \leqq 1 \\ & \qquad\; x_2 + x_3 \leqq 1 \\ & x_1 \qquad\; + x_3 \leqq 1 \end{array} \tag{4.19}$$

which is a linear program. The optimal solution for this problem is $(1/2, 1/2, 1/2)$ and the maximum value of the objective function is 6. The maximum value changes depending on whether the integrality condition is imposed or not, and this very fact shows that the integrality condition plays an essential role.

Consider another integer programming problem:

$$\begin{array}{ll} \text{Maximize} & 3x_1 + 4x_2 \\ \text{subject to} & x_1 + 2x_2 \leqq 4 \\ & x_1 + x_2 \;\;\, \leqq 3 \\ & x_1, x_2 \in \mathbb{Z} \end{array} \tag{4.20}$$

The feasible solutions lying in the nonnegative quadrant are indicated in Fig. 4.1 by the black circles (•). The optimal solution is $(x_1, x_2) = (2, 1)$

and the maximum value of the objective function is 10. If we remove the integrality condition we obtain a linear programming problem

$$\begin{array}{ll} \text{Maximize} & 3x_1 + 4x_2 \\ \text{subject to} & x_1 + 2x_2 \leqq 4 \\ & x_1 + x_2 \leqq 3 \end{array} \tag{4.21}$$

for which the same point $(x_1, x_2) = (2, 1)$ is the optimal solution. That is, in the problem of (4.20), the integrality condition does not really impose substantial constraints. Put otherwise, in the linear programming problem in (4.21), the integrality of the coefficients in the problem specification is inherited by the optimal solution.

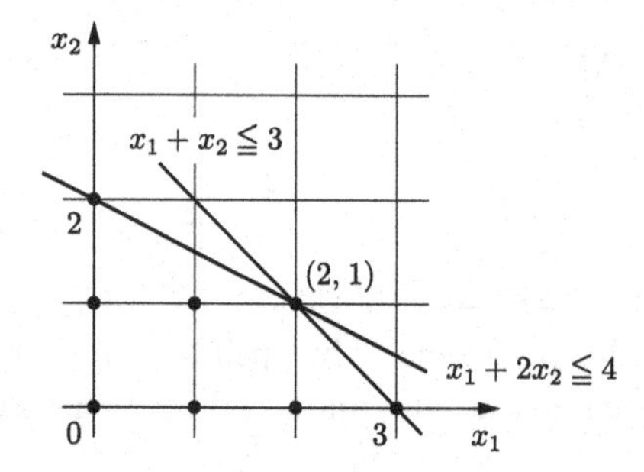


Fig. 4.1 An integer program reducible to linear programming.

As shown in the second example above, there are some classes of linear programming problems with integer coefficients whose optimal solutions can be chosen from integer vectors. The mathematical structures underlying such classes have a close relation with the unimodular matrices treated in Sec. 4.1. It is indeed the purpose of this section (Sec. 4.6) to explain such structures.

**Remark 4.5.** It is intuitively clear that the optimal solution to the problem (4.18) is given by $(x_1, x_2, x_3) = (0, 0, 1)$, but we will provide a rigorous proof here. Since the solution $(x_1, x_2, x_3) = (0, 0, 1)$ satisfies the constraints and the objective function has value 5 for this vector, we have:

$$\text{the optimal value of (4.18)} \geqq 5.$$

This value, 5, can be shown to be the optimal (maximum) value as follows. By adding the three inequality constraints we obtain $2(x_1 + x_2 + x_3) \leqq 3$, *i.e.*,

$$x_1 + x_2 + x_3 \leqq 3/2.$$

Since the left-hand side of this inequality is an integer, we can derive a stronger inequality

$$x_1 + x_2 + x_3 \leqq 1.$$

Using this, we can evaluate the objective function value from above as

$$3x_1 + 4x_2 + 5x_3 = 2(x_1 + x_2 + x_3) + (x_1 + x_3) + 2(x_2 + x_3) \leqq 2 + 1 + 2 = 5$$

for any $(x_1, x_2, x_3)$ that satisfies the constraints. This gives:

$$\text{the objective function of (4.18)} \leqq 5.$$

In this way, the maximum value of the objective function is proved to be 5.

In the above argument we have used the following reasoning:

$$x_1 + x_2 + x_3 \leqq \frac{3}{2} \ \ \& \ \ \text{integrality} \ \ \Longrightarrow x_1 + x_2 + x_3 \leqq 1.$$

Such a method of generating stronger inequalities is one of the important techniques in dealing with inequalities of integer variables. Adding the generated inequality to the linear program (4.19) yields another linear program

$$\begin{array}{ll} \text{Maximize} & 3x_1 + 4x_2 + 5x_3 \\ \text{subject to} & x_1 + x_2 \qquad \leqq 1 \\ & \qquad x_2 + x_3 \leqq 1 \\ & x_1 \qquad + x_3 \leqq 1 \\ & x_1 + x_2 + x_3 \leqq 1 \end{array} \tag{4.22}$$

of which the optimal solution is the same as that of the original integer programming problem (4.18). Such techniques are often used in the solution of integer programming problems. ■

**Remark 4.6.** The problem of testing for the existence of a real vector satisfying a given system of linear inequalities is a well-studied and relatively easy problem. A complete theory has been established (see Chap. 3) and algorithms based on linear programming are available for this problem. In contrast, the integer version of this problem, *i.e.*, the problem of testing for the existence of an integer vector satisfying a given system of linear inequalities, is an extremely difficult problem, and no efficient algorithms are known for this problem. ■

### 4.6.2 *Totally Unimodular Matrices*

#### 4.6.2.1 *Definition and examples*

An integer matrix (not necessarily square) is said to be a *totally unimodular matrix* if the value of each subdeterminant (minor) is 0, 1 or $-1$. Each entry of a totally unimodular matrix must be 0, 1 or $-1$, since it is equal to the value of the subdeterminant of order 1 consisting of that entry.

**Example 4.13.** Matrix $A = \begin{bmatrix} 1 & -1 \\ -1 & 1 \end{bmatrix}$ is a totally unimodular matrix. ■

**Example 4.14.** Each entry of the matrix $A = \begin{bmatrix} 1 & 1 & 1 \\ -1 & 1 & 0 \end{bmatrix}$ is 0, 1, or $-1$. But this is not a totally unimodular matrix. The value of the subdeterminant consisting of the first two columns is equal to 2. ■

**Example 4.15.** Matrix $A = \begin{bmatrix} 1 & 2 \\ 1 & 1 \end{bmatrix}$ is a unimodular matrix, but it is not totally unimodular. We have $\det A = -1$, but $a_{12} \notin \{0, 1, -1\}$. ■

**Example 4.16.** Matrix

$$A = \begin{bmatrix} 1 & 1 & 0 & 0 \\ 1 & 1 & 1 & 1 \\ 0 & 1 & 1 & 1 \end{bmatrix}$$

is a totally unimodular matrix. ■

The incidence matrix of a graph is an important example of a totally unimodular matrix. Given a directed graph, the *incidence matrix* of the graph is defined to be a matrix $A$ with its rows indexed by vertices and columns by edges, and $a_{ie} = 1$ and $a_{je} = -1$ for each edge $e = (v_i, v_j)$ (and the other entries being 0). It should be clear that the row set of the incidence matrix corresponds to the vertex set of the graph and the column set corresponds to the edge set.

**Example 4.17.** Consider the directed graph $G = (V, E)$ of Fig. 4.2. The vertex set is $V = \{v_1, v_2, v_3, v_4\}$ and the edge set is
$E = \{e_1 = (v_2, v_1), e_2 = (v_1, v_3), e_3 = (v_3, v_4), e_4 = (v_4, v_2), e_5 = (v_4, v_1)\}$.
The incidence matrix is given by

$$A = \begin{array}{c} \\ v_1 \\ v_2 \\ v_3 \\ v_4 \end{array} \begin{array}{c} \begin{array}{ccccc} e_1 & e_2 & e_3 & e_4 & e_5 \end{array} \\ \begin{bmatrix} -1 & 1 & 0 & 0 & -1 \\ 1 & 0 & 0 & -1 & 0 \\ 0 & -1 & 1 & 0 & 0 \\ 0 & 0 & -1 & 1 & 1 \end{bmatrix} \end{array}. \tag{4.23}$$

■

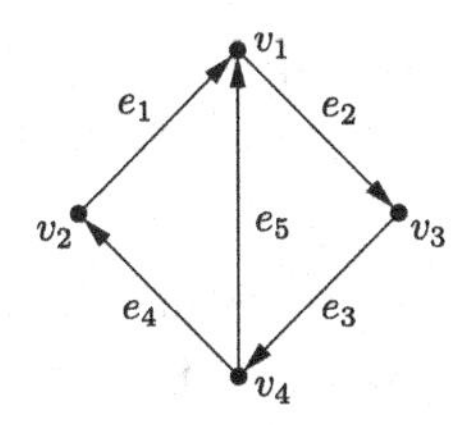


Fig. 4.2 A directed graph.

**Theorem 4.14.** *The incidence matrix of a directed graph is a totally unimodular matrix.*

**Proof.** We are to show that $\det B \in \{0, 1, -1\}$ for every square submatrix $B$ of an incidence matrix, which we prove by induction on the size of $B$. If the size of $B$ is 1, the claim is obviously true. When the size of $B$ is greater than or equal to 2, three cases can be distinguished:

(i) If there is a zero vector among the columns of $B$, then $\det B = 0$.

(ii) If there is a $(\pm)$ unit vector among the columns of $B$, then, by the Laplace expansion, $\det B$ is equal (up to a sign) to a subdeterminant that is smaller in size by one. The value of this smaller subdeterminant belongs to $\{0, 1, -1\}$ by the induction hypothesis.

(iii) If each column of $B$ contains exactly one 1 and one $-1$, then the sum of the row vectors of $B$ is a zero vector (*i.e.*, the row vectors are linearly dependent), which implies $\det B = 0$. □

**Remark 4.7.** If $G = (V, E)$ is an undirected graph, it is not natural to have $\pm$ signs in the entries of the incidence matrix. As a variant of the incidence matrix, we can define a matrix $\tilde{A}$ with $V$ as the row set and $E$ as the column set that has entries of 1 corresponding to edges, *i.e.*, $\tilde{a}_{ie} = \tilde{a}_{je} = 1$ for each edge $e = (v_i, v_j)$ (and the other entries being 0). For example, for a triangle-shaped graph, we have

$$\tilde{A} = \begin{bmatrix} 1 & 0 & 1 \\ 1 & 1 & 0 \\ 0 & 1 & 1 \end{bmatrix}.$$

As is seen from this example, the matrix $\tilde{A}$ is not necessarily a totally unimodular matrix. In the case of a bipartite graph $G = (U, V; E)$, however, $\tilde{A}$ is indeed a totally unimodular matrix. The reason is as follows. Direct all the edges from $U$ to $V$ to obtain a directed graph and let $A$ be its incidence matrix. The matrix $A$ is totally unimodular. On the other hand, a subdeterminant of $\tilde{A}$ is equal, up to a sign, to the determinant of the corresponding submatrix of $A$, since $\tilde{A}$ is obtained from $A$ by multiplying the rows corresponding to $V$ by $-1$. ■

Of the totally unimodular matrices in the examples at the beginning of this section, the matrix in Example 4.13 is the incidence matrix of a graph consisting of two vertices $v_1$ and $v_2$ and two edges $(v_1, v_2)$ and $(v_2, v_1)$. The matrix in Example 4.16 is not a graph incidence matrix, and therefore, does not fall into the category of Theorem 4.14. But it does possess the following characteristic property:

Each entry is 0 or 1, and the entries of 1 are consecutive in each row. (4.24)

We next show that, in general, this type of a matrix is a totally unimodular matrix.

**Theorem 4.15.** *A matrix $A$ with the property* (4.24) *is a totally unimodular matrix.*

**Proof.** If $A$ possesses the property (4.24), so does every submatrix of $A$. Therefore, it suffices to prove the following proposition:

If a square matrix $B$ satisfies (4.24), then $\det B \in \{0, 1, -1\}$. (4.25)

We prove (4.25) by induction on the size of $B$. If the size of $B$ is 1, then this is obviously true. If the size of $B$ is greater than or equal to 2, we examine the nonzero entries in the first column of $B$ and define the set of row indices $I = \{i \mid b_{i1} = 1\}$. For each $i \in I$, the $i$th row vector of $B$ is in the form of $(1, 1, \ldots, 1, 0, 0, \ldots, 0)$. We divide into two cases:

(i) If $I$ is an empty set, then $\det B = 0$.

(ii) If $I$ is nonempty, let $i_0 \in I$ be the index of the row that contains the smallest number of 1's among the rows with indices in $I$. Let $C$ be the matrix obtained from $B$ by subtracting the $i_0$th row from the $i$th row for each $i \in I$ $(i \neq i_0)$. The first column of $C$ contains only one entry of 1 at the $i_0$th row, whereas

$$\det B = \det C.$$

Let $D$ be the submatrix of $C$ avoiding the first column and the $i_0$th row. By the Laplace expansion, we have

$$\det C = \pm \det D.$$

By the choice of $i_0$, the matrix $D$ is endowed with the property (4.24), and is smaller than $B$. Hence $\det D \in \{0, 1, -1\}$ by the induction hypothesis. Therefore, $\det B = \det C = \pm \det D \in \{0, 1, -1\}$.

This proves (4.25), thereby proving the theorem as well. □

4.6.2.2 *Operations preserving total unimodularity*

For a totally unimodular matrix $A$, the following matrices are also totally unimodular:

$$\begin{gathered} A^\top, \quad [\, A \; I_m \,], \quad [\, A \; -A \,], \quad [\, A^\top \; -A^\top \; I_n \,], \\ \begin{bmatrix} A & I_m \\ I_n & O \end{bmatrix}, \quad \begin{bmatrix} A & O \\ I_n & I_n \end{bmatrix}, \quad \begin{bmatrix} A & -A \\ I_n & -I_n \end{bmatrix}, \end{gathered} \tag{4.26}$$

where $A$ is assumed to be an $m \times n$ matrix, and $I_m$ and $I_n$ denote the unit matrices of orders $m$ and $n$, respectively.

Next, we consider the relation between pivotal transformation and total unimodularity. Suppose that a matrix $A$ is partitioned into blocks as

$$A = \begin{bmatrix} B & C \\ D & E \end{bmatrix} \tag{4.27}$$

with $B$ being a (square) nonsingular matrix. Then the transformation of $A$ to

$$A_{\mathrm{piv}} = \begin{bmatrix} B^{-1} & B^{-1}C \\ -DB^{-1} & E - DB^{-1}C \end{bmatrix} \tag{4.28}$$

is called the *pivotal transformation* with respect to *pivot* $B$, and the resulting matrix $A_{\mathrm{piv}}$ is called a *pivotal transform* of $A$.

**Theorem 4.16.** *A pivotal transform of a totally unimodular matrix is totally unimodular. In particular, the inverse of a nonsingular totally unimodular matrix is totally unimodular.*

**Proof.** Suppose that the matrix $A$ in (4.27) is totally unimodular, and consider a subdeterminant $\det A_{\mathrm{piv}}[I, J]$ of the matrix $A_{\mathrm{piv}}$ in (4.28). The row set $I$ can be expressed as $I = I_1 \cup I_2$ with $I_1$ denoting the part of $I$ contained in the first block (the row set of $B^{-1}$) and $I_2$ denoting the rest in $I$. Likewise, the column set $J$ is expressed as $J = J_1 \cup J_2$. Then we have a formula

$$\det A_{\mathrm{piv}}[I_1 \cup I_2, J_1 \cup J_2] = \pm \det A[\overline{J_1} \cup I_2, \overline{I_1} \cup J_2] / \det B$$

for the subdeterminant (see Remark 4.8). Here, $\overline{I_1}$ denotes the complement of $I_1$ in the row set of $B^{-1}$ (= column set of $B$) and $\overline{J_1}$ denotes the complement of $J_1$ in the column set of $B^{-1}$ (= row set of $B$). Since $A$ is a totally unimodular matrix, we have

$$\det B \in \{1, -1\}, \qquad \det A[\overline{J_1} \cup I_2, \overline{I_1} \cup J_2] \in \{0, 1, -1\}.$$

Therefore, $\det A_{\rm piv}[I,J] \in \{0,1,-1\}$. The second assertion follows from the observation that the pivotal transform with respect to $A$ itself coincides with the inverse $A^{-1}$. □

**Example 4.18.** For the incidence matrix $A$ in (4.23), the pivotal transform with respect to the submatrix $B$ with row set $\{v_1, v_2, v_3\}$ and column set $\{e_1, e_2, e_3\}$ is given by

$$A_{\rm piv} = \left[\begin{array}{ccc|cc} 0 & 1 & 0 & -1 & 0 \\ 1 & 1 & 0 & -1 & -1 \\ 1 & 1 & 1 & -1 & -1 \\ \hline 1 & 1 & 1 & 0 & 0 \end{array}\right].$$

This matrix is totally unimodular. ■

**Remark 4.8.** We provide a proof of the formula

$$\det A_{\rm piv}[I_1 \cup I_2, J_1 \cup J_2] = \pm \det A[\overline{J_1} \cup I_2, \overline{I_1} \cup J_2] / \det B \tag{4.29}$$

for the subdeterminant of a pivotal transform (see Remark 4.2 for the notation of subdeterminants). Here $A_{\rm piv}$ is defined by (4.28) for an $m \times n$ matrix $A$ of the form of (4.27), and

- $I_1$ is a subset of the row set of $B^{-1}$ (= column set of $B$),
- $\overline{I_1}$ is the complement of $I_1$ in the row set of $B^{-1}$ (= column set of $B$),
- $J_1$ is a subset of the column set of $B^{-1}$ (= row set of $B$),
- $\overline{J_1}$ is the complement of $J_1$ in the column set of $B^{-1}$ (= row set of $B$),
- $I_2$ is a subset of the row set of $E$, and
- $J_2$ is a subset of the column set of $E$.

We mention that the formula (4.29) is not limited to integer matrices, but it is also valid for matrices of real or complex numbers.

First, we recall a fundamental fact about pivotal transformation. Denote the size of $B$ by $k$ and consider the $m \times (m+n)$ matrix

$$\tilde{A} = \left[\begin{array}{cc|cc} I_k & O & B & C \\ O & I_{m-k} & D & E \end{array}\right]$$

that is formed by putting the unit matrix to the left of the given $m \times n$ matrix $A$. By applying a row transformation to $\tilde{A}$ that transforms $\begin{bmatrix} B \\ D \end{bmatrix}$ to $\begin{bmatrix} I_k \\ O \end{bmatrix}$, we obtain

$$\begin{bmatrix} B^{-1} & O \\ -DB^{-1} & I_{m-k} \end{bmatrix} \left[\begin{array}{cc|cc} I_k & O & B & C \\ O & I_{m-k} & D & E \end{array}\right] = \left[\begin{array}{cc|cc} B^{-1} & O & I_k & B^{-1}C \\ -DB^{-1} & I_{m-k} & O & E - DB^{-1}C \end{array}\right].$$

That is, we have $Q\tilde{A} = \hat{A}$ for

$$Q = \begin{bmatrix} B^{-1} & O \\ -DB^{-1} & I_{m-k} \end{bmatrix}, \quad \hat{A} = \left[\begin{array}{cc|cc} B^{-1} & O & I_k & B^{-1}C \\ -DB^{-1} & I_{m-k} & O & E - DB^{-1}C \end{array}\right].$$

Then the $m \times n$ submatrix of $\hat{A}$ corresponding to the first and fourth column blocks coincides with the pivotal transform $A_{\text{piv}}$.

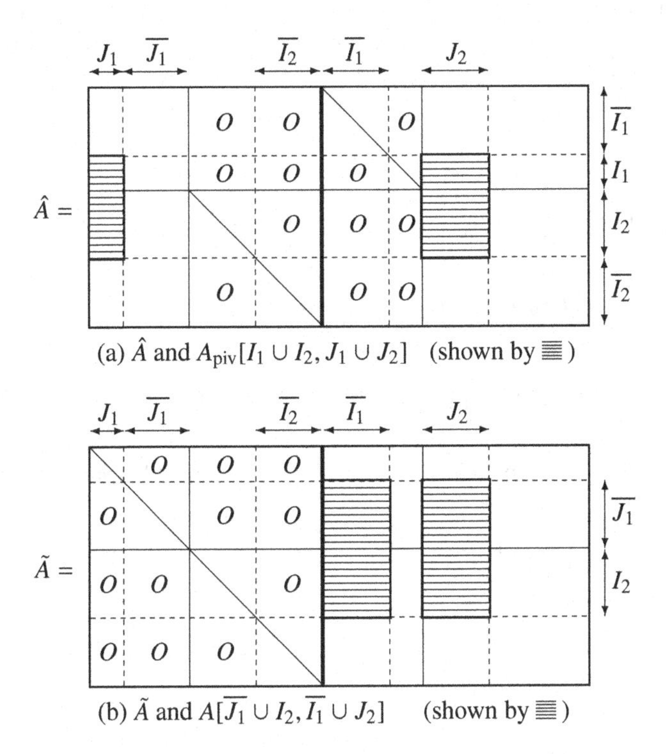


(a) $\hat{A}$ and $A_{\text{piv}}[I_1 \cup I_2, J_1 \cup J_2]$ (shown by ▤)

(b) $\tilde{A}$ and $A[\overline{J_1} \cup I_2, \overline{I_1} \cup J_2]$ (shown by ▤)

Fig. 4.3 Pivotal transformation.

As shown in Fig. 4.3(a), the set $\overline{I_1}$ is a subset of row indices of $\hat{A}$, but it can also be regarded as a subset of column indices of $\hat{A}$ via the correspondence through the unit matrix $I_k$ contained in $\hat{A}$. Denote the complement of $I_2$ within the row set of $E$ as $\overline{I_2}$. The set $\overline{I_2}$ can similarly be regarded as a subset of column indices of $\hat{A}$. With this understanding, we have

$$\det A_{\text{piv}}[I_1 \cup I_2, J_1 \cup J_2] = \pm \det \hat{A}[*, \overline{I_1} \cup \overline{I_2} \cup J_1 \cup J_2], \tag{4.30}$$

where the symbol "$*$" on the right-hand side means the whole row set, and accordingly, $\hat{A}[*, \overline{I_1} \cup \overline{I_2} \cup J_1 \cup J_2]$ denotes the submatrix of $\hat{A}$ of order

$m$ with column set $\overline{I_1} \cup \overline{I_2} \cup J_1 \cup J_2$. On the other hand, it follows from $\hat{A} = Q\tilde{A}$ that

$$\det \hat{A}[*, \overline{I_1} \cup \overline{I_2} \cup J_1 \cup J_2] = \det Q \cdot \det \tilde{A}[*, \overline{I_1} \cup \overline{I_2} \cup J_1 \cup J_2], \quad (4.31)$$

where $\det Q = (\det B)^{-1}$.

Similarly, the matrix $\tilde{A}$ is partitioned as in Fig. 4.3(b). The set $\overline{J_1}$, besides being a subset of column indices of $\tilde{A}$, can also be regarded as a subset of row indices of $\tilde{A}$ via the correspondence through the unit matrix $I_k$ contained in $\tilde{A}$. With this understanding, we have

$$\det \tilde{A}[*, \overline{I_1} \cup \overline{I_2} \cup J_1 \cup J_2] = \pm \det A[\overline{J_1} \cup I_2, \overline{I_1} \cup J_2]. \quad (4.32)$$

The formula (4.29) is obtained from (4.30), (4.31), and (4.32). ■

### 4.6.3 *Integrality of Linear Programs*

In Sec. 4.6.1 we have explained the significance of integrality of linear programs using a simple example. Here we explain the relation between total unimodularity and integrality.

Consider the following pair of linear programming problems (see Sec. 3.5.3):

| [Primal Problem P] | | [Dual Problem D] | |
|---|---|---|---|
| Minimize | $\boldsymbol{c}^\top \boldsymbol{x}$ | Maximize | $\boldsymbol{b}^\top \boldsymbol{y}$ |
| subject to | $A\boldsymbol{x} = \boldsymbol{b}$ | subject to | $A^\top \boldsymbol{y} \leqq \boldsymbol{c}$ |
| | $\boldsymbol{x} \geqq \boldsymbol{0}$ | | |

(4.33)

Here, $A$ is an $m \times n$ matrix, $\boldsymbol{b}$ is an $m$-dimensional vector, and $\boldsymbol{c}$ is an $n$-dimensional vector; we assume that $\operatorname{rank} A = m$.

If the matrix $A$ is totally unimodular, the integrality in the description of the problems is inherited by the solutions in the sense that *integer optimal solutions* exist.

**Theorem 4.17.** *Suppose that the linear programming problems in* (4.33) *have optimal solutions and the matrix $A$ is totally unimodular.*

(1) *If $\boldsymbol{b}$ is an integer vector, the problem* P *has an integer optimal solution* $\boldsymbol{x} \in \mathbb{Z}^n$.
(2) *If $\boldsymbol{c}$ is an integer vector, the problem* D *has an integer optimal solution* $\boldsymbol{y} \in \mathbb{Z}^m$.

**Proof.** (1) According to the theory of linear programming [37, 38, 40, 41], a basic solution exists among the optimal solutions to the problem P. A basic solution is a solution $\boldsymbol{x}$ given in the form[9] of $(B^{-1}\boldsymbol{b}, \boldsymbol{0})$ for an $m \times m$ nonsingular submatrix $B$ of $A$. Since $A$ is totally unimodular, the matrix $B$ is unimodular, and hence $B^{-1}$ is an integer matrix; therefore, $B^{-1}\boldsymbol{b}$ is an integer vector.

(2) Problem D is equivalent to a problem in the standard form P with $A$, $\boldsymbol{b}$, and $\boldsymbol{c}$ replaced by

$$\tilde{A} = \begin{bmatrix} A^\top & -A^\top & I \end{bmatrix}, \qquad \tilde{\boldsymbol{b}} = \boldsymbol{c}, \qquad \tilde{\boldsymbol{c}} = \begin{bmatrix} -\boldsymbol{b}^\top & \boldsymbol{b}^\top & \boldsymbol{0}^\top \end{bmatrix}^\top,$$

respectively (see Example 3.8). Here, $\tilde{A}$ is a totally unimodular matrix (see (4.26)) and $\tilde{\boldsymbol{b}}$ is an integer vector. Therefore, the assertion in (2) follows from that of (1). □

With integrality conditions added to linear programs, the following duality relation results:

$$\begin{aligned} &\inf\{\boldsymbol{c}^\top \boldsymbol{x} \mid \boldsymbol{x} \in P, \boldsymbol{x} \in \mathbb{Z}^n\} \geqq \inf\{\boldsymbol{c}^\top \boldsymbol{x} \mid \boldsymbol{x} \in P\} \\ &= \sup\{\boldsymbol{b}^\top \boldsymbol{y} \mid \boldsymbol{y} \in D\} \geqq \sup\{\boldsymbol{b}^\top \boldsymbol{y} \mid \boldsymbol{y} \in D, \boldsymbol{y} \in \mathbb{Z}^m\}, \end{aligned} \tag{4.34}$$

where

$$P = \{\boldsymbol{x} \mid A\boldsymbol{x} = \boldsymbol{b}, \boldsymbol{x} \geqq \boldsymbol{0}\}, \qquad D = \{\boldsymbol{y} \mid A^\top \boldsymbol{y} \leqq \boldsymbol{c}\}$$

represent the feasible regions of the problems P and D, respectively, and the equality "$=$" in the middle of (4.34) is due to the strong duality in linear programming (Theorem 3.12 in Sec. 3.5.3). In general, the two inequalities "$\geqq$" in (4.34) are more likely to be satisfied in strict inequalities "$>$" and hence we usually have

$$\inf\{\boldsymbol{c}^\top \boldsymbol{x} \mid \boldsymbol{x} \in P, \boldsymbol{x} \in \mathbb{Z}^n\} > \sup\{\boldsymbol{b}^\top \boldsymbol{y} \mid \boldsymbol{y} \in D, \boldsymbol{y} \in \mathbb{Z}^m\},$$

*i.e.*, a failure of the strong duality for integer programs. However, if the matrix $A$ is totally unimodular, we have an equality here, obtaining the strong duality under integrality constraints:

$$\inf\{\boldsymbol{c}^\top \boldsymbol{x} \mid \boldsymbol{x} \in P, \boldsymbol{x} \in \mathbb{Z}^n\} = \sup\{\boldsymbol{b}^\top \boldsymbol{y} \mid \boldsymbol{y} \in D, \boldsymbol{y} \in \mathbb{Z}^m\}. \tag{4.35}$$

Such strong duality is extremely convenient and useful in optimization (see Remark 3.8).

[9] More precisely, the *basic solution* determined by $B$ is the vector $\boldsymbol{x}$ with $x_j$ being equal to the $j$th component of $B^{-1}\boldsymbol{b}$ for $j \in J$ and $x_j = 0$ for $j \notin J$, where $J$ denotes the set of the column subset of $A$ that corresponds to $B$.

For example, in network-type optimization problems [27,28] (such as the maximum flow problem, the minimum cost flow problem, and the shortest path problem), the total unimodularity of graph incidence matrices (Theorem 4.14) guarantees the strong duality under integrality conditions, from which interesting combinatorial duality theorems follow.[10]

**Remark 4.9.** The proof of Theorem 4.17 shows that the total unimodularity of $A$ is not absolutely necessary to guarantee the integrality in the problem P, but a weaker condition: "every $m \times m$ subdeterminant of $A$ is 0, 1, or $-1$" suffices, where $m$ is the number of rows of $A$. The integrality property of linear programs that corresponds precisely to the total unimodularity of $A$ is as follows.

**Hoffman–Kruskal Theorem**: *Let $A$ be an $m \times n$ integer matrix. The polyhedron $\{\boldsymbol{x} \in \mathbb{R}^n \mid A\boldsymbol{x} \leqq \boldsymbol{b}, \boldsymbol{x} \geqq \boldsymbol{0}\}$ is an integer polyhedron for every vector $\boldsymbol{b} \in \mathbb{Z}^m$ if and only if $A$ is a totally unimodular matrix.*

Here, an *integer polyhedron* means a polyhedron (defined by a system of inequalities using rational coefficients) such that every face contains an integer vector. In the case of a bounded polyhedron, this condition is equivalent to all the vertices being integer vectors. ■

**Remark 4.10.** In considering integer solutions $\boldsymbol{x}$ for a system of linear equations $A\boldsymbol{x} = \boldsymbol{b}$, we can make effective use of the Smith normal form (Sec. 4.5) based on a rewriting of the equation:

$$A\boldsymbol{x} = \boldsymbol{b} \iff (UAV)(V^{-1}\boldsymbol{x}) = U\boldsymbol{b}$$

with unimodular matrices $U$ and $V$. For a system of inequalities, however, the Smith normal form is not that effective, since such a rewriting does not work, that is,

$$A\boldsymbol{x} \leqq \boldsymbol{b} \iff (UAV)(V^{-1}\boldsymbol{x}) \leqq U\boldsymbol{b}$$

is not true. Instead, we can make use of the following obvious equivalence:

$$A\boldsymbol{x} \leqq \boldsymbol{b} \iff (AV)(V^{-1}\boldsymbol{x}) \leqq \boldsymbol{b}.$$

This indicates that the Hermite normal form is a more adequate tool in considering integer solutions for a system of linear inequalities. ■

[10] Hall's theorem explained in Remark 1.6 is such an example.

# Chapter 5

# Polynomial Matrices

In this chapter we deal with matrices with entries of polynomials. Such matrices often appear in engineering as representations of linear systems in the frequency domain (Laplace transforms). Since division is not always possible between polynomials, consideration of divisibility is crucial. Fundamental facts on matrices with entries of polynomials, such as the Hermite normal form, the Smith normal form, and the Kronecker canonical form of matrix pencils, are discussed. Facts pertaining to elementary transformations, the Hermite normal form, and the Smith normal form are almost direct analogues of those for integer matrices in Chap. 4.

## 5.1 Polynomial Matrices and Their Examples

### 5.1.1 *Polynomial Matrices*

A *polynomial matrix* means a matrix whose entries are polynomials. Similarly, a vector whose components are polynomials is called a *polynomial vector*. In this chapter we deal with polynomials in a single variable, say, $s$. For example, the following expressions

$$s+2, \qquad 3s^2-4s-4, \qquad -s^9+s^3+s, \qquad 0$$

are polynomials, whereas

$$\frac{1}{s}, \qquad \frac{s+1}{s^2-4s-4}, \qquad \sqrt{s^9+s^3+3}, \qquad 1+s+\frac{1}{2!}s^2+\frac{1}{3!}s^3+\cdots$$

are not.

Let $A(s)$ be an $m \times n$ polynomial matrix, and denote its $(i,j)$ entry by $a_{ij}(s)$ $(i=1,\ldots,m;\ j=1,\ldots,n)$. The degree of $a_{ij}(s)$ as a polynomial is associated with each entry $a_{ij}(s)$, and the maximum degree over all entries is called the *degree* of the polynomial matrix $A(s)$, which we denote as $\delta(A(s))$. We can express $A(s)$ in the form of

$$A(s) = s^d A_d + s^{d-1} A_{d-1} + \cdots + s^2 A_2 + s A_1 + A_0 \tag{5.1}$$

with $m \times n$ constant matrices $A_d, A_{d-1}, \ldots, A_1, A_0$, called *coefficient matrices.* Usually we have $d = \delta(A(s))$, but there are also cases where $d > \delta(A(s))$ with $A_d = A_{d-1} = \cdots = A_{\delta(A(s))+1} = O$.

The determinant of a polynomial matrix $A(s)$ is a polynomial in the variable $s$. The rank of a polynomial matrix $A(s)$ is defined as the maximum value of $k$ such that a nonvanishing subdeterminant, which is nonzero as a polynomial in $s$, exists among the subdeterminants of $A(s)$ of order $k$.

### 5.1.2 *Examples in Engineering*

Using simple examples we explain how engineering systems can be described by polynomial matrices.

#### 5.1.2.1 *Electric circuit*

Figure 5.1 is an electric circuit consisting of a capacitor $C$, a resistor $R$, and an inductor $L$ connected in parallel to the current source $I_0(t)$. The currents $i_C$, $i_R$, $i_L$ and voltages $v_C$, $v_R$, $v_L$ of the elements are subject, respectively, to Kirchhoff's current conservation and voltage conservation laws

$$i_C + i_R + i_L = I_0, \qquad v_C = v_R = v_L,$$

and the physical characteristics of the elements are described as

$$i_C = C\frac{\mathrm{d}v_C}{\mathrm{d}t}, \qquad v_R = R i_R, \qquad v_L = L\frac{\mathrm{d}i_L}{\mathrm{d}t}.$$

In the frequency domain the element characteristics are expressed as

$$i_C = sC\, v_C, \qquad v_R = R\, i_R, \qquad v_L = sL\, i_L.$$

Here, $s$ is the variable for Laplace transformation (Remark 5.1), which can be regarded as a differential operator as well.

The relations above are described in the form of

$$A(s)\boldsymbol{x} = \boldsymbol{b} \tag{5.2}$$

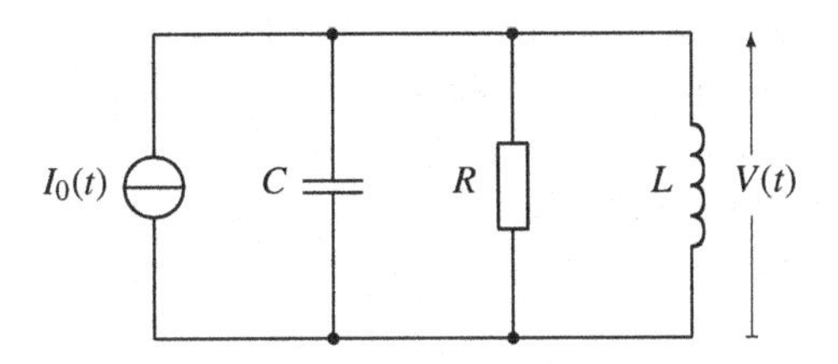


Fig. 5.1 An RLC circuit.

in terms of a vector of currents and voltages $\boldsymbol{x} = (i_C, i_R, i_L, v_C, v_R, v_L)^\top$ using a polynomial matrix

$$A(s) = \left[\begin{array}{ccc|ccc} 1 & 1 & 1 & & & \\ \hline & & & 1 & -1 & 0 \\ & & & 0 & 1 & -1 \\ \hline -1 & 0 & 0 & sC & 0 & 0 \\ 0 & R & 0 & 0 & -1 & 0 \\ 0 & 0 & sL & 0 & 0 & -1 \end{array}\right] \tag{5.3}$$

and a vector $\boldsymbol{b} = (I_0, 0, 0, 0, 0, 0)^\top$. Note that the coefficient matrix $A(s)$ is a polynomial matrix of degree one in variable $s$.

When the variable $s$ is regarded as a differential operator, the equation (5.2) is naturally recognized as a system of linear *differential equations.* However, it also contains relations (algebraic equations) for expressing the conservation laws. Such a system of equations is often called a *differential-algebraic equation* (DAE)[1] (to distinguish it from a pure differential equation). We will discuss DAE in more detail in Sec. 5.7.

**Remark 5.1.** For a function $x(t)$ of variable $t \in [0, \infty)$, the function $\hat{x}(s)$ defined by

$$\hat{x}(s) = \int_0^\infty x(t)\mathrm{e}^{-st}\,\mathrm{d}t \qquad (s \in \mathbb{C})$$

is called the *Laplace transform*[2] of $x(t)$. Usually, the variable $t$ represents time, and then $s$ corresponds to *complex frequency.* Under the zero initial condition $x(0) = 0$, the Laplace transform of the derivative $\dfrac{\mathrm{d}x}{\mathrm{d}t}(t)$ is equal to

[1]DAE stands for "differential-algebraic equation."

[2]Whereas "Laplace transform" refers to the transformed function $\hat{x}$, the term "Laplace transformation" is used to mean the operation $x \mapsto \hat{x}$ that transforms a function $x$ to another function $\hat{x}$. In the Japanese language, however, the same word is used to refer to both, and what it means should be understood from the context.

$s \cdot \hat{x}(s)$. In this sense, we can regard the variable $s$ as a differential operator and, accordingly, regard a polynomial of $s$ as a (higher-order) differential operator. For example, $s^2 + s + 3$ corresponds to $\mathrm{d}^2/\mathrm{d}t^2 + \mathrm{d}/\mathrm{d}t + 3$. We mention that, in the main text of this book, $\hat{x}(s)$ is identified with $x(t)$, both being denoted by the same symbol $x$. ■

#### 5.1.2.2 *Mechanical system*

Next we consider a simple mechanical vibration system, mass–spring–damper system, in Fig. 5.2. This system is composed of the following:

- two boxes (lumped masses) with masses $m_1$ and $m_2$,
- two springs with spring constants $k_1$ and $k_2$ (*i.e.*, force $= k_i \times$ displacement), and
- one viscous damper with damping coefficient $f$ (*i.e.*, force $= f \times$ velocity).

A rightward excitation force (input force) $u$ is applied to the right box.

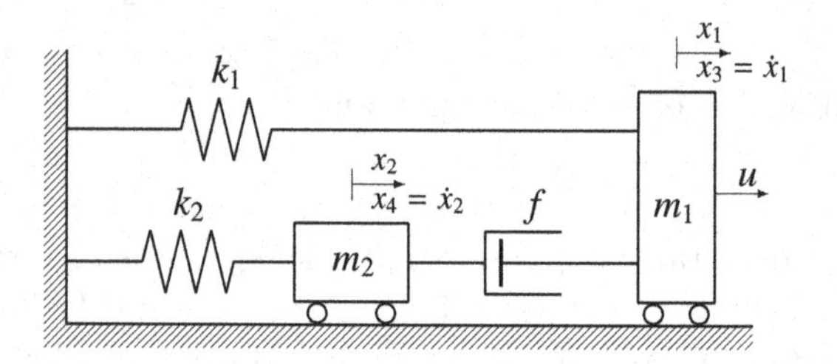


Fig. 5.2 A mechanical vibration system.

Among several possibilities, we can use the following four variables to describe this system:

$x_1 = x_1(t)$ : displacement (rightward) from the undeformed position of mass $m_1$,

$x_2 = x_2(t)$ : displacement (rightward) from the undeformed position of mass $m_2$,

$x_3 = x_3(t)$ : velocity (rightward) of mass $m_1$, and

$x_4 = x_4(t)$ : velocity (rightward) of mass $m_2$.

With the notation "$\cdot$" (read as dot) for the time derivative $\mathrm{d}/\mathrm{d}t$, the definition of velocity gives

$$\dot{x}_1 = x_3, \qquad \dot{x}_2 = x_4.$$

Since the force exerted by the damper is given by $f\cdot(x_3 - x_4)$, the equations of motion are

$$m_1\dot{x}_3 = -k_1x_1 - f(x_3 - x_4) + u,$$
$$m_2\dot{x}_4 = -k_2x_2 + f(x_3 - x_4).$$

By defining vectors $\boldsymbol{x} = (x_1, x_2, x_3, x_4)^\top$ and $\boldsymbol{u} = (u)$, we can put the above four equations into the form of

$$\frac{\mathrm{d}\boldsymbol{x}}{\mathrm{d}t}(t) = A\boldsymbol{x}(t) + B\boldsymbol{u}(t) \tag{5.4}$$

with

$$A = \begin{bmatrix} 0 & 0 & 1 & 0 \\ 0 & 0 & 0 & 1 \\ -k_1/m_1 & 0 & -f/m_1 & f/m_1 \\ 0 & -k_2/m_2 & f/m_2 & -f/m_2 \end{bmatrix}, \quad B = \begin{bmatrix} 0 \\ 0 \\ 1/m_1 \\ 0 \end{bmatrix}. \tag{5.5}$$

In control theory, an equation of the form of (5.4) is called a *state-space equation* in standard form. A state-space equation with outputs is as follows:

$$\frac{\mathrm{d}\boldsymbol{x}}{\mathrm{d}t}(t) = A\boldsymbol{x}(t) + B\boldsymbol{u}(t), \qquad \boldsymbol{y}(t) = C\boldsymbol{x}(t). \tag{5.6}$$

Here, $\boldsymbol{x}$ is referred to as the *state vector*, $\boldsymbol{u}$ as the *input vector*, and $\boldsymbol{y}$ as the *output vector*.

The frequency domain representation of the state-space equation (5.6) is given by[3]

$$(sI - A)\boldsymbol{x} = B\boldsymbol{u}, \qquad \boldsymbol{y} = C\boldsymbol{x}, \tag{5.7}$$

wherein a polynomial matrix $(sI - A)$ appears as a coefficient matrix. By eliminating the state vector $\boldsymbol{x}$ from (5.7), we can obtain the relation between the input $\boldsymbol{u}$ and the output $\boldsymbol{y}$:

$$\boldsymbol{y} = C(sI - A)^{-1}B\boldsymbol{u}. \tag{5.8}$$

The matrix

$$G(s) = C(sI - A)^{-1}B \tag{5.9}$$

that represents the input-output relation is called the *transfer function matrix*. The entries of the transfer function matrix $G(s)$ are not polynomial

[3]It is assumed that the initial value is $\boldsymbol{x}(0) = \boldsymbol{0}$.

but rational functions. If $C = (0,1,0,0)$ (*i.e.*, $y = x_2$) in the above example, we have

$$y = \frac{fs}{(m_1 s^2 + fs + k_1)(m_2 s^2 + fs + k_2) - f^2 s^2} \times u.$$

In a single-input single-output system, like the example here, the transfer function matrix $G(s)$ is a $1 \times 1$ matrix (scalar) and is called the *transfer function.*[4]

## 5.2 Properties of Polynomials

We explain some basic concepts and properties about univariate polynomials that we need in this chapter. In general, a *polynomial* $f(s)$ in variable $s$ can be represented as

$$f(s) = c_d s^d + c_{d-1} s^{d-1} + \cdots + c_2 s^2 + c_1 s + c_0 \quad (\text{with } c_d \neq 0) \qquad (5.10)$$

or else

$$f(s) = 0 \qquad (\text{zero polynomial}). \qquad (5.11)$$

In (5.10), the nonnegative integer $d$ is called the *degree* of $f(s)$, and the constants $c_d, c_{d-1}, \ldots, c_1, c_0$ are called the *coefficients*. In this book, the degree of $f(s)$ is denoted as $\delta(f(s))$ and the leading coefficient (the coefficient of $s^{\delta(f(s))}$) is denoted as[5] $\mathrm{lc}(f(s))$. For example:

$$\begin{array}{lll} \delta(s+2) = 1, & \delta(3s^2 - 4s - 4) = 2, & \delta(-s^9 + s^3 + s) = 9; \\ \mathrm{lc}(s+2) = 1, & \mathrm{lc}(3s^2 - 4s - 4) = 3, & \mathrm{lc}(-s^9 + s^3 + s) = -1. \end{array}$$

According to the definition above, the degree of a zero polynomial $f(s) = 0$ is left undefined, but it is often convenient to put $\delta(0) = -\infty$ to simplify the statements.

When discussing polynomials, it is important to specify what kind of numbers are used as coefficients $c_d, c_{d-1}, \ldots, c_1, c_0$. To formulate this mathematically, we fix a *field* $F$ (a set of numbers on which addition, subtraction, multiplication, and division are defined) as the field of coefficients and assume that all the coefficients are taken from $F$. In engineering applications, we often have $F = \mathbb{R}$ (field of real numbers) or $F = \mathbb{C}$ (field of complex

[4] Matrices with entries of rational functions are not treated in this book, but a theory for rational function matrices has also been developed. In particular, the Smith normal form for polynomial matrices (Sec. 5.5) has an analogue for rational function matrices, which is used in control theory (see Remark 5.3 in Sec. 5.5).

[5] "lc" stands for "leading coefficient."

numbers), but $F$ can also be a finite field, *e.g.*, in coding and cryptography. A polynomial with coefficients taken from $F$ is called a polynomial over $F$, and the set of all polynomials over $F$ in variable $s$ is denoted as $F[s]$.

The expression $p(s) \mid f(s)$ is used to mean that the first polynomial $p(s)$ is a divisor of the second polynomial $f(s)$, or in other words, $f(s)$ is divisible by $p(s)$. For a set of polynomials $f_1(s), \ldots, f_n(s)$, a polynomial $p(s)$ that satisfies $p(s) \mid f_i(s)$ for all $i = 1, \ldots, n$ is referred to as a *common divisor* of $f_1(s), \ldots, f_n(s)$. As long as not all $f_i(s)$ are 0, there exists a polynomial of maximum degree among the common divisors of $f_1(s), \ldots, f_n(s)$. The common divisor of maximum degree, which is determined uniquely up to a multiplicative constant factor, is referred to as the *greatest common divisor* of $f_1(s), \ldots, f_n(s)$. In this book, we denote by $\gcd(f_1(s), \ldots, f_n(s))$ the greatest common divisor with the leading coefficient being equal to 1.

Consider the division of a polynomial $f(s)$ by another polynomial $g(s)$, and let $q(s)$ be the quotient and $r(s)$ be the remainder. Then we have

$$f(s) = g(s)q(s) + r(s), \qquad \delta(g(s)) > \delta(r(s)), \tag{5.12}$$

where $\delta(r(s)) = -\infty$ if $r(s) = 0$. Since

$$\gcd(f(s), g(s)) = \gcd(r(s), g(s)) \tag{5.13}$$

holds, we can obtain the greatest common divisor $\gcd(f(s), g(s))$ for a given pair of polynomials $f(s)$ and $g(s)$ by repeating such a division. First, let us explain this using an example.

**Example 5.1.** The greatest common divisor of two polynomials

$$f(s) = s^4 + s^3 - s^2 + 1, \qquad g(s) = s^3 - 2s - 1$$

can be found as follows. First we divide $f(s)$ by $g(s)$ to obtain

$$\begin{array}{ccccccc} s^4 + s^3 - s^2 + 1 & = & (s^3 - 2s - 1) & \times & (s + 1) & + & (s^2 + 3s + 2) \\ \| & & \| & & \| & & \| \\ f(s) & & g(s) & & q_1(s) & & r_1(s) \end{array}$$

as in (5.12). Next we divide $g(s)$ by $r_1(s)$ to obtain

$$\begin{array}{ccccccc} s^3 - 2s - 1 & = & (s^2 + 3s + 2) & \times & (s - 3) & + & (5s + 5) \\ \| & & \| & & \| & & \| \\ g(s) & & r_1(s) & & q_2(s) & & r_2(s) \end{array}$$

as in (5.12). Then we divide $r_1(s)$ by $r_2(s)$ to obtain

$$\begin{array}{ccccccc} s^2 + 3s + 2 & = & (5s + 5) & \times & (\frac{1}{5}s + \frac{2}{5}) & + & 0 \\ \| & & \| & & \| & & \| \\ r_1(s) & & r_2(s) & & q_3(s) & & r_3(s) \end{array}$$

with remainder 0. By (5.13), the gcd remains invariant at each stage of division, and hence we have

$$\gcd(f(s), g(s)) = \gcd(g(s), r_1(s)) = \gcd(r_1(s), r_2(s)) = \frac{r_2(s)}{\mathrm{lc}(r_2(s))} = s + 1.$$

Indeed the given polynomials have factorizations

$$f(s) = (s+1)(s^3 - s + 1), \qquad g(s) = (s+1)(s^2 - s - 1)$$

that have $s + 1$ as a common factor (of maximum degree). ■

The *Euclidean algorithm* is a method of obtaining the greatest common divisor $\gcd(f(s), g(s))$ of a given pair of polynomials $f(s)$ and $g(s)$ by repeating division as in the example above. First, we can assume

$$\delta(f(s)) \geqq \delta(g(s)),$$

since otherwise we can exchange $f(s)$ and $g(s)$. We generate a sequence of polynomials $f_0(s), f_1(s), f_2(s), \ldots$ by setting $f_0(s) = f(s)$ and $f_1(s) = g(s)$, and defining $f_{i+1}(s)$ as the remainder in the division of $f_{i-1}(s)$ by $f_i(s)$ for $i = 1, 2, \ldots$. In the course of this process, the degree of $f_i(s)$ decreases strictly, and therefore, $f_{m+1}(s) = 0$ for some finite $m$. The resulting relations are

$$\begin{aligned} &f_0(s) = f(s), \quad f_1(s) = g(s), \\ &f_{i-1}(s) = f_i(s)q_i(s) + f_{i+1}(s), \quad \delta(f_i(s)) > \delta(f_{i+1}(s)) \quad (i = 1, 2, \ldots, m), \\ &f_{m+1}(s) = 0, \end{aligned} \tag{5.14}$$

from which follows

$$\gcd(f(s), g(s)) = f_m(s) \ / \ \mathrm{lc}(f_m(s)).$$

A scrutiny of the process of the Euclidean algorithm reveals the following important fact.

**Theorem 5.1.** *For given polynomials $f(s)$ and $g(s)$, there exist polynomials $u(s)$ and $v(s)$ that satisfy*

$$f(s)u(s) + g(s)v(s) = \gcd(f(s), g(s)). \tag{5.15}$$

**Proof.** We are to derive an equation of the form

$$f_m(s) = f_0(s)u_1(s) + f_1(s)v_1(s)$$

by eliminating $f_2(s), f_3(s), \ldots, f_{m-1}(s)$ from (5.14). To be more specific, we show, for $i = m-1, m-2, \ldots, 2, 1$ in this order, the existence of polynomials $u_i(s)$ and $v_i(s)$ that satisfy

$$f_m(s) = f_{i-1}(s)u_i(s) + f_i(s)v_i(s). \tag{5.16}$$

The equation (5.14) for $i = m - 1$ shows

$$f_m(s) = f_{m-2}(s) - f_{m-1}(s)q_{m-1}(s),$$

and therefore, (5.16) for $i = m - 1$ holds with

$$u_{m-1}(s) = 1, \qquad v_{m-1}(s) = -q_{m-1}(s). \tag{5.17}$$

Assume $i \leqq m - 2$. We have

$$f_i(s) = -f_{i-1}(s)q_{i-1}(s) + f_{i-2}(s)$$

from (5.14). By substituting this expression into (5.16), we obtain

$$\begin{aligned} f_m(s) &= f_{i-1}(s)u_i(s) + f_i(s)v_i(s) \\ &= f_{i-1}(s)u_i(s) + [-f_{i-1}(s)q_{i-1}(s) + f_{i-2}(s)]v_i(s) \\ &= f_{i-2}(s)v_i(s) + f_{i-1}(s)[u_i(s) - q_{i-1}(s)v_i(s)]. \end{aligned}$$

This shows that (5.16) is satisfied by the polynomials $u_i(s)$ and $v_i(s)$ determined by the recurrence relation

$$u_{i-1}(s) = v_i(s), \quad v_{i-1}(s) = u_i(s) - q_{i-1}(s)v_i(s) \quad (i = m-1, m-2, \dots, 2). \tag{5.18}$$

Finally, we define

$$u(s) = u_1(s) \;/\; \mathrm{lc}(f_m(s)), \qquad v(s) = v_1(s) \;/\; \mathrm{lc}(f_m(s)), \tag{5.19}$$

which satisfy (5.15). □

The above proof provides us with a concrete procedure, the *Euclidean algorithm*, for computing polynomials $u(s)$ and $v(s)$ that satisfy (5.15). That is, we start with the initial values in (5.17), proceed according to the recurrence formula (5.18), and finally normalize them as in (5.19).

**Example 5.2.** In the process of the Euclidean algorithm applied to

$$f(s) = s^4 + s^3 - s^2 + 1, \qquad g(s) = s^3 - 2s - 1$$

in Example 5.1, we have[6]

$$\begin{aligned} f(s) &= g(s)q_1(s) + r_1(s), & q_1(s) &= s + 1, \\ g(s) &= r_1(s)q_2(s) + r_2(s), & q_2(s) &= s - 3. \end{aligned}$$

The elimination of $r_1(s)$ results in

$$f(s)[-q_2(s)] + g(s)[1 + q_1(s)q_2(s)] = r_2(s) = 5 \cdot \mathrm{gcd}(f(s), g(s)),$$

[6] In the notation of the proof of Theorem 5.1, we have $f_2(s) = r_1(s)$, $f_3(s) = r_2(s)$, and $m = 3$.

showing that (5.15) holds for

$$u(s) = -\frac{1}{5}q_2(s) = \frac{1}{5}(-s+3),$$
$$v(s) = \frac{1}{5}(1 + q_1(s)q_2(s)) = \frac{1}{5}(s^2 - 2s - 2).$$

We can confirm this by calculating $f(s)u(s) + g(s)v(s)$ as

$$(s^4 + s^3 - s^2 + 1) \cdot \frac{1}{5}(-s+3) + (s^3 - 2s - 1) \cdot \frac{1}{5}(s^2 - 2s - 2)$$
$$= s + 1 = \gcd(f(s), g(s)).$$

■

## 5.3 Unimodular Matrices and Elementary Transformations

### 5.3.1 *Unimodular Matrices*

For polynomial matrices, nonsingularity, *i.e.*, having a nonzero determinant, does not imply that the inverse is a polynomial matrix. Generally, the inverse contains rational functions.

**Example 5.3.** Matrix $A(s) = \begin{bmatrix} s & -1 \\ 1 & 1 \end{bmatrix}$ is nonsingular, but its inverse

$$A(s)^{-1} = \begin{bmatrix} 1/(s+1) & 1/(s+1) \\ -1/(s+1) & s/(s+1) \end{bmatrix}$$

is not a polynomial matrix. ■

First, let us derive a necessary condition for a polynomial matrix to have a polynomial inverse matrix. Suppose that a polynomial matrix $A(s)$ is nonsingular and that its inverse $B(s)$ is a polynomial matrix. Since the determinant of a polynomial matrix is a polynomial, both $\det A(s)$ and $\det B(s)$ are polynomials, whereas

$$\det A(s) \cdot \det B(s) = \det(A(s)B(s)) = 1$$

by $A(s)B(s) = I$ (unit matrix). It then follows that $\det A(s)$ is a nonzero constant, which is a necessary condition for $A(s)$ to have a polynomial inverse matrix.

As is stated in Theorem 5.2 below, this necessary condition is, in fact, a sufficient condition. A (square) polynomial matrix is called a *unimodular matrix* if its determinant is a nonzero constant.

**Theorem 5.2.** *For a polynomial matrix $A(s)$, the inverse of $A(s)$ exists and is a polynomial matrix if and only if $A(s)$ is a unimodular matrix.*

**Proof.** The necessity has already been shown. To show the sufficiency, assume $A(s)$ to be a unimodular matrix. Then $A(s)$ is nonsingular and the inverse $A(s)^{-1}$ exists as a rational function matrix. Each entry of $A(s)^{-1}$ is expressed in the form of "cofactor of $A(s)$ / $\det A(s)$" where a cofactor of a polynomial matrix $A(s)$ is a polynomial and $\det A(s)$ is a nonzero constant. Therefore, each entry of $A(s)^{-1}$ is a polynomial. □

**Proposition 5.1.** *If a polynomial matrix $A(s)$ is unimodular, then its inverse $A(s)^{-1}$ is also unimodular.*

**Proof.** By Theorem 5.2, $A(s)^{-1}$ is a polynomial matrix. On the other hand, $\det A(s)^{-1} = 1/\det A(s)$ is a nonzero constant. Therefore, $A(s)^{-1}$ is a unimodular matrix. □

**Example 5.4.** Consider a polynomial matrix $A(s) = \begin{bmatrix} s+2 & s \\ 1 & 1 \end{bmatrix}$. Since the determinant $\det A(s) = 2$ is a nonzero constant, $A(s)$ is a unimodular matrix. Its inverse

$$A(s)^{-1} = \begin{bmatrix} 1/2 & -s/2 \\ -1/2 & (s+2)/2 \end{bmatrix}$$

is a polynomial matrix and, moreover, a unimodular matrix. ■

### 5.3.2 *Elementary Transformations*

Let us first recall elementary transformations for real matrices (matrices whose entries are real numbers). The elementary column transformations consist of the following three kinds of operations [6, 15]:

(1) exchanging two columns,
(2) multiplying a certain column by a nonzero number, and
(3) adding a scalar multiple of a certain column to another column.

For polynomial matrices, however, we need to preserve polynomiality in elementary transformations and their inverse transformations. With this in mind we introduce the following three kinds of operations:

(1) exchanging two columns,
(2) multiplying a certain column by a nonzero constant, and
(3) adding a polynomial multiple of a certain column to another column

as *elementary column transformations*, which are also referred to as *polynomial elementary column transformations*.

The above three types of polynomial elementary column transformations are represented, respectively, by matrices of the following forms:

$$E_1 = \left[\begin{array}{cc|ccc} & 1 & & & \\ 1 & & & & \\ \hline & & 1 & & \\ & & & \ddots & \\ & & & & 1 \end{array}\right], \quad E_2 = \left[\begin{array}{c|cccc} c & & & & \\ \hline & 1 & & & \\ & & 1 & & \\ & & & \ddots & \\ & & & & 1 \end{array}\right], \quad E_3 = \left[\begin{array}{cc|ccc} 1 & \beta(s) & & & \\ & 1 & & & \\ \hline & & 1 & & \\ & & & \ddots & \\ & & & & 1 \end{array}\right], \tag{5.20}$$

where $c$ stands for a nonzero constant and $\beta(s)$ is a polynomial.

Such matrices are called *polynomial elementary matrices.* More precisely, the matrices $E_1 = E_1(p,q)$, $E_2 = E_2(p;c)$, and $E_3 = E_3(p,q;\beta(s))$ are defined as follows ($p \neq q$):

(1) $E_1(p,q)$: The submatrix corresponding to row and column indices $\{p,q\}$ is $\begin{bmatrix} 0 & 1 \\ 1 & 0 \end{bmatrix}$, and the rest is a unit matrix,
(2) $E_2(p;c)$: The submatrix corresponding to row and column index $\{p\}$ is $[\ c\ ]$, and the rest is a unit matrix, and
(3) $E_3(p,q;\beta(s))$: The $(p,q)$ entry is $\beta(s)$, the other off-diagonal entries are all 0, and the diagonal entries are all 1. The submatrix corresponding to row and column indices $\{p,q\}$ is $\begin{bmatrix} 1 & \beta(s) \\ 0 & 1 \end{bmatrix}$ if $p < q$, and $\begin{bmatrix} 1 & 0 \\ \beta(s) & 1 \end{bmatrix}$ if $q < p$.

With the above definitions we have the following:

(1) If we multiply $A(s)$ by $E_1(p,q)$ from the right, the $p$th column and the $q$th column are exchanged.
(2) If we multiply $A(s)$ by $E_2(p;c)$ from the right, the $p$th column is multiplied by $c$.
(3) If we multiply $A(s)$ by $E_3(p,q;\beta(s))$ from the right, then $\beta(s)$ times the $p$th column is added to the $q$th column.

Polynomial elementary matrices are unimodular matrices.[7] Indeed,

$$\det E_1(p,q) = -1, \quad \det E_2(p;c) = c, \quad \det E_3(p,q;\beta(s)) = 1,$$

and moreover,

$$E_1(p,q)^{-1} = E_1(p,q), \quad E_2(p;c)^{-1} = E_2(p;1/c),$$
$$E_3(p,q;\beta(s))^{-1} = E_3(p,q;-\beta(s)).$$

[7] In Theorem 5.4 in Sec. 5.4, it will be shown that every unimodular matrix can be expressed as a product of polynomial elementary matrices.

Similarly, the following three kinds of operations are defined as *elementary row transformations* for polynomial matrices:

(1) exchanging two rows,
(2) multiplying a certain row by a nonzero constant, and
(3) adding a polynomial multiple of a certain row to another row.

These operations are also referred to as *polynomial elementary row transformations.* Polynomial elementary row transformations are equivalent to multiplying $A(s)$ by $E_1(p,q)$, $E_2(p;c)$, and $E_3(p,q;\beta(s))$ from the left of the matrix as follows:

(1) If we multiply $A(s)$ by $E_1(p,q)$ from the left, the $p$th row and $q$th row are exchanged.
(2) If we multiply $A(s)$ by $E_2(p;c)$ from the left, the $p$th row is multiplied by $c$.
(3) If we multiply $A(s)$ by $E_3(p,q;\beta(s))$ from the left, then $\beta(s)$ times the $q$th row is added to the $p$th row.

Polynomial elementary row and column transformations are together referred to as *polynomial elementary transformations.*

The significance of elementary transformations is summarized in the following proposition.[8] For polynomials $a_1(s), a_2(s), \ldots, a_n(s)$ $(n \geqq 1)$, in general, we denote their *greatest common divisor* (with the leading coefficient 1) as $\gcd(a_1(s), a_2(s), \ldots, a_n(s))$. In the exceptional case where all the $a_j(s)$ are 0, we set their greatest common divisor to be 0.

**Proposition 5.2.**

(1) *A row vector* $(a_1(s), a_2(s), \ldots, a_n(s)) \neq \mathbf{0}^\top$ *with polynomial components can be transformed to* $(b(s), 0, \ldots, 0)$ *through repeated elementary column transformations, where* $b(s) = \gcd(a_1(s), a_2(s), \ldots, a_n(s))$.
(2) *A column vector* $(a_1(s), a_2(s), \ldots, a_n(s))^\top \neq \mathbf{0}$ *with polynomial components can be transformed to* $(b(s), 0, \ldots, 0)^\top$ *through repeated elementary row transformations, where* $b(s) = \gcd(a_1(s), a_2(s), \ldots, a_n(s))$.

**Proof.** (1) With a suitable permutation of the components (or columns) we can bring the minimum degree component to the first component $a_1(s)$. For $j = 2, \ldots, n$, divide $a_j(s)$ by $a_1(s)$ to obtain quotient $q_j(s)$ and remainder

[8] Proposition 5.2 will be used in the proofs for the Hermite normal form (Theorem 5.3) and the Smith normal form (Theorem 5.7).

$r_j(s)$, *i.e.*,

$$a_j(s) = a_1(s)q_j(s) + r_j(s), \qquad \delta(a_1(s)) > \delta(r_j(s)), \tag{5.21}$$

where $\delta(r_j(s)) = -\infty$ if $r_j(s) = 0$. By adding $-q_j(s)$ times the first component to the $j$th component for each $j = 2, \ldots, n$, the row vector $(a_1(s), a_2(s), \ldots, a_n(s))$ is transformed to another row vector $(a_1(s), r_2(s), \ldots, r_n(s))$. If $r_j(s) = 0$ for all $j = 2, \ldots, n$, the assertion of (1) holds with $b(s) = a_1(s)/\mathrm{lc}(a_1(s))$. If this is not the case, we repeat the same transformation. In the course of such transformations, the minimum degree of the nonzero components decreases strictly since $\delta(a_1(s)) > \delta(r_j(s)) \geqq 0$ $(j = 2, \ldots, n)$, and so the process ends after a finite number of iterations. Note an important fact that

$$\gcd(a_1(s), r_2(s), \ldots, r_n(s)) = \gcd(a_1(s), a_2(s), \ldots, a_n(s)),$$

which follows from (5.21).

The proof for (2) is similar to that for (1). □

**Example 5.5.** Row vector $(s^3 + s^2, s^2 + 2s + 1, s^2 - 1)$ is transformed through elementary column transformations as follows:

$$\begin{aligned}
&(s^3 + s^2, s^2 + 2s + 1, s^2 - 1) \Rightarrow (s^2 + 2s + 1, s^3 + s^2, s^2 - 1) \\
&\to (s^2 + 2s + 1, s + 1, s^2 - 1) \to (s^2 + 2s + 1, s + 1, -2s - 2) \\
&\Rightarrow (s + 1, s^2 + 2s + 1, -2s - 2) \to (s + 1, 0, -2s - 2) \\
&\to (s + 1, 0, 0).
\end{aligned}$$

Here, "$\Rightarrow$" stands for a swap of components and "$\to$" for a transformation via division (5.21). The transformation above is the result of multiplying the given row vector by the matrix

$$\begin{bmatrix} 0 & 1 & 0 \\ 1 & 0 & 0 \\ 0 & 0 & 1 \end{bmatrix} \begin{bmatrix} 1 & -s+1 & 0 \\ 0 & 1 & 0 \\ 0 & 0 & 1 \end{bmatrix} \begin{bmatrix} 1 & 0 & -1 \\ 0 & 1 & 0 \\ 0 & 0 & 1 \end{bmatrix} \begin{bmatrix} 0 & 1 & 0 \\ 1 & 0 & 0 \\ 0 & 0 & 1 \end{bmatrix} \begin{bmatrix} 1 & -s-1 & 0 \\ 0 & 1 & 0 \\ 0 & 0 & 1 \end{bmatrix} \begin{bmatrix} 1 & 0 & 2 \\ 0 & 1 & 0 \\ 0 & 0 & 1 \end{bmatrix}$$

$$= \begin{bmatrix} 1 & -s-1 & 2 \\ -s+1 & s^2 & -2s+1 \\ 0 & 0 & 1 \end{bmatrix} \tag{5.22}$$

from the right. ■

### 5.3.3 *Determinantal Divisors*

The proof of Proposition 5.2 shows that the greatest common divisor of the components of a vector remains invariant under elementary transformations. Such a relation between elementary transformations and the greatest common divisor is extended to matrices as follows.

For an $m \times n$ polynomial matrix $A(s)$ and a natural number $k$, the greatest common divisor of all minors (subdeterminants) of order $k$ is called the $k$th *determinantal divisor* of $A(s)$, which we denote by $d_k(A(s))$. That is,

$$d_k(A(s)) = \gcd\{\det A[I,J] \mid |I| = |J| = k\}, \tag{5.23}$$

where the notation $A[I,J]$ for the submatrix with row subset $I$ and column subset $J$ is explained in Remark 4.2. On the right-hand side above, $I$ runs over all subsets of size $k$ of the row set of $A(s)$, and $J$ runs over all subsets of size $k$ of the column set of $A(s)$. The number of such $I$ is equal to $\binom{m}{k}$ (binomial coefficient) and the number of such $J$ is to $\binom{n}{k}$, and therefore, the number of submatrices $A[I,J]$ in (5.23) is $\binom{m}{k} \times \binom{n}{k}$. The index $k$ stays in the range of $1 \leqq k \leqq \min(m,n)$, but it is often convenient to define $d_0(A(s)) = 1$ by allowing $k = 0$.

**Example 5.6.** For polynomial matrix

$$A(s) = \begin{bmatrix} s^3+s^2 & s^2+2s+1 & s^2-1 \\ s^5+2s^4-s^3-s^2 & s^4+3s^3+s^2-s-1 & 2s^3-s^2-s+1 \end{bmatrix}$$

we have

$$\begin{aligned} d_1(A(s)) &= \gcd(s^3+s^2,\ s^2+2s+1,\ s^2-1,\ s^5+2s^4-s^3-s^2, \\ &\qquad s^4+3s^3+s^2-s-1,\ 2s^3-s^2-s+1) \\ &= 1, \\ d_2(A(s)) &= \gcd\left( \det \begin{bmatrix} s^3+s^2 & s^2+2s+1 \\ s^5+2s^4-s^3-s^2 & s^4+3s^3+s^2-s-1 \end{bmatrix}, \right. \\ &\qquad \det \begin{bmatrix} s^3+s^2 & s^2-1 \\ s^5+2s^4-s^3-s^2 & 2s^3-s^2-s+1 \end{bmatrix}, \\ &\qquad \left. \det \begin{bmatrix} s^2+2s+1 & s^2-1 \\ s^4+3s^3+s^2-s-1 & 2s^3-s^2-s+1 \end{bmatrix} \right) \\ &= s^3(s+1). \end{aligned}$$

■

The determinantal divisors remain invariant under polynomial elementary transformations.

**Proposition 5.3.** *Let $A(s)$ be an $m \times n$ polynomial matrix. For any elementary matrices $E(s)$ and $F(s)$ of orders $m$ and $n$, respectively, we have*

$$d_k(E(s)A(s)F(s)) = d_k(A(s)) \qquad (1 \leqq k \leqq \min(m,n)). \tag{5.24}$$

**Proof.** It suffices to prove the invariance under column transformations only; so we assume $E(s)$ to be the identity matrix. The equation (5.24) holds obviously for $F(s) = E_1(p,q)$ or $E_2(p;c)$. To consider the remaining case of $F(s) = E_3(p,q;\beta(s))$, we put

$$B(s) = A(s)\, E_3(p,q;\beta(s))$$

and compare $B[I,J]$ and $A[I,J]$, where $|I| = |J| = k$. There are four cases depending on the inclusion relation of $p$ and $q$ in the subset $J$: (i) $p \in J$, $q \in J$, (ii) $p \in J$, $q \notin J$, (iii) $p \notin J$, $q \notin J$, and (iv) $p \notin J$, $q \in J$. In the cases of (i), (ii), and (iii), we have $\det B[I,J] = \det A[I,J]$, which is a multiple of $d_k(A(s))$. In the case of (iv), we have[9]

$$\det B[I,J] = \det A[I,J] \pm \beta(s) \cdot \det A[I,(J \setminus \{q\}) \cup \{p\}],$$

which shows that $\det B[I,J]$ is a multiple of $d_k(A(s))$. Therefore, $d_k(B(s))$ is a multiple of $d_k(A(s))$. Since the inverse of an elementary transformation is an elementary transformation, we also have that $d_k(A(s))$ is a multiple of $d_k(B(s))$. Therefore, $d_k(B(s)) = d_k(A(s))$. □

## 5.4 Hermite Normal Form

We consider transforming a given polynomial matrix to a simplest possible form by means of elementary column transformations. Let $A(s)$ be an $m \times n$ polynomial matrix of full row rank ($\operatorname{rank} A(s) = m$). Since $m \leqq n$, $A(s)$ is either a square matrix or a rectangular matrix with more columns than rows.

The targeted "simplest form" is an $m \times n$ polynomial matrix $B(s) =$

[9] The notation $(J \setminus \{q\}) \cup \{p\}$ means the set (of column indices) obtained from $J$ by deleting $q$ and adding $p$. The sign of $\pm\beta(s)$ is determined from the sign of the permutation which reorders the sequence formed by replacing $q$ with $p$ in the ascending sequence of members of $J$, to be again ascending. See also Remark 4.2.

$(b_{ij}(s))$ that satisfies the following conditions:[10]

$$b_{ii}(s) \neq 0, \quad \mathrm{lc}(b_{ii}(s)) = 1 \qquad (1 \leqq i \leqq m), \tag{5.25}$$

$$\delta(b_{ij}(s)) < \delta(b_{ii}(s)) \qquad (1 \leqq j < i \leqq m), \tag{5.26}$$

$$b_{ij}(s) = 0 \qquad (1 \leqq i \leqq m;\ i < j \leqq n). \tag{5.27}$$

That is, the targeted form is a lower (left) triangular matrix with polynomial entries such that the diagonal entries have leading coefficients 1 and strictly higher degrees than the off-diagonal entries in the same row. A matrix that satisfies these conditions is said to be in *Hermite normal form.*

The following important theorem, Theorem 5.3, states that every polynomial matrix can be brought to a matrix of Hermite normal form via elementary column transformations. Recall Proposition 5.2(1), which states that every row vector can be brought to the form of $(b(s), 0, \ldots, 0)$ via elementary column transformations. This proposition is certainly a special case ($m = 1$) of Theorem 5.3, but at the same time, it plays the essential role in the proof of Theorem 5.3.

**Theorem 5.3.** *Every polynomial matrix of full row rank can be transformed to Hermite normal form through repeated elementary column transformations.*

**Proof.** Let $A(s) = (a_{ij}(s))$ be a polynomial matrix of full row rank. By Proposition 5.2(1), we can transform the first row vector $(a_{11}(s), a_{12}(s), \ldots, a_{1n}(s))$ of $A(s)$, which is nonzero by the rank assumption, to a vector of the form of $(b_{11}(s), 0, \ldots, 0)$ with $\mathrm{lc}(b_{11}(s)) = 1$ through repeated elementary column transformations. Denote by $A^{(1)}(s) = (a_{ij}^{(1)}(s))$ the matrix resulting from these column transformations applied to the matrix $A(s)$. Then

$$a_{11}^{(1)}(s) = b_{11}(s), \qquad a_{1j}^{(1)}(s) = 0 \quad (j \geqq 2).$$

Next, consider the row vector $(a_{22}^{(1)}(s), a_{23}^{(1)}(s), \ldots, a_{2n}^{(1)}(s))$ consisting of the entries of $A^{(1)}(s)$ in the second row with column indices greater than or equal to 2. This row vector is nonzero, again by the rank assumption, and we can transform it to the form of $(b_{22}(s), 0, \ldots, 0)$ with $\mathrm{lc}(b_{22}(s)) = 1$.

[10] In (5.25), lc( ) denotes the leading coefficient (coefficient of the highest degree term). Since the degree of the zero polynomial is assumed to be $-\infty$, the condition in (5.26) is always satisfied if $b_{ij}(s) = 0$. Therefore, (5.26) is equivalent to the condition that $\delta(b_{ij}(s)) < \delta(b_{ii}(s))$ if $b_{ij}(s) \neq 0$.

Moreover, by division of $a_{21}^{(1)}(s)$ by $b_{22}(s)$, we can find polynomials $q(s)$ and $r(s)$ such that

$$a_{21}^{(1)}(s) = b_{22}(s)q(s) + r(s), \qquad \delta(b_{22}(s)) > \delta(r(s)).$$

Let $b_{21}(s) = r(s)$. By adding $-q(s)$ times the second column to the first column we can transform the second row vector to $(b_{21}(s), b_{22}(s), 0, \ldots, 0)$ with $\delta(b_{21}(s)) < \delta(b_{22}(s))$. Denote by $A^{(2)}(s) = (a_{ij}^{(2)}(s))$ the resulting matrix. Then

$$\begin{aligned} &a_{11}^{(2)}(s) = b_{11}(s), \quad a_{1j}^{(2)}(s) = 0 \ \ (j \geqq 2), \\ &a_{21}^{(2)}(s) = b_{21}(s), \quad a_{22}^{(2)}(s) = b_{22}(s), \quad a_{2j}^{(2)} = 0 \ \ (j \geqq 3). \end{aligned}$$

By repeating this process we arrive at $A^{(m)}(s)$, which is a matrix in Hermite normal form. □

For a polynomial matrix $A(s)$, the matrix in Hermite normal form given in Theorem 5.3 is referred to as the Hermite normal form of $A(s)$. (The uniqueness of this normal form will be shown later in Theorem 5.6.)

**Example 5.7.** We consider again the polynomial matrix in Example 5.6:

$$A(s) = \begin{bmatrix} s^3 + s^2 & s^2 + 2s + 1 & s^2 - 1 \\ s^5 + 2s^4 - s^3 - s^2 & s^4 + 3s^3 + s^2 - s - 1 & 2s^3 - s^2 - s + 1 \end{bmatrix},$$

whose first row is the row vector treated in Example 5.5. By applying the column transformation of Example 5.5 to $A(s)$, *i.e.*, by multiplying $A(s)$ by the matrix of (5.22) from the right, we obtain

$$A^{(1)}(s) = \begin{bmatrix} s+1 & 0 & 0 \\ s^3 + s^2 - 1 & s^3 & -s^4 + s^3 \end{bmatrix}.$$

To transform the second row vector, we multiply this matrix $A^{(1)}(s)$ by the following product of polynomial elementary matrices

$$\begin{bmatrix} 1 & 0 & 0 \\ 0 & 1 & s-1 \\ 0 & 0 & 1 \end{bmatrix} \begin{bmatrix} 1 & 0 & 0 \\ -1 & 1 & 0 \\ 0 & 0 & 1 \end{bmatrix} = \begin{bmatrix} 1 & 0 & 0 \\ -1 & 1 & s-1 \\ 0 & 0 & 1 \end{bmatrix} \tag{5.28}$$

from the right. We thus obtain the Hermite normal form

$$A^{(2)}(s) = \begin{bmatrix} s+1 & 0 & 0 \\ s^2 - 1 & s^3 & 0 \end{bmatrix} = B(s).$$

■

As a corollary to Theorem 5.3 (existence of Hermite normal form) we obtain a characterization of unimodular matrices.

**Theorem 5.4.** *The following three conditions* (a) *to* (c) *are equivalent for a square polynomial matrix* $A(s)$*:*

(a) $A(s)$ *is a unimodular matrix.*
(b) $A(s)$ *is nonsingular and its Hermite normal form is the unit matrix* $I$*.*
(c) $A(s)$ *is expressed as a product of polynomial elementary matrices.*

**Proof.** Let $A(s)$ be an $n \times n$ polynomial matrix. In each case of (a), (b), and (c), $A(s)$ is a nonsingular matrix, for which the Hermite normal form $B(s) = (b_{ij}(s))$ exists by Theorem 5.3. According to the construction of the Hermite normal form, there exist polynomial elementary matrices $E_1(s), E_2(s), \ldots, E_k(s)$ such that $B(s) = A(s)E_1(s)E_2(s) \cdots E_k(s)$.

[(a) $\Rightarrow$ (b)] Since the matrix $B(s)$ is a lower-triangular matrix and $\det E_i(s)$ is a nonzero constant for $i = 1, \ldots, k$, we have

$$b_{11}(s)\ b_{22}(s)\ \cdots\ b_{nn}(s) = \det B(s) = c \cdot \det A(s) \in F \setminus \{0\}$$

for some $c \in F \setminus \{0\}$. This shows that $b_{11}(s), b_{22}(s), \ldots, b_{nn}(s)$ are constants. Furthermore, it follows from this and $\delta(b_{ij}(s)) < \delta(b_{ii}(s))$ $(j < i)$ that $b_{ij}(s) = 0$ $(j < i)$. Therefore, $B(s) = I$ (unit matrix).

[(b) $\Rightarrow$ (c)] It follows from $I = B(s) = A(s)E_1(s)E_2(s) \cdots E_k(s)$ that

$$A(s) = E_k(s)^{-1} \cdots E_2(s)^{-1} E_1(s)^{-1},$$

where each $E_i(s)^{-1}$ is a polynomial elementary matrix.

[(c) $\Rightarrow$ (a)] This is obvious, since an elementary matrix is unimodular. □

Theorem 5.4 enables us to rephrase Theorem 5.3 as follows.

**Theorem 5.5.** *For any polynomial matrix* $A(s)$ *of full row rank, there exists a unimodular matrix* $V(s)$ *such that the matrix* $A(s)V(s)$ *is in Hermite normal form.*

**Example 5.8.** The transformation of the polynomial matrix $A(s)$ in Example 5.7 to its Hermite normal form $B(s)$ is represented as $A(s)V(s) = B(s)$ with a unimodular matrix $V(s)$. The matrix $V(s)$ is given by the product of the matrices in (5.22) and (5.28) as

$$V(s) = (5.22) \times (5.28) = \begin{bmatrix} s+2 & -s-1 & -s^2+3 \\ -s^2-s+1 & s^2 & s^3-s^2-2s+1 \\ 0 & 0 & 1 \end{bmatrix}.$$

The unimodularity of $V(s)$ can also be verified by a straightforward calculation of the determinant. ■

The existence of the Hermite normal form has been established in Theorem 5.3 (or Theorem 5.5). The Hermite normal form is, in fact, uniquely determined.

**Theorem 5.6.** *The Hermite normal form of a polynomial matrix of full row rank is uniquely determined.*

**Proof.** We prove the uniqueness by contradiction. Suppose that $B(s) = (b_{ij}(s))$ and $B'(s) = (b'_{ij}(s))$ are two different matrices in Hermite normal form that are obtained from a given $m \times n$ polynomial matrix $A(s)$ of rank $m$ by unimodular column transformations. Define

$$\Lambda(A(s)) = \{A(s)\boldsymbol{x}(s) \mid \boldsymbol{x}(s) \in F[s]^n\},$$

where $F[s]^n$ denotes the set of all $n$-dimensional vectors whose components are polynomials in $s$ over $F$. Note that $\Lambda(A(s))$ is a subset of $F[s]^m$. For any unimodular matrix $V(s)$ of order $n$ we have

$$\begin{aligned}\Lambda(A(s)V(s)) &= \{A(s)V(s)\boldsymbol{x}(s) \mid \boldsymbol{x}(s) \in F[s]^n\}\\ &= \{A(s)\boldsymbol{y}(s) \mid \boldsymbol{y}(s) = V(s)\boldsymbol{x}(s),\ \boldsymbol{x}(s) \in F[s]^n\}\\ &= \{A(s)\boldsymbol{y}(s) \mid \boldsymbol{y}(s) \in F[s]^n\}\\ &= \Lambda(A(s)),\end{aligned}$$

from which it follows that $\Lambda(A(s)) = \Lambda(B(s)) = \Lambda(B'(s))$.

Since $B(s)$ and $B'(s)$ are distinct, there exists $(i,j)$ with $b_{ij}(s) \neq b'_{ij}(s)$, where $i \geqq j$. Take such $(i,j)$ with $i$ minimum. We may assume $\delta(b_{ii}(s)) \geqq \delta(b'_{ii}(s))$; otherwise exchange $B(s)$ and $B'(s)$.

We denote the column vectors of the matrix $B(s)$ as $\boldsymbol{b}_1(s), \boldsymbol{b}_2(s), \ldots, \boldsymbol{b}_n(s)$ and those of $B'(s)$ as $\boldsymbol{b}'_1(s), \boldsymbol{b}'_2(s), \ldots, \boldsymbol{b}'_n(s)$, where $\boldsymbol{b}_k(s) = \boldsymbol{b}'_k(s) = \boldsymbol{0}$ $(k > m)$. We now examine the $j$th column vectors for the $j$ chosen above. Since

$$\boldsymbol{b}'_j(s) \in \Lambda(B'(s)) = \Lambda(B(s)), \qquad \boldsymbol{b}_j(s) \in \Lambda(B(s)),$$

we have $\boldsymbol{b}'_j(s) - \boldsymbol{b}_j(s) \in \Lambda(B(s))$, and therefore, there exists $\boldsymbol{x}(s) \in F[s]^n$ such that

$$\begin{aligned}\boldsymbol{b}'_j(s) - \boldsymbol{b}_j(s) &= B(s)\boldsymbol{x}(s)\\ &= x_1(s)\boldsymbol{b}_1(s) + x_2(s)\boldsymbol{b}_2(s) + \cdots + x_m(s)\boldsymbol{b}_m(s).\end{aligned}$$

By the choice of the row index $i$, the first $(i-1)$ components of the vector $\boldsymbol{b}'_j(s) - \boldsymbol{b}_j(s)$ are all 0. Since $B(s)$ is a lower (left) triangular matrix, this implies that $x_1(s) = x_2(s) = \cdots = x_{i-1}(s) = 0$. Then, the $i$th component in the above equation shows that

$$b'_{ij}(s) - b_{ij}(s) = x_i(s) b_{ii}(s). \tag{5.29}$$

Here, $x_i(s)$ is a nonzero polynomial, since $b_{ij}(s) \neq b'_{ij}(s)$.

First suppose that $i \neq j$. By the conditions of Hermite normal form and our assumption $\delta(b'_{ii}(s)) \leqq \delta(b_{ii}(s))$, we have

$$\delta(b'_{ij}(s)) < \delta(b'_{ii}(s)) \leqq \delta(b_{ii}(s)), \qquad \delta(b_{ij}(s)) < \delta(b_{ii}(s)),$$

from which follows $\delta(b'_{ij}(s) - b_{ij}(s)) < \delta(b_{ii}(s))$. But this contradicts (5.29).

We next consider the case of $i = j$. The equation (5.29) with $i = j$ yields $0 \neq b'_{ii}(s) = (1 + x_i(s)) b_{ii}(s)$. Since $\delta(b_{ii}(s)) \geqq \delta(b'_{ii}(s))$ by our assumption, $x_i(s)$ must be a constant, say, $c_i$, and $\delta(b_{ii}(s)) = \delta(b'_{ii}(s))$ holds. Then it follows from $\mathrm{lc}(b_{ii}(s)) = \mathrm{lc}(b'_{ii}(s)) = 1$ that $c_i = 0$, which contradicts the above-mentioned fact that $x_i(s)$ is a nonzero polynomial.

In either case, a contradiction has been derived from the assumed existence of two different Hermite normal forms. This proves the uniqueness of the Hermite normal form. □

In this section, we have discussed Hermite normal forms for polynomial matrices of full row rank. For the general case without the rank assumption, see [2, Chap. VI, Sec. 2].

## 5.5 Smith Normal Form

Hermite normal form is a triangularization realized by column transformations. By using both column and row elementary transformations, we can obtain a diagonal form. Since repeated elementary transformations are equivalent to multiplying with a unimodular matrix (Theorem 5.4), we can alternatively say that a given matrix can be transformed to a diagonal form by multiplying unimodular matrices from the left and the right.

**Example 5.9.** Polynomial matrix

$$A(s) = \begin{bmatrix} s+1 & 0 \\ s^2 - 1 & s^3 \end{bmatrix}$$

is already in Hermite normal form and cannot be transformed to a simpler form with column transformations only. However, if we employ a row transformation of adding $(-s+1)$ times the first row to the second row, we can transform it to a diagonal matrix as follows:

$$\begin{bmatrix} 1 & 0 \\ -s+1 & 1 \end{bmatrix} \begin{bmatrix} s+1 & 0 \\ s^2-1 & s^3 \end{bmatrix} = \begin{bmatrix} s+1 & 0 \\ 0 & s^3 \end{bmatrix}.$$

■

**Proposition 5.4.** *For any $m \times n$ polynomial matrix $A(s)$, there exist a unimodular matrix $U(s)$ of order $m$ and a unimodular matrix $V(s)$ of order $n$ such that*

$$U(s)A(s)V(s) = \left[\begin{array}{ccc|c} \alpha_1(s) & & 0 & \\ & \ddots & & O_{r,n-r} \\ 0 & & \alpha_r(s) & \\ \hline & O_{m-r,r} & & O_{m-r,n-r} \end{array}\right]. \tag{5.30}$$

*Here, $r = \operatorname{rank} A(s)$, and $\alpha_1(s), \alpha_2(s), \ldots, \alpha_r(s)$ are polynomials satisfying*

$$\textit{Normalization}: \quad \operatorname{lc}(\alpha_1(s)) = \operatorname{lc}(\alpha_2(s)) = \cdots = \operatorname{lc}(\alpha_r(s)) = 1, \tag{5.31}$$

$$\textit{Degree condition}: \quad \delta(\alpha_1(s)) \leqq \delta(\alpha_2(s)) \leqq \cdots \leqq \delta(\alpha_r(s)). \tag{5.32}$$

**Proof.** Set $A^{(0)}(s) = A(s)$. If $A^{(0)}(s) = O$, we already have the form of (5.30) with $r = 0$. If $A^{(0)}(s) \neq O$, we choose a nonzero entry with the lowest degree and move it to the $(1,1)$ entry by permuting the rows and columns. If necessary, we multiply the first column by a nonzero constant to obtain $\operatorname{lc}(a_{11}^{(0)}(s)) = 1$.

We focus on the first row. For each $j = 2, \ldots, n$ we take polynomials $q_j(s)$ and $r_j(s)$ such that

$$a_{1j}^{(0)}(s) = a_{11}^{(0)}(s)q_j(s) + r_j(s), \qquad \delta(a_{11}^{(0)}(s)) > \delta(r_j(s))$$

and add $-q_j(s)$ times the first column to the $j$th column to change the $(1,j)$ entry to $r_j(s)$. Next, we focus on the first column. For each $i = 2, \ldots, m$ we take polynomials $q_i'(s)$ and $r_i'(s)$ such that

$$a_{i1}^{(0)}(s) = a_{11}^{(0)}(s)q_i'(s) + r_i'(s), \qquad \delta(a_{11}^{(0)}(s)) > \delta(r_i'(s))$$

and add $-q_i'(s)$ times the first row to the $i$th row to change the $(i,1)$ entry to $r_i'(s)$.

If, as a result of the above operations, the first row is $(a_{11}^{(0)}(s), 0, \ldots, 0)$ and the first column is $(a_{11}^{(0)}(s), 0, \ldots, 0)^\top$, then we define

$$A^{(1)}(s) = (a_{ij}^{(1)}(s) \mid 2 \leqq i \leqq m,\ 2 \leqq j \leqq n)$$

to be the $(m-1) \times (n-1)$ matrix consisting of the second row to the $m$th row and the second column to the $n$th column, thereby completing the first stage with

$$\left[\begin{array}{c|ccc} a_{11}^{(0)}(s) & 0 & \cdots & 0 \\ \hline 0 & & & \\ \vdots & & A^{(1)}(s) & \\ 0 & & & \end{array}\right].$$

Otherwise, there is a nonzero polynomial, other than $a_{11}^{(0)}(s)$, in the first row or in the first column, and its degree is strictly lower than that of $a_{11}^{(0)}(s)$. We move this to the $(1,1)$ entry and repeat the above process. In so doing, after a finite number of iterations, the first stage is completed with a polynomial $a_{11}^{(0)}(s)$ and an $(m-1) \times (n-1)$ matrix $A^{(1)}(s)$.

Then we apply the same procedure as above to this matrix $A^{(1)}(s)$ and obtain the first diagonal entry $a_{22}^{(1)}(s)$ and an $(m-2) \times (n-2)$ matrix

$$A^{(2)}(s) = (a_{ij}^{(2)}(s) \mid 3 \leqq i \leqq m,\ 3 \leqq j \leqq n),$$

to complete the second stage. Continuing in this way, we end up with diagonal entries $a_{11}^{(0)}(s), a_{22}^{(1)}(s), \ldots, a_{rr}^{(r-1)}(s)$ and $A^{(r)}(s) = O$. Finally, we permute the rows and columns to rearrange the diagonal entries to meet the degree condition (5.32). □

In the diagonal form (5.30), we can impose a further condition, a divisibility condition among the diagonal entries $\alpha_1(s), \alpha_2(s), \ldots, \alpha_r(s)$:

$$\alpha_1(s) \mid \alpha_2(s) \mid \cdots \mid \alpha_r(s), \tag{5.33}$$

where the notation $a(s) \mid b(s)$ means that $a(s)$ is a factor of $b(s)$. The condition (5.33) says that $\alpha_i(s)$ divides $\alpha_{i+1}(s)$ for $i = 1, \ldots, r-1$. A diagonal matrix (5.30) that satisfies the divisibility condition (5.33) is said to be in *Smith normal form*.

As shown in Theorem 5.7 below, we can transform each polynomial matrix $A(s)$ to Smith normal form by multiplication with unimodular matrices from the left and the right. Furthermore, the Smith normal form is uniquely determined by $A(s)$. We note that the diagonal matrix in Proposition 5.4 (without the divisibility condition) is not uniquely determined, as demonstrated in Example 5.10 below.

**Example 5.10.** The diagonal matrix

$$A(s) = \begin{bmatrix} s+1 & 0 \\ 0 & s^3 \end{bmatrix}$$

is not in Smith normal form, since it does not satisfy the divisibility condition (5.33). The Smith normal form of this matrix can be obtained by polynomial elementary (row and column) transformations as follows:

$$\begin{bmatrix} s+1 & 0 \\ 0 & s^3 \end{bmatrix} \xrightarrow[E_3(1,2;s^2-s+1)]{\text{column}} \begin{bmatrix} s+1 & s^3+1 \\ 0 & s^3 \end{bmatrix}$$

$$\xrightarrow[E_3(1,2;-1)]{\text{row}} \begin{bmatrix} s+1 & 1 \\ 0 & s^3 \end{bmatrix} \xrightarrow[E_1(1,2)]{\text{column}} \begin{bmatrix} 1 & s+1 \\ s^3 & 0 \end{bmatrix}$$

$$\xrightarrow[E_3(1,2;-s-1)]{\text{column}} \begin{bmatrix} 1 & 0 \\ s^3 & -s^3(s+1) \end{bmatrix} \xrightarrow[E_3(2,1;-s^3)]{\text{row}} \begin{bmatrix} 1 & 0 \\ 0 & -s^3(s+1) \end{bmatrix}$$

$$\xrightarrow[E_2(2;-1)]{\text{column}} \begin{bmatrix} 1 & 0 \\ 0 & s^3(s+1) \end{bmatrix} \quad \text{(Smith normal form)}.$$

A key ingredient in the above transformation is the fact (Theorem 5.1 in Sec. 5.2) that, for two polynomials $\alpha(s)$ and $\beta(s)$ in general, there exist polynomials $v(s)$ and $u(s)$ that satisfy

$$\alpha(s)v(s) + \beta(s)u(s) = \gcd(\alpha(s), \beta(s)).$$

In the above example, $\alpha(s) = s+1$ and $\beta(s) = s^3$ with $\gcd(\alpha(s), \beta(s)) = \gcd(s+1, s^3) = 1$, and

$$\alpha(s)v(s) + \beta(s)u(s) = (s+1)(s^2-s+1) + s^3(-1) = 1$$

for $v(s) = s^2 - s + 1$ and $u(s) = -1$. In the transformation shown above, we have employed a column transformation using $E_3(1,2;v) = E_3(1,2;s^2 - s+1)$ first and then a row transformation using $E_3(1,2;u) = E_3(1,2;-1)$ to create $1 = \gcd(\alpha, \beta)$ as the $(1,2)$ entry. ■

**Theorem 5.7.** *Every polynomial matrix can be transformed to Smith normal form through unimodular row and column transformations. That is, for any $m \times n$ polynomial matrix $A(s)$, there exist a unimodular matrix $U(s)$ of order $m$ and a unimodular matrix $V(s)$ of order $n$ such that*

$$U(s)A(s)V(s) = \left[\begin{array}{ccc|c} \alpha_1(s) & & O & \\ & \ddots & & O_{r,n-r} \\ O & & \alpha_r(s) & \\ \hline & O_{m-r,r} & & O_{m-r,n-r} \end{array}\right]. \tag{5.34}$$

*Here, $r = \operatorname{rank} A$, and $\alpha_1(s), \alpha_2(s), \ldots, \alpha_r(s)$ are polynomials satisfying*

$$\textit{Normalization } (5.31): \quad \operatorname{lc}(\alpha_1(s)) = \operatorname{lc}(\alpha_2(s)) = \cdots = \operatorname{lc}(\alpha_r(s)) = 1,$$

$$\textit{Divisibility } (5.33): \quad \alpha_1(s) \mid \alpha_2(s) \mid \cdots \mid \alpha_r(s)$$

*and are uniquely determined by $A(s)$.*

**Proof.** By Proposition 5.4 we can assume that $A(s)$ is a diagonal matrix (5.30) that satisfies the normalization condition (5.31) and the degree condition (5.32). Let $\alpha_1(s), \alpha_2(s), \ldots, \alpha_r(s)$ be the nonzero diagonal entries. If the divisibility condition (5.33) is satisfied, we are done. Otherwise, let $i$ be the minimum index for which $\alpha_i(s) \mid \alpha_{i+1}(s)$ fails.[11] Consider the submatrix $\mathrm{diag}\,(\alpha_i(s), \alpha_{i+1}(s))$ that corresponds to rows and columns in $\{i, i+1\}$. Put $\alpha(s) = \alpha_i(s)$, $\beta(s) = \alpha_{i+1}(s)$, and $g(s) = \gcd(\alpha(s), \beta(s))$, and take polynomials $v(s)$ and $u(s)$ that satisfy

$$\alpha(s)v(s) + \beta(s)u(s) = g(s)$$

to transform the matrix as

$$\begin{bmatrix} 1 & u(s) \\ 0 & 1 \end{bmatrix} \begin{bmatrix} \alpha(s) & 0 \\ 0 & \beta(s) \end{bmatrix} \begin{bmatrix} 1 & v(s) \\ 0 & 1 \end{bmatrix}$$

$$= \begin{bmatrix} \alpha(s) & \alpha(s)v(s) + \beta(s)u(s) \\ 0 & \beta(s) \end{bmatrix} = \begin{bmatrix} \alpha(s) & g(s) \\ 0 & \beta(s) \end{bmatrix}.$$

Next, we exchange the columns and transform the resulting matrix as

$$\begin{bmatrix} 1 & 0 \\ -\beta(s)/g(s) & 1 \end{bmatrix} \begin{bmatrix} g(s) & \alpha(s) \\ \beta(s) & 0 \end{bmatrix} \begin{bmatrix} 1 & -\alpha(s)/g(s) \\ 0 & 1 \end{bmatrix} = \begin{bmatrix} g(s) & 0 \\ 0 & -\alpha(s)\beta(s)/g(s) \end{bmatrix}.$$

Finally, by changing the sign of the second column, we can obtain $\mathrm{diag}\,(g(s), \alpha(s)\beta(s)/g(s))$.

The diagonal entries are thus changed from $(\alpha_1(s), \alpha_2(s), \ldots, \alpha_r(s))$ to

$$(\alpha_1(s), \ldots, \alpha_{i-1}(s);\ g(s),\ \alpha_i(s)\alpha_{i+1}(s)/g(s);\ \alpha_{i+2}(s), \ldots, \alpha_r(s)).$$

Let $(\delta_1, \delta_2, \ldots, \delta_r)$ and $(\delta'_1, \delta'_2, \ldots, \delta'_r)$ be the degrees of the above sequences rearranged in ascending order, where

$$(\delta_1, \delta_2, \ldots, \delta_r) = (\delta(\alpha_1(s)), \delta(\alpha_2(s)), \ldots, \delta(\alpha_r(s)))$$

as a consequence of the degree condition (5.32). Since $\delta(g(s)) < \delta(\alpha_i(s))$, the sequence $(\delta'_1, \delta'_2, \ldots, \delta'_r)$ is strictly smaller in lexicographical order than $(\delta_1, \delta_2, \ldots, \delta_r)$. After a finite number of such transformations, the divisibility condition (5.33) is eventually satisfied, and then the Smith normal form is obtained.

The uniqueness can be shown as follows. In the Smith normal form $U(s)A(s)V(s)$, the $k$th determinantal divisor $d_k(U(s)A(s)V(s))$ is given as

$$d_k(U(s)A(s)V(s)) = \alpha_1(s)\alpha_2(s)\cdots\alpha_k(s)$$

[11]What follows is a generalization of the argument in Example 5.10.

as a consequence of the divisibility condition (5.33). On the other hand, $d_k(U(s)A(s)V(s))$ is equal to $d_k(A(s))$, since the determinantal divisors remain invariant under elementary transformations (Proposition 5.3). Therefore,

$$d_k(A(s)) = \alpha_1(s)\alpha_2(s)\cdots\alpha_k(s) \qquad (k = 1, \ldots, r),$$

which shows

$$\alpha_k(s) = d_k(A(s)) \;/\; d_{k-1}(A(s)) \qquad (k = 1, \ldots, r).$$

The right-hand side is determined by $A(s)$, independently of $U(s)$ and $V(s)$. □

For a polynomial matrix $A(s)$, the matrix in (5.34) in Theorem 5.7 is called the Smith normal form of $A(s)$. The diagonal entries

$$\alpha_1(s), \alpha_2(s), \ldots, \alpha_r(s)$$

in the Smith normal form (5.34) are referred to as the *elementary divisors*[12] of $A(s)$, and are denoted as

$$e_1(A(s)), e_2(A(s)), \ldots, e_r(A(s)).$$

The following relations hold between elementary divisors and determinantal divisors:

$$d_k(A(s)) = e_1(A(s))e_2(A(s))\cdots e_k(A(s)) \qquad (k = 1, \ldots, r), \tag{5.35}$$

$$e_k(A(s)) = \frac{d_k(A(s))}{d_{k-1}(A(s))} \qquad (k = 1, \ldots, r). \tag{5.36}$$

Smith normal form is also called (in Japan) "elementary divisor normal form."

**Example 5.11.** Recall the polynomial matrix

$$A(s) = \begin{bmatrix} s^3 + s^2 & s^2 + 2s + 1 & s^2 - 1 \\ s^5 + 2s^4 - s^3 - s^2 & s^4 + 3s^3 + s^2 - s - 1 & 2s^3 - s^2 - s + 1 \end{bmatrix}$$

introduced in Example 5.6 and investigated in Examples 5.7, 5.9, and 5.10. The elementary divisors are $e_1(A(s)) = 1$ and $e_2(A(s)) = s^3(s + 1)$, and the determinantal divisors are $d_1(A(s)) = 1$ and $d_2(A(s)) = s^3(s+1)$. The Smith and Hermite normal forms are given as follows:

$$\text{Smith normal form} = \begin{bmatrix} 1 & 0 & 0 \\ 0 & s^3(s+1) & 0 \end{bmatrix},$$

$$\text{Hermite normal form} = \begin{bmatrix} s+1 & 0 & 0 \\ s^2 - 1 & s^3 & 0 \end{bmatrix}.$$

■

[12] In some textbooks, *e.g.*, in [2], $\alpha_i(s)$'s are called *invariant polynomials*, while "elementary divisors" mean other polynomials derived from $\alpha_i(s)$'s.

**Remark 5.2.** A polynomial matrix $A(s)$ is said to be *equivalent* to another polynomial matrix $B(s)$ if there exist unimodular matrices $U(s)$ and $V(s)$ such that

$$U(s)\,A(s)\,V(s) = B(s). \tag{5.37}$$

If (5.37) holds, we also have

$$A(s) = U(s)^{-1}\,B(s)\,V(s)^{-1},$$

where $U(s)^{-1}$ and $V(s)^{-1}$ are unimodular matrices. Therefore, if $A(s)$ is equivalent to $B(s)$, then $B(s)$ is equivalent to $A(s)$. The relation of being equivalent is an equivalence relation (in the sense of Remark 1.3 in Sec.1.1.2), according to which the set of polynomial matrices are partitioned into equivalence classes. A representative of an equivalence class is given by the Smith normal form. A necessary and sufficient condition for two polynomial matrices to be equivalent is that they share the same elementary divisors. ■

**Remark 5.3.** Hermite and Smith normal forms have been defined also for integer matrices (in Secs. 4.3 and 4.4) through a construction that is similar to the one for polynomial matrices presented in this section. The fact underlying this similarity is that the set $\mathbb{Z}$ of integers and the set $F[s]$ of polynomials are endowed with a common algebraic structure.

In general, a set $R$ is called a *ring* if addition, subtraction, and multiplication are defined for elements of $R$. A ring $R$ is called an *integral domain*, if the multiplication is commutative ($ab = ba$ for all $a, b \in R$), the unit element for multiplication exists (there exists $e \in R$ for which $ea = ae = a$ for all $a \in R$), and no zero divisor exists ($a \neq 0$ and $b \neq 0$ imply $ab \neq 0$). The set of integers $\mathbb{Z}$ and the set of univariate polynomials $F[s]$ are both integral domains.

Furthermore, "division" (with remainder) can be carried out in $\mathbb{Z}$ and $F[s]$. As an abstraction of this, an integral domain $R$ is called a *Euclidean domain*, if it admits a function $\varphi : R \setminus \{0\} \to \mathbb{Z}$ that satisfies the following conditions:

(i) $\varphi(a) \geqq 0 \quad (a \in R \setminus \{0\})$,
(ii) $\varphi(ab) \geqq \varphi(a) \quad (a, b \in R \setminus \{0\})$,
(iii) For any $a \in R$ and $b \in R \setminus \{0\}$, there exist $q \in R$ and $r \in R$ such that $a = bq + r$ and either $r = 0$ or $r \neq 0$, $\varphi(r) < \varphi(b)$.

For example, $\varphi(a) = |a|$ serves as such a function for $R = \mathbb{Z}$, and $\varphi(a) = \delta(a)$ (the degree of polynomial $a$) for $R = F[s]$, and therefore, $\mathbb{Z}$ and $F[s]$ are both Euclidean domains.

We can consider the Hermite and Smith normal forms for matrices over a Euclidean domain $R$ in general. However, rings of multivariate polynomials (polynomials in more than one variable) are not Euclidean domains, although they are certainly integral domains. Accordingly, the Hermite and Smith normal forms cannot be defined for matrices of multivariate polynomials.

As yet another example of a Euclidean domain, we mention the set of proper rational functions, which is defined as

$$R = \{a/b \mid a, b \in F[s],\ b \neq 0,\ \delta(a) \leqq \delta(b)\}, \quad \varphi(a/b) = \delta(b) - \delta(a).$$

Here, a rational function is said to be *proper*, if

$$[\text{degree of denominator}] \geqq [\text{degree of numerator}].$$

Also, the set of all proper and stable rational functions with real coefficients, often denoted by $\mathrm{RH}_\infty$, forms a Euclidean domain. Here, a rational function with real coefficients is said to be *stable* if it does not possess poles in

$$\mathbb{C}_{+\mathrm{e}} = \{z \in \mathbb{C} \mid \mathrm{Re}(z) \geqq 0\} \cup \{\infty\},$$

the closed right half of the extended complex plane. The function representing the number of zeros contained in $\mathbb{C}_{+\mathrm{e}}$ (with the multiplicities counted) serves as the function $\varphi$. The Smith normal form of matrices over $\mathrm{RH}_\infty$ plays an important role in control theory [48, 50, 51]. ■

## 5.6 Solutions to Systems of Linear Equations

In this section, we consider a system of linear equations $A(s)\boldsymbol{x}(s) = \boldsymbol{b}(s)$ with particular interest in the existence of a polynomial vector solution $\boldsymbol{x}(s)$. It is assumed that $A(s)$ is an $m \times n$ polynomial matrix and $\boldsymbol{b}(s)$ is an $m$-dimensional polynomial vector. The inverse of a nonsingular polynomial matrix is not necessarily a polynomial matrix, but possibly involves rational functions. As can be imagined easily from this fact, divisibility conditions need to be taken into account as the key issue in considering polynomial solutions. The argument for polynomial solutions goes in parallel with that for integer solutions in Sec. 4.5.

First, a fundamental fact on transformation between polynomial vectors is stated.

**Proposition 5.5.** *Let $V(s)$ be a unimodular matrix of order $n$ and $\boldsymbol{x}(s)$ be an $n$-dimensional vector of rational functions. Then $V(s)\boldsymbol{x}(s)$ is a polynomial vector if and only if $\boldsymbol{x}(s)$ is a polynomial vector.*

**Proof.** Set $\boldsymbol{y}(s) = V(s)\boldsymbol{x}(s)$. If $\boldsymbol{x}(s)$ is a polynomial vector, then the vector $\boldsymbol{y}(s)$, which is the product of a polynomial matrix and a polynomial vector, is a polynomial vector. Conversely, if $\boldsymbol{y}(s)$ is a polynomial vector, then $\boldsymbol{x}(s) = V(s)^{-1}\boldsymbol{y}(s)$ is a polynomial vector, since the inverse of a unimodular matrix $V(s)$ is a polynomial matrix (Theorem 5.2). □

In considering a polynomial solution $\boldsymbol{x}(s)$ for a system of equations $A(s)\boldsymbol{x}(s) = \boldsymbol{b}(s)$, we can rewrite the equation $A(s)\boldsymbol{x}(s) = \boldsymbol{b}(s)$ to

$$(U(s)A(s)V(s))(V(s)^{-1}\boldsymbol{x}(s)) = U(s)\boldsymbol{b}(s)$$

using unimodular matrices $U(s)$ and $V(s)$. By Proposition 5.5, $\boldsymbol{x}(s)$ is a polynomial vector if and only if $\tilde{\boldsymbol{x}}(s) = V(s)^{-1}\boldsymbol{x}(s)$ is a polynomial vector. With suitable unimodular matrices $U(s)$ and $V(s)$, we can transform the coefficient matrix $A(s)$ to the Smith normal form (5.34) (or the diagonal form in (5.30)) $\tilde{A}(s) = U(s)A(s)V(s)$, thereby making the analysis easier. The transformation of the variables is summarized here for easy reference:

$$\tilde{A}(s) = U(s)A(s)V(s), \quad \tilde{\boldsymbol{x}}(s) = V(s)^{-1}\boldsymbol{x}(s), \quad \tilde{\boldsymbol{b}}(s) = U(s)\boldsymbol{b}(s). \tag{5.38}$$

Then $A(s)\boldsymbol{x}(s) = \boldsymbol{b}(s)$ is equivalent to

$$\tilde{A}(s)\tilde{\boldsymbol{x}}(s) = \tilde{\boldsymbol{b}}(s).$$

The above discussion leads naturally to the following theorem.

**Theorem 5.8.** *The following three conditions* (a) *to* (c) *are equivalent for an $m \times n$ polynomial matrix $A(s)$ and an $m$-dimensional polynomial vector $\boldsymbol{b}(s)$:*

(a) *$A(s)\boldsymbol{x}(s) = \boldsymbol{b}(s)$ has a polynomial vector solution $\boldsymbol{x}(s)$.*
(b) *If $\boldsymbol{y}(s)^\top A(s)$ is a polynomial vector, then $\boldsymbol{y}(s)^\top \boldsymbol{b}(s)$ is a polynomial, where $\boldsymbol{y}(s)$ is taken from rational function vectors.*
(c) *$A(s)$ and $[A(s) \mid \boldsymbol{b}(s)]$ have the same elementary divisors.*[13]

**Proof.** Using the Smith normal form, the proof goes in parallel with the proof of Theorem 4.11. □

The solvability conditions for varying right-hand side vector $\boldsymbol{b}(s)$ are as follows.

**Theorem 5.9.** *The following three conditions* (a) *to* (c) *are equivalent for an $m \times n$ polynomial matrix $A(s)$ with* $\operatorname{rank} A(s) = m$*:*

[13]Since the number of elementary divisors is equal to the rank, condition (c) contains the condition that $A(s)$ and $[A(s) \mid \boldsymbol{b}(s)]$ have the same rank.

(a) *For every $m$-dimensional polynomial vector $\boldsymbol{b}(s)$, $A(s)\boldsymbol{x}(s) = \boldsymbol{b}(s)$ has a polynomial vector solution $\boldsymbol{x}(s)$.*
(b) *If $\boldsymbol{y}(s)^\top A(s)$ is a polynomial vector, then $\boldsymbol{y}(s)$ is a polynomial vector, where $\boldsymbol{y}(s)$ is taken from rational function vectors.*
(c) *The elementary divisors $e_1(A(s)), e_2(A(s)), \ldots, e_m(A(s))$ are all equal to one.*[14]

**Proof.** The condition (a) above says (literally) that the condition (a) in Theorem 5.8 holds for all $m$-dimensional polynomial vectors $\boldsymbol{b}(s)$. The conditions (b) and (c) above are equivalent to the corresponding conditions in Theorem 5.8 holding for all $m$-dimensional polynomial vectors $\boldsymbol{b}(s)$ (in particular, for all the unit vectors $\boldsymbol{e}_1, \ldots, \boldsymbol{e}_m$). □

## 5.7 Matrix Pencils

### 5.7.1 *Definition*

Polynomial matrices of degree at most one are called *matrix pencils*, or simply *pencils*. Matrix pencils are represented in the form of

$$A(s) = sA_1 + A_0 \tag{5.39}$$

using constant matrices $A_1$ and $A_0$. In this section, $A(s)$ denotes an $m \times n$ matrix pencil of rank $r$, unless otherwise stated. We continue to assume that the coefficients are taken from a field $F$ (that is, $A_1$ and $A_0$ are matrices over $F$). In applications, we usually have $F = \mathbb{R}$ (field of real numbers) or $F = \mathbb{C}$ (field of complex numbers), and accordingly, we assume in this section that $F$ includes the field of rational numbers $\mathbb{Q}$.

A matrix pencil $A(s)$ is called a *regular pencil* if it is square ($m = n$) and the determinant $\det A(s)$ is nonzero as a polynomial in $s$. A matrix pencil that is not regular is called a *singular pencil.* That is, a singular pencil $A(s)$ is either nonsquare ($m \neq n$) or square with $\det A(s) = 0$.

**Example 5.12.** $A(s) = \begin{bmatrix} s & 1 \\ 1 & s \end{bmatrix}$ is a regular pencil. Indeed we have $\det A(s) = s^2 - 1 \neq 0$. The degree of the determinant is equal to the matrix size $n = 2$. The coefficient matrices $A_1 = \begin{bmatrix} 1 & 0 \\ 0 & 1 \end{bmatrix}$ and $A_0 = \begin{bmatrix} 0 & 1 \\ 1 & 0 \end{bmatrix}$ are both nonsingular. ■

[14]This condition is equivalent to $e_m(A(s)) = 1$ and, also to $d_m(A(s)) = 1$.

**Example 5.13.** $A(s) = \begin{bmatrix} 1 & s \\ 1 & 1 \end{bmatrix}$ is a regular pencil. Indeed we have $\det A(s) = -s + 1 \neq 0$. The coefficient matrix $A_0 = \begin{bmatrix} 1 & 0 \\ 1 & 1 \end{bmatrix}$ is nonsingular, but $A_1 = \begin{bmatrix} 0 & 1 \\ 0 & 0 \end{bmatrix}$ is singular. As a result, the degree of the determinant $\det A(s) = -s + 1$ is lower than the matrix size $n = 2$. ■

**Example 5.14.** $A(s) = \begin{bmatrix} s & 1 & 0 \\ 0 & s & 1 \end{bmatrix}$ is a singular pencil. ■

### 5.7.2 *Strict Equivalence*

Two matrix pencils

$$A(s) = sA_1 + A_0,$$
$$B(s) = sB_1 + B_0$$

are said to be *strictly equivalent*, if there exist nonsingular constant matrices $P$ and $Q$ such that

$$P\,A(s)\,Q = B(s). \tag{5.40}$$

The condition (5.40) is equivalent to

$$P\,A_1\,Q = B_1, \qquad P\,A_0\,Q = B_0. \tag{5.41}$$

Let us explain the significance of this concept in the context of differential equations. When a matrix pencil $A(s) = sA_1 + A_0$ corresponds to a differential equation

$$A_1 \frac{\mathrm{d}\boldsymbol{x}}{\mathrm{d}t}(t) + A_0 \boldsymbol{x}(t) = \boldsymbol{f}(t), \tag{5.42}$$

the strict equivalence (5.40) corresponds to a transformation of variables

$$\boldsymbol{y} = Q^{-1}\boldsymbol{x}, \qquad \boldsymbol{g} = P\,\boldsymbol{f} \tag{5.43}$$

to rewrite (5.42) to

$$B_1 \frac{\mathrm{d}\boldsymbol{y}}{\mathrm{d}t}(t) + B_0 \boldsymbol{y}(t) = \boldsymbol{g}(t).$$

In this way, the concept of strict equivalence matches the requirement that the derivatives should not be involved in the transformation (5.43). In contrast, the equivalence based on the relation $U(s)A(s)V(s) = B(s)$ with unimodular matrices $U(s)$ and $V(s)$ (see Remark 5.2 in Sec. 5.5) is not appropriate for such a transformation of differential equations, since a transformation of the form of

$$\boldsymbol{y} = V(s)^{-1}\boldsymbol{x}, \qquad \boldsymbol{g} = U(s)\,\boldsymbol{f} \tag{5.44}$$

may possibly result in a new variable $\boldsymbol{y}$ that contains some derivatives of the original variable $\boldsymbol{x}$.

In Sec. 5.7.3 below, we will discuss the Kronecker canonical form as a normal form under strict equivalence. In addition, in Sec. 5.7.4, we will clarify the relationship between equivalence and strict equivalence.

### 5.7.3 *Kronecker Canonical Form*

#### 5.7.3.1 *Theorem*

When given a matrix pencil $A(s)$, we can choose appropriate nonsingular constant matrices $P$ and $Q$ to put

$$PA(s)Q$$

into a block-diagonal form, called the *Kronecker canonical form.* Kronecker canonical form can be recognized as the matrix pencil having the simplest possible form among those strictly equivalent to the given matrix pencil $A(s)$.

Before introducing general notations, let us first take a look at an example of a Kronecker canonical form.

**Example 5.15.** The following matrix pencil is in Kronecker canonical form:[15]

$$\left[\begin{array}{cc|cc|ccc|ccc|cc|cc}
s+2 & 1 & \leftarrow H & & & & & & & & & & & \\
0 & s+2 & & & & & & & & & & & & \\
\hline
 & & s & 1 & & & & & & & & & & \\
 & K_2 \rightarrow & 0 & s & & & & & & & & & & \\
\hline
 & & & & 1 & s & 0 & & & & & & & \\
 & & & N_3 \rightarrow & 0 & 1 & s & & & & & & & \\
 & & & & 0 & 0 & 1 & & & & & & & \\
\hline
 & & & & & & & s & 1 & 0 & & & & \\
 & & & & & & L_2 \rightarrow & 0 & s & 1 & & & & \\
\hline
 & & & & & & & & & L_1 \rightarrow & s & 1 & & \\
\hline
 & & & & & & & & & & & & s & 0 \\
 & & & & & & & & & & & U_2 \rightarrow & 1 & s \\
 & & & & & & & & & & & & 0 & 1
\end{array}\right].$$

This is a block-diagonal matrix with diagonal blocks $H$, $K_2$, $N_3$, $L_2$, $L_1$, and $U_2$, where these notations are introduced below. ■

[15]In the notation of Theorem 5.10, we have $\nu = 2$, $c = 1$ ($\rho_1 = 2$), $d = 1$ ($\mu_1 = 3$), $p = 2$ ($\varepsilon_1 = 2$, $\varepsilon_2 = 1$), and $q = 1$ ($\eta_1 = 2$).

We introduce notations and then state the theorem. For positive integers $\rho$ and $\mu$, we define a $\rho \times \rho$ matrix pencil $K_\rho(s)$ and a $\mu \times \mu$ matrix pencil $N_\mu(s)$ as

$$K_\rho(s) = \begin{bmatrix} s & 1 & 0 & \cdots & 0 \\ 0 & s & 1 & \ddots & \vdots \\ \vdots & \ddots & \ddots & \ddots & 0 \\ \vdots & & \ddots & s & 1 \\ 0 & \cdots & \cdots & 0 & s \end{bmatrix}, \qquad N_\mu(s) = \begin{bmatrix} 1 & s & 0 & \cdots & 0 \\ 0 & 1 & s & \ddots & \vdots \\ \vdots & \ddots & \ddots & \ddots & 0 \\ \vdots & & \ddots & 1 & s \\ 0 & \cdots & \cdots & 0 & 1 \end{bmatrix}.$$

For positive integers $\varepsilon$ and $\eta$, we define an $\varepsilon \times (\varepsilon + 1)$ matrix pencil $L_\varepsilon(s)$ and an $(\eta + 1) \times \eta$ matrix pencil $U_\eta(s)$ as

$$L_\varepsilon(s) = \begin{bmatrix} s & 1 & 0 & \cdots & 0 \\ 0 & s & 1 & \ddots & \vdots \\ \vdots & \ddots & \ddots & \ddots & 0 \\ 0 & \cdots & 0 & s & 1 \end{bmatrix}, \qquad U_\eta(s) = \begin{bmatrix} s & 0 & \cdots & 0 \\ 1 & s & \ddots & \vdots \\ 0 & 1 & \ddots & 0 \\ \vdots & \ddots & \ddots & s \\ 0 & \cdots & 0 & 1 \end{bmatrix}.$$

**Theorem 5.10.** *For any matrix pencil $A(s)$ over a field $F$, there exist nonsingular matrices $P$ and $Q$ over $F$ such that*

$$\begin{aligned} PA(s)Q = \operatorname{diag}\,(&H(s); K_{\rho_1}(s), \ldots, K_{\rho_c}(s); N_{\mu_1}(s), \ldots, N_{\mu_d}(s); \\ &L_{\varepsilon_1}(s), \ldots, L_{\varepsilon_p}(s); U_{\eta_1}(s), \ldots, U_{\eta_q}(s); O), \end{aligned} \tag{5.45}$$

*where*

$$\rho_1 \geqq \cdots \geqq \rho_c \geqq 1, \quad \mu_1 \geqq \cdots \geqq \mu_d \geqq 1,$$
$$\varepsilon_1 \geqq \cdots \geqq \varepsilon_p \geqq 1, \quad \eta_1 \geqq \cdots \geqq \eta_q \geqq 1,$$

*and $H(s) = sH_1 + H_0$ is a $\nu \times \nu$ matrix pencil with nonsingular matrices $H_1$ and $H_0$ over $F$. The values of the parameters $\nu$, $c$, $d$, $p$, $q$, $\rho_1, \ldots, \rho_c$, $\mu_1, \ldots, \mu_d$, $\varepsilon_1, \ldots, \varepsilon_p$, and $\eta_1, \ldots, \eta_q$ are uniquely determined by $A(s)$.*

**Proof.** The proof is given later in Sec. 5.7.3.3. □

**Remark 5.4.** We can choose $H_1 = I$ or $H_0 = I$ in the Kronecker canonical form in Theorem 5.10. ■

**Remark 5.5.** For a regular pencil $A(s)$, the horizontal blocks $L_\varepsilon(s)$ and the vertical blocks $U_\eta(s)$ do not appear (*i.e.*, $p = q = 0$). The zero matrix $O$ at the end of the right-hand side of (5.45) does not appear, either (being a $0 \times 0$ matrix). ■

### 5.7.3.2 *Applications and examples*

Kronecker canonical form is useful for the analysis [49] of *differential-algebraic equations* (DAE). Consider a constant-coefficient linear differential equation:[16]

$$C \frac{\mathrm{d}\boldsymbol{x}}{\mathrm{d}t}(t) = A\boldsymbol{x}(t) + \boldsymbol{f}(t). \tag{5.46}$$

Here, $\boldsymbol{f}(t)$ denotes a known function expressing the input from an external source and $\boldsymbol{x}(t)$ is the unknown function to be determined. It is assumed that $A$ and $C$ are square constant matrices such that $A - sC$ is a regular matrix pencil.

By Theorem 5.10, Remark 5.4, and Remark 5.5, a suitable choice of square constant matrices $P$ and $Q$ provides us with a transformation:[17]

$$P(A - sC)Q = \mathrm{diag}\,(B - sI_{\mu_0}; I_{\mu_1} - sT_{\mu_1}, \ldots, I_{\mu_d} - sT_{\mu_d}), \tag{5.47}$$

where $B$ is a constant matrix and $T_\mu$ is a $\mu \times \mu$ matrix in the form of

$$T_\mu = \begin{bmatrix} 0 & 1 & 0 & \cdots & 0 \\ 0 & 0 & 1 & \ddots & \vdots \\ \vdots & \ddots & \ddots & \ddots & 0 \\ \vdots & & \ddots & 0 & 1 \\ 0 & \cdots & \cdots & 0 & 0 \end{bmatrix},$$

with which we have $N_\mu(s) = I_\mu + sT_\mu$. Accordingly, by transforming the variables to

$$\begin{bmatrix} \boldsymbol{y}_0 \\ \boldsymbol{y}_1 \\ \vdots \\ \boldsymbol{y}_d \end{bmatrix} = Q^{-1}\boldsymbol{x}, \qquad \begin{bmatrix} \boldsymbol{g}_0 \\ \boldsymbol{g}_1 \\ \vdots \\ \boldsymbol{g}_d \end{bmatrix} = P\boldsymbol{f},$$

we can rewrite the equation (5.46) as

$$\frac{\mathrm{d}\boldsymbol{y}_0}{\mathrm{d}t}(t) = B\boldsymbol{y}_0(t) + \boldsymbol{g}_0(t), \tag{5.48}$$

$$T_{\mu_k}\frac{\mathrm{d}\boldsymbol{y}_k}{\mathrm{d}t}(t) = \boldsymbol{y}_k(t) + \boldsymbol{g}_k(t) \qquad (k = 1, \ldots, d). \tag{5.49}$$

Here, $\boldsymbol{y}_0(t), \boldsymbol{y}_1(t), \ldots, \boldsymbol{y}_d(t)$ are the unknown functions to be determined.

[16]This is the case of $A_1 = C$ and $A_0 = -A$ in (5.42).
[17]Apply Theorem 5.10 to $A + sC$ and change $s$ to $-s$. We have $\mu_0 = \nu + \rho_1 + \cdots + \rho_c$.

The equation (5.48) is a linear differential equation in *normal form*, and the solution $\boldsymbol{y}_0(t)$ satisfying the initial condition $\boldsymbol{y}_0(0) = \boldsymbol{\eta}_0$ for an arbitrarily specified vector $\boldsymbol{\eta}_0 \in \mathbb{R}^{\mu_0}$ is uniquely determined. To be specific, it is given as

$$\boldsymbol{y}_0(t) = \exp(Bt)\,\boldsymbol{\eta}_0 + \int_0^t \exp[B(t-\tau)]\,\boldsymbol{g}_0(\tau)\,\mathrm{d}\tau, \tag{5.50}$$

where $\exp(Bt)$ is the exponential function in a matrix variable defined by

$$\exp(Bt) = \sum_{j=0}^{\infty} \frac{1}{j!}(Bt)^j.$$

In contrast, the equation (5.49) contains singularity. In the following argument, we fix $k$ in the range of $1 \leqq k \leqq d$.

If $\mu_k = 1$ in (5.49), by writing $\boldsymbol{y}_k(t) = y(t)$ and $\boldsymbol{g}_k(t) = g(t)$, we obtain

$$0 \cdot y'(t) = y(t) + g(t),$$

which can be solved as

$$y(t) = -g(t).$$

The initial value $y(0)$ cannot be specified arbitrarily.

If $\mu_k = 2$ in (5.49), on expressing $\boldsymbol{y}_k(t) = (y(t), z(t))^\top$ and $\boldsymbol{g}_k(t) = (g(t), h(t))^\top$, we obtain

$$\begin{bmatrix} 0 & 1 \\ 0 & 0 \end{bmatrix} \begin{bmatrix} y'(t) \\ z'(t) \end{bmatrix} = \begin{bmatrix} y(t) \\ z(t) \end{bmatrix} + \begin{bmatrix} g(t) \\ h(t) \end{bmatrix},$$

which can be solved as

$$y(t) = -g(t) - h'(t), \tag{5.51}$$

$$z(t) = -h(t). \tag{5.52}$$

Note that the initial value cannot be specified arbitrarily and that the solution contains the derivative of the external term.

The solution of (5.49) for general $\mu_k \geqq 1$ is given as

$$\boldsymbol{y}_k(t) = -\sum_{p=0}^{\mu_k - 1} {T_{\mu_k}}^p \boldsymbol{g}_k^{(p)}(t). \tag{5.53}$$

Here, $\boldsymbol{g}_k^{(p)}(t)$ denotes the $p$th-order derivative of function $\boldsymbol{g}_k(t)$, and ${T_{\mu_k}}^p = I_{\mu_k}$ for $p = 0$. The equation (5.53) shows that, if $\mu_k \geqq 2$, the solution contains the $(\mu_k - 1)$st-order derivative of the external term, and hence, the continuity of the solution is not guaranteed even if the external term is continuous.

In this way, the Kronecker canonical form reveals the structural properties of a system of differential-algebraic equations. For example, if $\mu_1 \geqq 2$ in (5.47), it is often the case that the system described by (5.46) is, in some sense or other, unstable or inconsistent.

**Example 5.16.** The matrix

$$A(s) = \left[\begin{array}{ccc|ccc} 1 & 1 & 1 & & & \\ \hline & & & 1 & -1 & 0 \\ & & & 0 & 1 & -1 \\ \hline -1 & 0 & 0 & sC & 0 & 0 \\ 0 & R & 0 & 0 & -1 & 0 \\ 0 & 0 & sL & 0 & 0 & -1 \end{array}\right] \tag{5.54}$$

in (5.3), which describes the RLC circuit (Fig. 5.1) in Sec. 5.1.2, is a regular pencil. Its Kronecker canonical form is given by

$$B(s) = \left[\begin{array}{cc|cccc} s & -1/L & & & & \\ 1/C & s+1/(RC) & & & & \\ \hline & & 1 & & & \\ & & & 1 & & \\ & & & & 1 & \\ & & & & & 1 \end{array}\right] \tag{5.55}$$

with the parameter values in Theorem 5.10 being $\nu = 2$, $c = 0$, $d = 4$ ($\mu_1 = \mu_2 = \mu_3 = \mu_4 = 1$), $p = 0$, and $q = 0$. The system is thus decomposed into a second-order dynamical system (differential equation) and four accompanying algebraic relations. The nonexistence of blocks with $\mu_k \geqq 2$ shows that this system is free from singularity. The transformation matrices $P$ and $Q$ in $PA(s)Q = B(s)$ are given, for example, by

$$P = \begin{bmatrix} 0 & -1/L & -1/L & 0 & 0 & 1/L \\ 1/C & 1/(RC) & 0 & 1/C & -1/(RC) & 0 \\ 1 & 1/R & 0 & 0 & -1/R & 0 \\ 0 & -1/R & 0 & 0 & 1/R & 0 \\ 0 & -1 & 0 & 0 & 0 & 0 \\ 0 & -1 & -1 & 0 & 0 & 0 \end{bmatrix},$$

$$Q = \begin{bmatrix} -1 & -1/R & 1 & 0 & 0 & 0 \\ 0 & 1/R & 0 & 1 & 0 & 0 \\ 1 & 0 & 0 & 0 & 0 & 0 \\ 0 & 1 & 0 & 0 & 0 & 0 \\ 0 & 1 & 0 & 0 & 1 & 0 \\ 0 & 1 & 0 & 0 & 0 & 1 \end{bmatrix},$$

where the choice of $P$ and $Q$ is not unique. ■

**Example 5.17.** Consider the RLC circuit [46] of Fig. 5.3. This circuit is composed of a voltage source $V_0(t)$ (branch 1), a resistor $R_1$ (branch 2), another resistor $R_2$ (branch 3), an inductor $L$ (branch 4), and a capacitor $C$ (branch 5). By choosing the vector $\boldsymbol{x} = (i_1, \dots, i_5, v_1, \dots, v_5)^\top$ consisting of currents $i_1, \dots, i_5$ and voltages $v_1, \dots, v_5$ as the state variables, we can describe this circuit as $A(s)\boldsymbol{x} = \boldsymbol{b}$ using a matrix pencil

$$A(s) = \left[\begin{array}{ccccc|ccccc} 1 & -1 & 0 & 0 & -1 & & & & & \\ -1 & 0 & 1 & 1 & 1 & & & & & \\ \hline & & & & & -1 & 0 & 0 & 0 & -1 \\ & & & & & 0 & 1 & 1 & 0 & -1 \\ & & & & & 0 & 0 & -1 & 1 & 0 \\ \hline 0 & 0 & 0 & 0 & 0 & -1 & 0 & 0 & 0 & 0 \\ 0 & R_1 & 0 & 0 & 0 & 0 & -1 & 0 & 0 & 0 \\ 0 & 0 & R_2 & 0 & 0 & 0 & 0 & -1 & 0 & 0 \\ 0 & 0 & 0 & sL & 0 & 0 & 0 & 0 & -1 & 0 \\ 0 & 0 & 0 & 0 & -1 & 0 & 0 & 0 & 0 & sC \end{array}\right] \tag{5.56}$$

and a vector $\boldsymbol{b} = (0, 0, 0, 0, 0; V_0, 0, 0, 0, 0)^\top$. This matrix $A(s)$ is a regular pencil, and its Kronecker canonical form is given by

$$B(s) = \mathrm{diag}\left(\left[ s + \frac{R_1 R_2}{L(R_1 + R_2)} \right], \begin{bmatrix} 1 & s \\ 0 & 1 \end{bmatrix}, [1], [1], [1], [1], [1], [1], [1]\right)$$

with the parameter values in Theorem 5.10 being $\nu = 1$, $c = 0$, $d = 8$ ($\mu_1 = 2$, $\mu_2 = \cdots = \mu_8 = 1$), $p = 0$, and $q = 0$. Since $\mu_1 = 2$, a solution of the form of (5.51) and (5.52) exists, and therefore, the derivative of the voltage source $V_0(t)$ is reflected in the internal state. ■

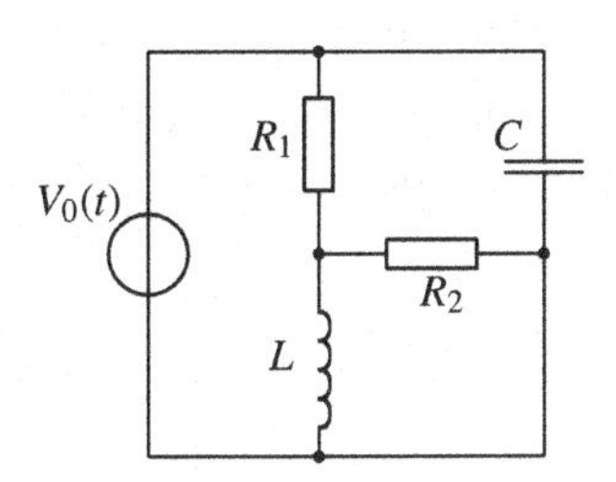


Fig. 5.3 An electric circuit.

#### 5.7.3.3 *Proof of Theorem 5.10*

The proof of Theorem 5.10 for Kronecker canonical form is too long to be included in this book. Here we will provide only the proof for a regular pencil over $\mathbb{C}$, which is the most important case in application, and refer the reader to [2, Chap. XII] for a complete proof. See also [29, Sec. 5.1.3] for an alternative proof.

Since the matrix pencil $A(s) = sA_1 + A_0$ is regular, there exists $c \in \mathbb{C}$ such that $cA_1 + A_0$ is a nonsingular (constant) matrix. From the expression

$$A(s) = sA_1 + A_0 = (s-c)A_1 + (cA_1 + A_0)$$

we see that

$$A(s) \approx (cA_1 + A_0)^{-1}A(s) = (s-c)(cA_1 + A_0)^{-1}A_1 + I, \tag{5.57}$$

where $\approx$ means strict equivalence. In view of the Jordan normal form of the matrix $(cA_1 + A_0)^{-1}A_1$, there exists a nonsingular matrix $S_1$ such that

$$S_1^{-1} \cdot (cA_1 + A_0)^{-1}A_1 \cdot S_1 = \begin{bmatrix} J_0 & O \\ O & J_1 \end{bmatrix}, \tag{5.58}$$

where $J_0$ is an aggregation of the Jordan blocks (cells) that correspond to eigenvalue 0 and $J_1$ is a matrix composed of the Jordan blocks that correspond to nonzero eigenvalues.[18] By (5.57) and (5.58) we have

$$A(s) \approx \begin{bmatrix} (s-c)J_0 + I & O \\ O & (s-c)J_1 + I \end{bmatrix}. \tag{5.59}$$

We transform the first block of the matrix on the right-hand side of (5.59) as

$$(s-c)J_0 + I = sJ_0 + (I - cJ_0) \approx s(I - cJ_0)^{-1}J_0 + I. \tag{5.60}$$

Since the eigenvalues of the matrix $(I - cJ_0)^{-1}J_0$ are all 0, there exists a nonsingular matrix $S_2$ that brings this matrix to its Jordan normal form:

$$S_2^{-1} \cdot (I - cJ_0)^{-1}J_0 \cdot S_2 = \mathrm{diag}\,(J(0,\mu_1), \ldots, J(0,\mu_d)), \tag{5.61}$$

where $J(\alpha,\mu)$ denotes[19] the Jordan block of size $\mu$ for eigenvalue $\alpha$. From (5.60) and (5.61) as well as the identity $sJ(0,\mu) + I = N_\mu(s)$, we obtain

$$(s-c)J_0 + I \approx \mathrm{diag}\,(N_{\mu_1}(s), \ldots, N_{\mu_d}(s)). \tag{5.62}$$

---

[18] For example, we may have $J_0 = \mathrm{diag}\left([0], \begin{bmatrix} 0 & 1 \\ 0 & 0 \end{bmatrix}, \begin{bmatrix} 0 & 1 \\ 0 & 0 \end{bmatrix}, \begin{bmatrix} 0 & 1 & 0 \\ 0 & 0 & 1 \\ 0 & 0 & 0 \end{bmatrix}\right)$ and $J_1 = \mathrm{diag}\left([2], [-2], \begin{bmatrix} 3 & 1 \\ 0 & 3 \end{bmatrix}, \begin{bmatrix} 4 & 1 \\ 0 & 4 \end{bmatrix}\right)$.

[19] We have $J(\alpha,\mu) = K_\mu(\alpha)$ using the notation in Theorem 5.10.

Next we transform the second block of the matrix on the right-hand side of (5.59) as

$$(s-c)J_1 + I = sJ_1 + (I - cJ_1) \approx sI + J_1^{-1}(I - cJ_1). \tag{5.63}$$

Consider the Jordan normal form of the matrix $J_1^{-1}(I - cJ_1)$, let $J(0,\rho_1), J(0,\rho_2), \ldots, J(0,\rho_c)$ be the Jordan blocks corresponding to eigenvalue 0, and let $H_0$ be the matrix consisting of the other Jordan blocks; $H_0$ being nonsingular. Then there exists a nonsingular matrix $S_3$ such that

$$S_3^{-1} \cdot J_1^{-1}(I - cJ_1) \cdot S_3 = \mathrm{diag}\,(H_0, J(0,\rho_1), \ldots, J(0,\rho_c)). \tag{5.64}$$

On setting $H(s) = sI + H_0$ and noting $sI + J(0,\rho) = K_\rho(s)$, we obtain

$$(s-c)J_1 + I \approx \mathrm{diag}\,(H(s);\ K_{\rho_1}(s), \ldots, K_{\rho_c}(s)). \tag{5.65}$$

By substituting (5.62) and (5.65) into (5.59), we obtain an expression of the form of (5.45) without $L_\varepsilon(s)$, $U_\eta(s)$, and $O$.

To prove the uniqueness, let

$$B(s) = \mathrm{diag}\,(H(s); K_{\rho_1}(s), \ldots, K_{\rho_c}(s); N_{\mu_1}(s), \ldots, N_{\mu_d}(s))$$

be a matrix pencil in Kronecker canonical form that is strictly equivalent to $A(s)$, where $\rho_1 \geqq \cdots \geqq \rho_c \geqq 1$ and $\mu_1 \geqq \cdots \geqq \mu_d \geqq 1$. Since $H(s) = sH_1 + H_0$ is a $\nu \times \nu$ matrix pencil with $H_1$ and $H_0$ being nonsingular, we have

$$\det H(s) = s^\nu \det H_1 + \cdots + \det H_0,$$

in which $\det H_1$ and $\det H_0$ are both nonzero. Therefore, $\nu$ is equal to the difference between the highest degree and the lowest degree of the terms contained in

$$\det B(s) = s^{\rho_1 + \cdots + \rho_c} \det H(s).$$

On the other hand, $\det B(s)$ is equal to $\det A(s)$ up to a constant factor. Therefore, $\nu$ is uniquely determined from $A(s)$.

Next, we derive the uniqueness of $c$, $d$, $\rho_1, \ldots, \rho_c$, and $\mu_1, \ldots, \mu_d$ from the fact that $B(s)$ has the same elementary divisors as $A(s)$. The elementary divisors of $K_\rho(s)$ are given by 1 (repeated $\rho - 1$ times) and $s^\rho$, whereas those of $N_\mu(s)$ are all 1. Let $e_1(s), e_2(s), \ldots, e_\nu(s)$ denote the elementary divisors of $H(s)$, where $e_1(s) \,|\, e_2(s) \,|\cdots|\, e_\nu(s)$. Since $\det H(s)$ has a nonzero constant term, we have $\gcd(\det H(s), s^\rho) = 1$, which is equivalent to saying that $\gcd(e_k(s), s^\rho) = 1$ for $k = 1, \ldots, \nu$. Therefore, the elementary divisors of $B(s)$ are given as follows.[20]

If $c \leqq \nu$: $1, 1, \ldots, 1;\ e_1(s), \ldots, e_{\nu-c}(s);\ s^{\rho_c} e_{\nu-c+1}(s), \ldots, s^{\rho_1} e_\nu(s)$.

If $c \geqq \nu$: $1, 1, \ldots, 1;\ s^{\rho_c}, \ldots, s^{\rho_{\nu+1}};\ s^{\rho_\nu} e_1(s), \ldots, s^{\rho_1} e_\nu(s)$.

Since these polynomials are equal to the elementary divisors of $A(s)$, the structural indices $c, \rho_1, \ldots, \rho_c$ are uniquely determined by $A(s)$. The same argument applied to $A^\circ(s) = sA_0 + A_1$ and its Kronecker canonical form shows that $d, \mu_1, \ldots, \mu_d$ are also determined uniquely by $A(s)$.

### 5.7.4 *Equivalence and Strict Equivalence*

In this section we compare the concepts of equivalence and strict equivalence for matrix pencils. Let us recall that two polynomial matrices $A(s)$ and $B(s)$ are said be *equivalent* if there exist unimodular matrices $U(s)$ and $V(s)$ such that

$$U(s)\, A(s)\, V(s) = B(s), \tag{5.66}$$

which is explained in Remark 5.2 in Sec. 5.5. By Theorem 5.7 for Smith normal form, a necessary and sufficient condition for equivalence is that all the elementary divisors of the two polynomial matrices are the same.

Two matrix pencils that are strictly equivalent to each other are obviously equivalent. The converse, however, is not true in general.

**Example 5.18.** Matrix pencil $A(s) = \begin{bmatrix} 1 & s \\ 0 & 1 \end{bmatrix}$ is equivalent to $B(s) = \begin{bmatrix} 1 & 0 \\ 0 & 1 \end{bmatrix}$. Indeed, we have the relation $U(s)A(s)V(s) = B(s)$ for unimodular matrices

$$U(s) = \begin{bmatrix} 1 & -s \\ 0 & 1 \end{bmatrix}, \qquad V(s) = \begin{bmatrix} 1 & 0 \\ 0 & 1 \end{bmatrix}.$$

But $A(s)$ is not strictly equivalent to $B(s)$. This is obvious from the fact that $A(s)$ includes $s$ and $B(s)$ does not, which cannot be true if $PA(s)Q = B(s)$ for nonsingular constant matrices $P$ and $Q$. ∎

**Example 5.19.** Matrix pencil[21]

$$A(s) = sA_1 + A_0 = s \begin{bmatrix} 1 & 1 & 2 \\ 1 & 1 & 2 \\ 1 & 1 & 3 \end{bmatrix} + \begin{bmatrix} 2 & 1 & 3 \\ 3 & 2 & 5 \\ 3 & 2 & 6 \end{bmatrix} = \begin{bmatrix} s+2 & s+1 & 2s+3 \\ s+3 & s+2 & 2s+5 \\ s+3 & s+2 & 3s+6 \end{bmatrix}$$

[20] That is, $B(s)$ has $\max(\nu, c)$ elementary divisors that are distinct from 1. With the convention of $\rho_j = 0$ ($j \geqq c+1$) and $e_j(s) = 1$ ($j \leqq 0$), they are expressed as $s^{\rho_{\max(\nu,c)+1-i}} e_{i+\nu-\max(\nu,c)}(s)$ ($i = 1, \ldots, \max(\nu, c)$).

[21] This is an example given in Sec. 2, Chap. XII of [2].

is equivalent to

$$B(s) = sB_1 + B_0 = s \begin{bmatrix} 0 & 0 & 0 \\ 0 & 0 & 0 \\ 0 & 0 & 1 \end{bmatrix} + \begin{bmatrix} 1 & 0 & 0 \\ 0 & 1 & 0 \\ 0 & 0 & 1 \end{bmatrix} = \begin{bmatrix} 1 & 0 & 0 \\ 0 & 1 & 0 \\ 0 & 0 & s+1 \end{bmatrix}.$$

Indeed, $U(s)A(s)V(s) = B(s)$ is true for unimodular matrices

$$U(s) = \begin{bmatrix} 1 & 0 & 0 \\ -1 & 1 & 0 \\ 0 & -1 & 1 \end{bmatrix}, \qquad V(s) = \begin{bmatrix} 1 & -s-1 & -1 \\ -1 & s+2 & -1 \\ 0 & 0 & 1 \end{bmatrix}.$$

But, since $\operatorname{rank} A_1 = 2$ and $\operatorname{rank} B_1 = 1$, there are no nonsingular constant matrices $P$ and $Q$ that satisfy $PA(s)Q = B(s)$. ■

As shown in the above examples, there is a difference, in general, between equivalence and strict equivalence. Their relationship can be made clear by introducing another matrix pencil

$$A^\circ(s) = sA_0 + A_1 = A_1 + sA_0 \tag{5.67}$$

with the coefficient matrices swapped in the given matrix pencil $A(s) = sA_1 + A_0$. By (5.41), $A(s)$ and $B(s)$ are strictly equivalent if and only if $A^\circ(s)$ and $B^\circ(s)$ are strictly equivalent. The relationship between strict equivalence (denoted as $\approx$) and equivalence (denoted as $\sim$) can be summarized into the following diagram:

$$\begin{array}{ccc} \boxed{A(s) \approx B(s)} & \Longleftrightarrow & \boxed{A^\circ(s) \approx B^\circ(s)} \\ \Downarrow & & \Downarrow \\ \boxed{A(s) \sim B(s)} & & \boxed{A^\circ(s) \sim B^\circ(s)} \end{array} \tag{5.68}$$

For regular pencils, the following theorem holds.

**Theorem 5.11.** *Two regular pencils $A(s)$ and $B(s)$ are strictly equivalent if and only if $A(s)$ and $B(s)$ are equivalent and $A^\circ(s)$ and $B^\circ(s)$ are equivalent.*

**Proof.** The proof is briefly sketched in Remark 5.6 below. □

As a consequence of the above theorem, the diagram (5.68) is strengthened to

$$\begin{array}{ccc} \boxed{A(s) \approx B(s)} & \Longleftrightarrow & \boxed{A^\circ(s) \approx B^\circ(s)} \\ \Updownarrow & & \Updownarrow \\ \multicolumn{3}{c}{\boxed{A(s) \sim B(s) \quad \text{and} \quad A^\circ(s) \sim B^\circ(s)}} \end{array} \tag{5.69}$$

for regular pencils.

**Example 5.20.** Let us apply Theorem 5.11 to the matrix pencils in Example 5.18. The two pencils $A(s)$ and $B(s)$ have been shown to be equivalent, but

$$A^\circ(s) = \begin{bmatrix} s & 1 \\ 0 & s \end{bmatrix}, \qquad B^\circ(s) = \begin{bmatrix} s & 0 \\ 0 & s \end{bmatrix}$$

are not equivalent. This is because $A^\circ(s)$ has elementary divisors 1 and $s^2$, whereas $B^\circ(s)$ has $s$ and $s$. From this fact we can conclude, by Theorem 5.11, that $A(s)$ and $B(s)$ are not strictly equivalent. ■

**Example 5.21.** Let us apply Theorem 5.11 to the matrix pencils in Example 5.19. The elementary divisors of[22]

$$A^\circ(s) = \begin{bmatrix} 1 & 1 & 2 \\ 1 & 1 & 2 \\ 1 & 1 & 3 \end{bmatrix} + s \begin{bmatrix} 2 & 1 & 3 \\ 3 & 2 & 5 \\ 3 & 2 & 6 \end{bmatrix} = \begin{bmatrix} 1+2s & 1+s & 2+3s \\ 1+3s & 1+2s & 2+5s \\ 1+3s & 1+2s & 3+6s \end{bmatrix}$$

are 1, 1, and $s^2(s+1)$, and the elementary divisors of

$$B^\circ(s) = \begin{bmatrix} 0 & 0 & 0 \\ 0 & 0 & 0 \\ 0 & 0 & 1 \end{bmatrix} + s \begin{bmatrix} 1 & 0 & 0 \\ 0 & 1 & 0 \\ 0 & 0 & 1 \end{bmatrix} = \begin{bmatrix} s & 0 & 0 \\ 0 & s & 0 \\ 0 & 0 & 1+s \end{bmatrix}$$

are 1, $s$, and $s(s+1)$. Therefore, $A^\circ(s)$ and $B^\circ(s)$ are not equivalent and, by Theorem 5.11, $A(s)$ and $B(s)$ are not strictly equivalent. ■

**Remark 5.6.** Theorem 5.11 is a reformulation of Theorem 2 in Chap. XII of [2] in a form suitable for applications. Here, we briefly explain the proof of Theorem 5.11, with reference to the arguments in Sec. 2 of Chap. XII of [2].

We introduce another variable $t$ to turn the matrix pencil $A(s) = sA_1 + A_0$ to a *homogeneous form* $A(s,t) = sA_1 + tA_0$. For $A(s,t)$ in homogeneous form, we define the determinantal divisors $d_k(A(s,t))$ $(k = 1, \ldots, r)$ similarly as in (5.23), and define the elementary divisors by

$$e_k(A(s,t)) = \frac{d_k(A(s,t))}{d_{k-1}(A(s,t))} \qquad (k = 1, \ldots, r),$$

which is consistent with (5.36). Since the irreducible factors of a homogeneous polynomial are homogeneous, the polynomials $d_k(A(s,t))$ and

[22]After the transformation $\begin{bmatrix} 1 & 0 & 0 \\ -1 & 1 & 0 \\ -1 & 0 & 1 \end{bmatrix} A^\circ(s) \begin{bmatrix} 1 & 0 & 0 \\ -1 & 1 & -2 \\ 0 & 0 & 1 \end{bmatrix} = \begin{bmatrix} s & 1+s & s \\ 0 & s & 0 \\ 0 & s & 1+s \end{bmatrix}$ we calculate the determinantal divisors to obtain $d_1(A^\circ(s)) = 1$, $d_2(A^\circ(s)) = 1$, and $d_3(A^\circ(s)) = s^2(s+1)$. Hence $e_1(A^\circ(s)) = 1$, $e_2(A^\circ(s)) = 1$, and $e_3(A^\circ(s)) = s^2(s+1)$.

$e_k(A(s,t))$ introduced above are homogeneous polynomials in $(s,t)$. The following theorem is known [2, Chap. XII, Theorem 2]: *Two regular pencils $A(s)$ and $B(s)$ are strictly equivalent if and only if their homogeneous forms share the same elementary divisors, i.e., $e_k(A(s,t)) = e_k(B(s,t))$ for $k = 1,\ldots,r$.*

This theorem can be translated into Theorem 5.11 as follows. First note that $A(s,1) = A(s)$ and $A(1,t) = A^\circ(t)$. As for the determinantal divisors, $d_k(A(s,t))|_{t=1}$, which denotes $d_k(A(s,t))$ with $t = 1$, coincides with $d_k(A(s))$; symmetrically, $d_k(A(s,t))|_{s=1}$, which denotes $d_k(A(s,t))$ with $s = 1$, coincides with $d_k(A^\circ(t))$. As a consequence, we have

$$e_k(A(s,t))|_{t=1} = e_k(A(s)), \qquad e_k(A(s,t))|_{s=1} = e_k(A^\circ(t))$$

for the elementary divisors. For homogeneous polynomials $f(s,t)$ and $g(s,t)$, in general, it holds that

$$f(s,t) = g(s,t) \iff f(s,1) = g(s,1) \text{ and } f(1,t) = g(1,t).$$

Using this general fact for $f(s,t) = e_k(A(s,t))$ and $g(s,t) = e_k(B(s,t))$ we then obtain

$$\begin{aligned} & e_k(A(s,t)) = e_k(B(s,t)) \\ & \iff e_k(A(s)) = e_k(B(s)) \text{ and } e_k(A^\circ(t)) = e_k(B^\circ(t)). \end{aligned}$$

Thus, Theorem 5.11 is equivalent to Theorem 2 of Chap. XII of [2]. ∎

# Chapter 6

# Generalized Inverses

For nonsingular square matrices the inverse matrices exist and are uniquely determined, but this is not the case with singular square matrices or rectangular matrices. Extending the concept of inverse matrices to arbitrary matrices, including rectangular matrices, turns out to be useful in applications. In this chapter, fundamental facts about generalized inverses are presented. The derivation and characterization of several types of generalized inverses are shown, and an application to Newton's method is illustrated.

## 6.1 Generalized Inverse

### 6.1.1 *Definition and Construction*

How should we extend the concept of inverse matrices for singular square matrices or rectangular matrices? The inverse $A^{-1}$ of a nonsingular square matrix $A$ is a matrix such that the solution of a system of linear equations $A\boldsymbol{x} = \boldsymbol{y}$ is given by $\boldsymbol{x} = A^{-1}\boldsymbol{y}$. For a general $m \times n$ matrix $A$, it would be natural to regard an $n \times m$ matrix $G$ as an "inverse" of $A$ (in some extended sense) if, for each $\boldsymbol{y}$ for which the system of equations

$$A\boldsymbol{x} = \boldsymbol{y} \tag{6.1}$$

has solutions, the vector

$$\boldsymbol{x} = G\boldsymbol{y}$$

gives one of the solutions. Such a matrix $G$ is referred to as the *generalized inverse*[1] of $A$ and denoted as $A^{-}$. We consider real matrices below.[2]

[1]There are a variety of concepts called generalized inverses or pseudo-inverses in many fields. Not all of them, however, are a generalized inverse in the sense defined here. For example, the *Drazin inverse* and the *Bott–Duffin inverse* are not [54].

[2]Complex matrices can also be treated in a similar manner.

The generalized inverse can be constructed as follows. We denote the rank of $A$ by $r$, *i.e.*,

$$r = \operatorname{rank} A,$$

and transform the matrix $A$ to its rank normal form [6, 15]:

$$SAT = \begin{bmatrix} I_r & O_{r,n-r} \\ O_{m-r,r} & O_{m-r,n-r} \end{bmatrix}, \tag{6.2}$$

where $S$ is a nonsingular matrix of order $m$ and $T$ is a nonsingular matrix of order $n$; the matrices $S$ and $T$ are not uniquely determined. With the change of variables

$$\tilde{\boldsymbol{x}} = T^{-1}\boldsymbol{x}, \qquad \tilde{\boldsymbol{y}} = S\boldsymbol{y},$$

we can rewrite the equation $A\boldsymbol{x} = \boldsymbol{y}$ in (6.1) as

$$(SAT)\tilde{\boldsymbol{x}} = \tilde{\boldsymbol{y}}.$$

This equation is solvable if and only if

$$\tilde{y}_{r+1} = 0,\ \tilde{y}_{r+2} = 0,\ \ldots,\ \tilde{y}_m = 0, \tag{6.3}$$

and then the solution is given as

$$\tilde{x}_1 = \tilde{y}_1,\ \ldots,\ \tilde{x}_r = \tilde{y}_r;\quad \tilde{x}_{r+1},\ \ldots,\ \tilde{x}_n\text{: arbitrary real numbers.} \tag{6.4}$$

Let us determine an $n \times m$ matrix $\tilde{G}$ such that, for each vector $\tilde{\boldsymbol{y}}$ that satisfies the condition (6.3), the vector

$$\tilde{\boldsymbol{x}} = \tilde{G}\tilde{\boldsymbol{y}}$$

satisfies (6.4). Such a matrix $\tilde{G}$ is given as

$$\tilde{G} = \begin{bmatrix} I_r & B \\ C & D \end{bmatrix},$$

where $B$ is an arbitrary $r \times (m - r)$ matrix, $C$ is an arbitrary $(n - r) \times r$ matrix, and $D$ is an arbitrary $(n - r) \times (m - r)$ matrix. Then we have

$$\tilde{\boldsymbol{x}} = \begin{bmatrix} I_r & B \\ C & D \end{bmatrix} \tilde{\boldsymbol{y}}.$$

The correspondence $\tilde{\boldsymbol{y}} \mapsto \tilde{\boldsymbol{x}}$ is expressed in terms of the original variables as

$$\boldsymbol{x} = T \begin{bmatrix} I_r & B \\ C & D \end{bmatrix} S\boldsymbol{y},$$

and therefore, the generalized inverse $A^-$ of $A$ is given as

$$A^- = T \begin{bmatrix} I_r & B \\ C & D \end{bmatrix} S. \tag{6.5}$$

The generalized inverse $A^-$ is determined from a relatively weak requirement that it should give any solution $\boldsymbol{x}$ for each $\boldsymbol{y}$ for which the equation $A\boldsymbol{x} = \boldsymbol{y}$ is solvable. This means, in particular, that no constraints are imposed from those $\boldsymbol{y}$ for which no solutions $\boldsymbol{x}$ exist. The degrees of freedom inherent in the generalized inverse are reflected in the arbitrariness of the matrices $B$, $C$, and $D$ in (6.5). In the extreme case of $A = O_{m,n}$ (zero matrix), every $n \times m$ matrix is qualified as $A^-$.

**Example 6.1.** Let us obtain the generalized inverse $A^-$ of a $4 \times 3$ matrix

$$A = \begin{bmatrix} -2 & 1 & 1 \\ 1 & -2 & 1 \\ 1 & 1 & -2 \\ -2 & 1 & 1 \end{bmatrix}.$$

The matrix $A$, of rank 2, is transformed to the rank normal form (6.2) by

$$S = \left[\begin{array}{cccc} 1 & 0 & 0 & 0 \\ 0 & 1 & 0 & 0 \\ \hline 1 & 1 & 1 & 0 \\ -1 & 0 & 0 & 1 \end{array}\right], \qquad T = \left[\begin{array}{cc|c} -2/3 & -1/3 & 1 \\ -1/3 & -2/3 & 1 \\ 0 & 0 & 1 \end{array}\right]. \tag{6.6}$$

Putting

$$B = \begin{bmatrix} b_{11} & b_{12} \\ b_{21} & b_{22} \end{bmatrix}, \qquad C = \begin{bmatrix} c_1 & c_2 \end{bmatrix}, \qquad D = \begin{bmatrix} d_1 & d_2 \end{bmatrix}$$

in (6.5), we can calculate $A^-$ as

$$A^- = \left[\begin{array}{cc|c} -2/3 & -1/3 & 1 \\ -1/3 & -2/3 & 1 \\ 0 & 0 & 1 \end{array}\right] \left[\begin{array}{cc|cc} 1 & 0 & b_{11} & b_{12} \\ 0 & 1 & b_{21} & b_{22} \\ \hline c_1 & c_2 & d_1 & d_2 \end{array}\right] \left[\begin{array}{cccc} 1 & 0 & 0 & 0 \\ 0 & 1 & 0 & 0 \\ \hline 1 & 1 & 1 & 0 \\ -1 & 0 & 0 & 1 \end{array}\right]$$

$$
= \begin{bmatrix} -2/3 & -1/3 & 0 & 0 \\ -1/3 & -2/3 & 0 & 0 \\ 0 & 0 & 0 & 0 \end{bmatrix}
$$

$$
+ b_{11} \begin{bmatrix} -2/3 & -2/3 & -2/3 & 0 \\ -1/3 & -1/3 & -1/3 & 0 \\ 0 & 0 & 0 & 0 \end{bmatrix} + b_{12} \begin{bmatrix} 2/3 & 0 & 0 & -2/3 \\ 1/3 & 0 & 0 & -1/3 \\ 0 & 0 & 0 & 0 \end{bmatrix}
$$

$$
+ b_{21} \begin{bmatrix} -1/3 & -1/3 & -1/3 & 0 \\ -2/3 & -2/3 & -2/3 & 0 \\ 0 & 0 & 0 & 0 \end{bmatrix} + b_{22} \begin{bmatrix} 1/3 & 0 & 0 & -1/3 \\ 2/3 & 0 & 0 & -2/3 \\ 0 & 0 & 0 & 0 \end{bmatrix}
$$

$$
+ c_1 \begin{bmatrix} 1 & 0 & 0 & 0 \\ 1 & 0 & 0 & 0 \\ 1 & 0 & 0 & 0 \end{bmatrix} + c_2 \begin{bmatrix} 0 & 1 & 0 & 0 \\ 0 & 1 & 0 & 0 \\ 0 & 1 & 0 & 0 \end{bmatrix}
$$

$$
+ d_1 \begin{bmatrix} 1 & 1 & 1 & 0 \\ 1 & 1 & 1 & 0 \\ 1 & 1 & 1 & 0 \end{bmatrix} + d_2 \begin{bmatrix} -1 & 0 & 0 & 1 \\ -1 & 0 & 0 & 1 \\ -1 & 0 & 0 & 1 \end{bmatrix}.
$$

Eight free parameters $b_{11}$, $b_{12}$, $b_{21}$, $b_{22}$, $c_1$, $c_2$, $d_1$, and $d_2$ are contained in the generalized inverse $A^-$. ■

### 6.1.2 *Characterization*

Next, we give an alternative definition of the generalized inverse in a somewhat abstract form. The original definition of $A^-$ introduced above is:

$$
A^-\boldsymbol{y} \text{ is a solution to } A\boldsymbol{x} = \boldsymbol{y} \text{ for every } \boldsymbol{y} \in \mathrm{Im}(A). \tag{6.7}
$$

Here, two simple facts are observed:

(i) $\boldsymbol{y} \in \mathrm{Im}(A)$ if and only if there exists $\boldsymbol{x} \in \mathbb{R}^n$ that satisfies $\boldsymbol{y} = A\boldsymbol{x}$,

(ii) $A^-\boldsymbol{y}$ is a solution to $A\boldsymbol{x} = \boldsymbol{y}$ if and only if $AA^-\boldsymbol{y} = \boldsymbol{y}$.

With these observations, the definition (6.7) can be rewritten as:

$$
AA^-A\boldsymbol{x} = A\boldsymbol{x} \text{ for every } \boldsymbol{x} \in \mathbb{R}^n. \tag{6.8}
$$

Therefore, we can alternatively define $A^-$ to be the generalize inverse of $A$, if

$$
AA^-A = A. \tag{6.9}
$$

As is expected from the above argument, the matrix $A^-$ as defined in (6.5) satisfies the condition (6.9) and conversely, we can derive the expression (6.5) for $A^-$ from the condition (6.9) with the aid of the rank normal form in (6.2).

### 6.1.3 *Expression of General Solutions*

The generalized inverse $A^-$ enables us to explicitly represent a general solution $\boldsymbol{x}$ for a solvable system of equations $A\boldsymbol{x} = \boldsymbol{y}$ as

$$\boldsymbol{x} = A^-\boldsymbol{y} + (I - A^-A)\boldsymbol{s}, \qquad \boldsymbol{s} \in \mathbb{R}^n. \tag{6.10}$$

This expression is derived as follows. As is well known, a general solution can be represented as the sum of a particular solution and an arbitrary solution of the homogeneous equation, and accordingly we have an expression

$$\boldsymbol{x} = A^-\boldsymbol{y} + \boldsymbol{z}, \qquad \boldsymbol{z} \in \mathrm{Ker}(A).$$

Here, as shown in Remark 6.1 below, $\mathrm{Ker}(A)$ can be expressed, by using the generalized inverse $A^-$, as

$$\mathrm{Ker}(A) = \mathrm{Ker}(A^-A) = \mathrm{Im}(I - A^-A) \tag{6.11}$$

and therefore, $\boldsymbol{z}$ belongs to $\mathrm{Ker}(A)$ if and only if $\boldsymbol{z}$ can be represented as

$$\boldsymbol{z} = (I - A^-A)\boldsymbol{s}$$

for some $\boldsymbol{s}$. The expression (6.10) is thus derived.

**Remark 6.1.** We prove (6.11) here.
(i) Proof of $\mathrm{Ker}(A) = \mathrm{Ker}(A^-A)$: If $A\boldsymbol{x} = \boldsymbol{0}$, then $A^-A\boldsymbol{x} = \boldsymbol{0}$; therefore, $\mathrm{Ker}(A) \subseteqq \mathrm{Ker}(A^-A)$ holds. Conversely, if $A^-A\boldsymbol{x} = \boldsymbol{0}$, then $AA^-A\boldsymbol{x} = A\boldsymbol{x} = \boldsymbol{0}$ by (6.9); therefore, $\mathrm{Ker}(A) \supseteqq \mathrm{Ker}(A^-A)$ is also true.
(ii) Proof of $\mathrm{Ker}(A^-A) = \mathrm{Im}(I - A^-A)$: If $A^-A\boldsymbol{x} = \boldsymbol{0}$, then $\boldsymbol{x} = (I - A^-A)\boldsymbol{x}$; therefore, $\mathrm{Ker}(A^-A) \subseteqq \mathrm{Im}(I - A^-A)$. On the other hand, since $A^-AA^-A = A^-A$ by (6.9), we have $A^-A(I - A^-A) = O$ and therefore $\mathrm{Ker}(A^-A) \supseteqq \mathrm{Im}(I - A^-A)$. ■

## 6.2 Minimum-Norm Generalized Inverse

### 6.2.1 *Definition and Construction*

As we have already seen, the generalized inverse provides one of the possible solutions for solvable systems of equations. Here we consider the generalized inverse that gives a solution of minimum Euclidean norm.[3] Such a generalized inverse is referred to as the *minimum-norm generalized inverse.*

To derive a representation of the minimum-norm generalized inverse, it is convenient, just as in Sec. 6.1, to transform the given matrix $A$ to

[3]For a real vector $\boldsymbol{x} = (x_i)$, $\|\boldsymbol{x}\|_2 = (\sum_i {x_i}^2)^{1/2}$ is called the *Euclidean norm.*

a suitable normal form (or the like). Whereas we used a transformation of the form $SAT$ with nonsingular matrices $S$ and $T$ in Sec. 6.1, we now use a transformation of the form $SAQ$ with an orthogonal matrix $Q$ and a nonsingular matrix $S$ so as to keep the norm $\|\boldsymbol{x}\|_2$ of solution $\boldsymbol{x}$ unchanged (see below).

We first note that, when the equation $A\boldsymbol{x} = \boldsymbol{y}$ is transformed to

$$(AQ)(Q^\top \boldsymbol{x}) = \boldsymbol{y}$$

with an orthogonal matrix $Q$, the Euclidean norm of the solution is unchanged, *i.e.*,

$$\|Q^\top \boldsymbol{x}\|_2 = \|\boldsymbol{x}\|_2.$$

We choose an orthogonal matrix $Q$ such that $AQ$ takes a simple form. Specifically, we make use of the QR decomposition [3, 15, 30] of $A^\top$:

$$A^\top P = QR, \tag{6.12}$$

where $Q$ is an orthogonal matrix of order $n$, $R$ is an $n \times m$ upper-triangular matrix,[4] and $P$ is a permutation matrix of order $m$. From this decomposition, we can obtain

$$P^\top AQ = R^\top = \begin{bmatrix} \tilde{R}^\top & O_{r,n-r} \\ * & O_{m-r,n-r} \end{bmatrix}, \tag{6.13}$$

where $r = \operatorname{rank} A$, $\tilde{R}^\top$ is a nonsingular lower-triangular matrix of order $r$, and $*$ is an $(m-r) \times r$ matrix. Furthermore, by repeated elementary row transformations the matrix is transformed to

$$SAQ = \begin{bmatrix} I_r & O_{r,n-r} \\ O_{m-r,r} & O_{m-r,n-r} \end{bmatrix}, \tag{6.14}$$

where $S$ is a nonsingular matrix of order $m$.

With the change of variables $\tilde{\boldsymbol{x}} = Q^\top \boldsymbol{x}$ and $\tilde{\boldsymbol{y}} = S\boldsymbol{y}$, the equation $A\boldsymbol{x} = \boldsymbol{y}$ is transformed to

$$(SAQ)\tilde{\boldsymbol{x}} = \tilde{\boldsymbol{y}}. \tag{6.15}$$

Since $\|\tilde{\boldsymbol{x}}\|_2 = \|\boldsymbol{x}\|_2$, finding a solution $\boldsymbol{x}$ with the minimum norm for $A\boldsymbol{x} = \boldsymbol{y}$ is equivalent to finding a solution $\tilde{\boldsymbol{x}}$ with the minimum norm for the equation (6.15).

The transformed equation (6.15) has a solution if and only if

$$\tilde{y}_{r+1} = 0, \ \tilde{y}_{r+2} = 0, \ \ldots, \ \tilde{y}_m = 0, \tag{6.16}$$

[4] A matrix $R = (r_{ij})$ is called upper-triangular if $r_{ij} = 0$ for $i > j$.

and in this case, the minimum norm solution $\tilde{\boldsymbol{x}}$ is given by

$$\tilde{x}_1 = \tilde{y}_1, \ \ldots, \ \tilde{x}_r = \tilde{y}_r; \quad \tilde{x}_{r+1} = \cdots = \tilde{x}_n = 0.$$

This relation can be written in a matrix form as

$$\tilde{\boldsymbol{x}} = \begin{bmatrix} I_r & B \\ O & D \end{bmatrix} \tilde{\boldsymbol{y}},$$

where $B$ is an arbitrary $r \times (m-r)$ matrix and $D$ is an arbitrary $(n-r) \times (m-r)$ matrix. In the original variables we have

$$\boldsymbol{x} = Q \begin{bmatrix} I_r & B \\ O & D \end{bmatrix} S\boldsymbol{y}.$$

Therefore, the generalized inverse that gives a solution of the minimum Euclidean norm (the minimum-norm generalized inverse) is given as

$$A^{\vee} = Q \begin{bmatrix} I_r & B \\ O & D \end{bmatrix} S. \tag{6.17}$$

In this book, the minimum-norm generalized inverse of $A$ is denoted as $A^{\vee}$.

The minimum-norm generalized inverse $A^{\vee}$ is determined by the condition that it should give the minimum norm solution for solvable equations. For a solvable system of equations, the minimum norm solution is uniquely determined, whereas no constraints are imposed on the value of $A^{\vee}\boldsymbol{y}$ for those $\boldsymbol{y}$ for which the equation $A\boldsymbol{x} = \boldsymbol{y}$ is unsolvable. Such degrees of freedom are reflected in the arbitrariness of the matrices $B$ and $D$ in (6.17).

**Example 6.2.** Let us obtain the minimum-norm generalized inverse $A^{\vee}$ of the $4 \times 3$ matrix $A$ of Example 6.1. We can take

$$Q = \frac{1}{\sqrt{6}} \left[\begin{array}{cc|c} -2 & 0 & \sqrt{2} \\ 1 & -\sqrt{3} & \sqrt{2} \\ 1 & \sqrt{3} & \sqrt{2} \end{array}\right], \qquad R = \left[\begin{array}{cc|cc} \sqrt{6} & -\sqrt{6}/2 & -\sqrt{6}/2 & \sqrt{6} \\ 0 & 3/\sqrt{2} & -3/\sqrt{2} & 0 \\ \hline 0 & 0 & 0 & 0 \end{array}\right],$$

and $P = I$ for $A^{\top}P = QR$ in (6.12), and

$$S = \left[\begin{array}{cccc} 1/\sqrt{6} & 0 & 0 & 0 \\ \sqrt{2}/6 & \sqrt{2}/3 & 0 & 0 \\ \hline 1 & 1 & 1 & 0 \\ -1 & 0 & 0 & 1 \end{array}\right]$$

for $SAQ = \begin{bmatrix} I & O \\ O & O \end{bmatrix}$ in (6.14). Putting

$$B = \begin{bmatrix} \check{b}_{11} & \check{b}_{12} \\ \check{b}_{21} & \check{b}_{22} \end{bmatrix}, \qquad D = \begin{bmatrix} \check{d}_1 & \check{d}_2 \end{bmatrix}$$

in (6.17), we can calculate $A^\vee$ as

$$A^\vee = \frac{1}{\sqrt{6}} \left[\begin{array}{cc|c} -2 & 0 & \sqrt{2} \\ 1 & -\sqrt{3} & \sqrt{2} \\ 1 & \sqrt{3} & \sqrt{2} \end{array}\right] \left[\begin{array}{cc|cc} 1 & 0 & \check{b}_{11} & \check{b}_{12} \\ 0 & 1 & \check{b}_{21} & \check{b}_{22} \\ \hline 0 & 0 & \check{d}_1 & \check{d}_2 \end{array}\right] \left[\begin{array}{cccc} 1/\sqrt{6} & 0 & 0 & 0 \\ \sqrt{2}/6 & \sqrt{2}/3 & 0 & 0 \\ \hline 1 & 1 & 1 & 0 \\ -1 & 0 & 0 & 1 \end{array}\right]$$

$$= \begin{bmatrix} -1/3 & 0 & 0 & 0 \\ 0 & -1/3 & 0 & 0 \\ 1/3 & 1/3 & 0 & 0 \end{bmatrix}$$

$$+ \frac{\check{b}_{11}}{\sqrt{6}} \begin{bmatrix} -2 & -2 & -2 & 0 \\ 1 & 1 & 1 & 0 \\ 1 & 1 & 1 & 0 \end{bmatrix} + \frac{\check{b}_{12}}{\sqrt{6}} \begin{bmatrix} 2 & 0 & 0 & -2 \\ -1 & 0 & 0 & 1 \\ -1 & 0 & 0 & 1 \end{bmatrix}$$

$$+ \frac{\check{b}_{21}}{\sqrt{2}} \begin{bmatrix} 0 & 0 & 0 & 0 \\ -1 & -1 & -1 & 0 \\ 1 & 1 & 1 & 0 \end{bmatrix} + \frac{\check{b}_{22}}{\sqrt{2}} \begin{bmatrix} 0 & 0 & 0 & 0 \\ 1 & 0 & 0 & -1 \\ -1 & 0 & 0 & 1 \end{bmatrix}$$

$$+ \frac{\check{d}_1}{\sqrt{3}} \begin{bmatrix} 1 & 1 & 1 & 0 \\ 1 & 1 & 1 & 0 \\ 1 & 1 & 1 & 0 \end{bmatrix} + \frac{\check{d}_2}{\sqrt{3}} \begin{bmatrix} -1 & 0 & 0 & 1 \\ -1 & 0 & 0 & 1 \\ -1 & 0 & 0 & 1 \end{bmatrix}.$$

Six free parameters $\check{b}_{11}$, $\check{b}_{12}$, $\check{b}_{21}$, $\check{b}_{22}$, $\check{d}_1$, and $\check{d}_2$ are contained in the minimum-norm generalized inverse $A^\vee$. ■

### 6.2.2 *Characterization*

Similarly to the (general) generalized inverse, the minimum-norm generalized inverse $A^\vee$ can be characterized in an abstract form. As shown in (6.10), the solutions to $A\boldsymbol{x} = \boldsymbol{y}$ are expressed as

$$\boldsymbol{x} = A^\vee \boldsymbol{y} + (I - A^\vee A)\boldsymbol{s}, \qquad \boldsymbol{s} \in \mathbb{R}^n. \tag{6.18}$$

Therefore, $A^\vee \boldsymbol{y}$ has the minimum norm if and only if

$$\|A^\vee \boldsymbol{y}\|_2{}^2 \leqq \|A^\vee \boldsymbol{y} + (I - A^\vee A)\boldsymbol{s}\|_2{}^2$$

holds for all $\boldsymbol{s} \in \mathbb{R}^n$, which is equivalent (see Remark 6.2) to

$$(I - A^\vee A)^\top A^\vee \boldsymbol{y} = \boldsymbol{0}.$$

This condition should hold for all $\boldsymbol{y} \in \operatorname{Im}(A)$, *i.e.*, for all $\boldsymbol{y}$ that can be written as $\boldsymbol{y} = A\boldsymbol{x}$ for some $\boldsymbol{x} \in \mathbb{R}^n$, and therefore we must have

$$(I - A^\vee A)^\top A^\vee A = O.$$

This condition is equivalent to

$$(A^{\vee}A)^{\top} = A^{\vee}A \tag{6.19}$$

under the condition $AA^{\vee}A = A$ in (6.9) characterizing a generalized inverse. Thus, we can say that the minimum-norm generalized inverse $A^{\vee}$ is characterized by

$$AA^{\vee}A = A, \qquad (A^{\vee}A)^{\top} = A^{\vee}A. \tag{6.20}$$

As is expected from the above argument, the matrix $A^{\vee}$ as defined in (6.17) satisfies the condition (6.20) and conversely, we can derive the expression (6.17) for $A^{\vee}$ from the condition (6.20) by using the expression (6.14).

**Remark 6.2.** For a vector $\boldsymbol{a}$ and a matrix $B$, the inequality

$$\|\boldsymbol{a}\|_2{}^2 \leqq \|\boldsymbol{a} + B\boldsymbol{s}\|_2{}^2$$

holds for all $\boldsymbol{s}$ if and only if $B^{\top}\boldsymbol{a} = \boldsymbol{0}$. This is derived as the condition that, in the expression

$$\|\boldsymbol{a} + B\boldsymbol{s}\|_2{}^2 - \|\boldsymbol{a}\|_2{}^2 = 2\boldsymbol{s}^{\top}B^{\top}\boldsymbol{a} + \boldsymbol{s}^{\top}B^{\top}B\boldsymbol{s},$$

the linear term in $\boldsymbol{s}$ should vanish. ■

**Remark 6.3.** According to the general form (6.5) of a generalized inverse, the matrix $A^{\vee}$ can be expressed as

$$A^{\vee} = T \begin{bmatrix} I & B \\ C & D \end{bmatrix} S$$

with some nonsingular matrices $S$ and $T$ such that

$$A = S^{-1} \begin{bmatrix} I & O \\ O & O \end{bmatrix} T^{-1}.$$

Let $T^{\top}T$ be partitioned into blocks as

$$T^{\top}T = \begin{bmatrix} V_{11} & V_{12} \\ V_{21} & V_{22} \end{bmatrix}.$$

Then $A^{\vee}$ satisfies the condition (6.20) if and only if $C = -V_{22}^{-1}V_{21}$. Note that $T^{\top}T$ is a positive-definite symmetric matrix, and therefore $V_{22}$ is nonsingular, $V_{22}^{\top} = V_{22}$, and $V_{21}^{\top} = V_{12}$. ■

**Example 6.3.** Let us obtain the minimum-norm generalized inverse $A^{\vee}$ of the $4 \times 3$ matrix $A$ of Example 6.1 by using the fact stated in Remark 6.3. For the matrix $T$ in (6.6) we have

$$T^{\top}T = \left[\begin{array}{cc|c} 5/9 & 4/9 & -1 \\ 4/9 & 5/9 & -1 \\ \hline -1 & -1 & 3 \end{array}\right],$$

from which follows $(c_1, c_2) = -3^{-1}(-1, -1) = (1/3, 1/3)$. Therefore,

$$\begin{aligned} A^{\vee} = & \begin{bmatrix} -1/3 & 0 & 0 & 0 \\ 0 & -1/3 & 0 & 0 \\ 1/3 & 1/3 & 0 & 0 \end{bmatrix} \\ & + b_{11} \begin{bmatrix} -2/3 & -2/3 & -2/3 & 0 \\ -1/3 & -1/3 & -1/3 & 0 \\ 0 & 0 & 0 & 0 \end{bmatrix} + b_{12} \begin{bmatrix} 2/3 & 0 & 0 & -2/3 \\ 1/3 & 0 & 0 & -1/3 \\ 0 & 0 & 0 & 0 \end{bmatrix} \\ & + b_{21} \begin{bmatrix} -1/3 & -1/3 & -1/3 & 0 \\ -2/3 & -2/3 & -2/3 & 0 \\ 0 & 0 & 0 & 0 \end{bmatrix} + b_{22} \begin{bmatrix} 1/3 & 0 & 0 & -1/3 \\ 2/3 & 0 & 0 & -2/3 \\ 0 & 0 & 0 & 0 \end{bmatrix} \\ & + d_1 \begin{bmatrix} 1 & 1 & 1 & 0 \\ 1 & 1 & 1 & 0 \\ 1 & 1 & 1 & 0 \end{bmatrix} + d_2 \begin{bmatrix} -1 & 0 & 0 & 1 \\ -1 & 0 & 0 & 1 \\ -1 & 0 & 0 & 1 \end{bmatrix}. \end{aligned}$$

There is a one-to-one correspondence, through a linear transformation, between these parameters $(b_{11}, b_{12}, b_{21}, b_{22}, d_1, d_2)$ and the parameters $(\breve{b}_{11}, \breve{b}_{12}, \breve{b}_{21}, \breve{b}_{22}, \breve{d}_1, \breve{d}_2)$ in Example 6.2. ■

**Remark 6.4.** We mention here a formula

$$A^{\vee} = A^{\top}(AA^{\top})^{-} \tag{6.21}$$

for the minimum-norm generalized inverse $A^{\vee}$. Indeed we can show that the matrix $X = A^{\top}(AA^{\top})^{-}$ on the right-hand side above satisfies the conditions $AXA = A$ and $(XA)^{\top} = XA$ in (6.20) for the minimum-norm generalized inverse, as follows. First, by $\mathrm{Im}(AA^{\top}) = \mathrm{Im}(A)$, we have $AA^{\top}C = A$ for some matrix $C$, and then

$$AXA = AA^{\top}(AA^{\top})^{-}AA^{\top}C = AA^{\top}C = A.$$

Second, $(XA)^{\top} = XA$ follows from

$$XA = A^{\top}(AA^{\top})^{-}A = C^{\top}AA^{\top}(AA^{\top})^{-}AA^{\top}C = C^{\top}AA^{\top}C.$$

Although the expression (6.21) is certainly a convenient formula, it should be noted that not every $A^{\vee}$ can be expressed in the form of (6.21).

This fact is understood by considering the ranks. The rank of the matrix $A^\vee$ can be any integer ranging from $r$ to $\min(m, n)$; see (6.17). On the other hand, $\operatorname{rank} A^\top(AA^\top)^- = r$, since $\operatorname{rank} A^\top(AA^\top)^- \geqq r$ by the fact that $A^\top(AA^\top)^-$ is a generalized inverse of $A$, and since $\operatorname{rank} A^\top(AA^\top)^- \leqq \operatorname{rank} A^\top = r$. What the formula (6.21) really means is that $A^\top(AA^\top)^-$ is a possible choice of $A^\vee$, but this formula does not exhaust all possibilities of $A^\vee$ even if $(AA^\top)^-$ runs over all possible generalized inverses of $AA^\top$. ■

## 6.3 Least-Square Generalized Inverse

### 6.3.1 *Definition and Construction*

As shown in Sec. 6.1, the generalized inverse is defined in terms of the requirement that it should give solutions to solvable systems of equations. According to this definition, no constraint arises from unsolvable systems of equations. In this section, we take into consideration those systems of equations that have no solutions, and consider a generalized inverse that gives a solution $\boldsymbol{x}$ (*least-square solution*) to $A\boldsymbol{x} = \boldsymbol{y}$, which gives the minimum Euclidean norm of the residual error $\|A\boldsymbol{x} - \boldsymbol{y}\|_2$. Such a generalized inverse is called the *least-square generalized inverse.*

We first note that the Euclidean norm of the residual error $\|A\boldsymbol{x} - \boldsymbol{y}\|_2$ remains invariant under an orthogonal transformation of the equation $A\boldsymbol{x} = \boldsymbol{y}$, that is,

$$\|Q^\top(A\boldsymbol{x} - \boldsymbol{y})\|_2 = \|A\boldsymbol{x} - \boldsymbol{y}\|_2$$

for any orthogonal matrix $Q$. With this fact in mind we consider transforming the equation $A\boldsymbol{x} = \boldsymbol{y}$ to

$$Q^\top A\boldsymbol{x} = Q^\top \boldsymbol{y},$$

thereby bringing $Q^\top A$ into a simpler form. Specifically, we make use of the QR decomposition [3, 15, 30] of $A$:

$$AP = QR, \tag{6.22}$$

where $Q$ is an orthogonal matrix of order $m$, $R$ is an $m \times n$ upper-triangular matrix, and $P$ is a permutation matrix of order $n$. From this decomposition, we can obtain

$$Q^\top AP = R = \begin{bmatrix} \tilde{R} & * \\ O_{m-r,r} & O_{m-r,n-r} \end{bmatrix}, \tag{6.23}$$

where $r = \operatorname{rank} A$, $\tilde{R}$ is a nonsingular upper-triangular matrix of order $r$, and $*$ is an $r \times (n-r)$ matrix. Furthermore, by repeated elementary column transformations, the matrix is transformed to

$$Q^\top AT = \begin{bmatrix} I_r & O_{r,n-r} \\ O_{m-r,r} & O_{m-r,n-r} \end{bmatrix}, \tag{6.24}$$

where $T$ is a nonsingular matrix of order $n$.

With the change of variables $\tilde{\boldsymbol{x}} = T^{-1}\boldsymbol{x}$ and $\tilde{\boldsymbol{y}} = Q^\top \boldsymbol{y}$, the equation $A\boldsymbol{x} = \boldsymbol{y}$ is transformed to

$$(Q^\top AT)\tilde{\boldsymbol{x}} = \tilde{\boldsymbol{y}}. \tag{6.25}$$

Since

$$\|(Q^\top AT)\tilde{\boldsymbol{x}} - \tilde{\boldsymbol{y}}\|_2 = \|A\boldsymbol{x} - \boldsymbol{y}\|_2,$$

finding a least-square solution $\boldsymbol{x}$ for $A\boldsymbol{x} = \boldsymbol{y}$ is equivalent to finding a least-square solution $\tilde{\boldsymbol{x}}$ for the equation (6.25).

By (6.24), the least-square solution $\tilde{\boldsymbol{x}}$ for the transformed equation (6.25) is obviously given as

$$\tilde{x}_1 = \tilde{y}_1, \ \ldots, \ \tilde{x}_r = \tilde{y}_r; \quad \tilde{x}_{r+1}, \ \ldots, \ \tilde{x}_n\text{: arbitrary real numbers.} \tag{6.26}$$

This relation can be written in a matrix form as

$$\tilde{\boldsymbol{x}} = \begin{bmatrix} I_r & O \\ C & D \end{bmatrix} \tilde{\boldsymbol{y}},$$

where $C$ is an arbitrary $(n-r) \times r$ matrix and $D$ is an arbitrary $(n-r) \times (m-r)$ matrix. In the original variables we have

$$\boldsymbol{x} = T \begin{bmatrix} I_r & O \\ C & D \end{bmatrix} Q^\top \boldsymbol{y}.$$

Therefore, the least-square generalized inverse of $A$, to be denoted as $A^\wedge$ in this book, is given by

$$A^\wedge = T \begin{bmatrix} I_r & O \\ C & D \end{bmatrix} Q^\top. \tag{6.27}$$

The least-square generalized inverse $A^\wedge$ is determined by the condition that it should give the least-square solution for systems of equations, whether they are solvable or not. In general, however, the least-square solution is not uniquely determined. Such degrees of freedom are reflected in the arbitrariness of matrices $C$ and $D$ of in (6.27).

**Example 6.4.** Let us obtain the least-square generalized inverse $A^\wedge$ of the $4 \times 3$ matrix $A$ of Example 6.1. We can take

$$Q = \frac{1}{\sqrt{10}} \left[\begin{array}{cc|cc} -2 & 0 & \sqrt{5} & 1 \\ 1 & -\sqrt{5} & 0 & 2 \\ 1 & \sqrt{5} & 0 & 2 \\ -2 & 0 & -\sqrt{5} & 1 \end{array}\right], \quad R = \left[\begin{array}{cc|c} \sqrt{10} & -\sqrt{10}/2 & -\sqrt{10}/2 \\ 0 & 3/\sqrt{2} & -3/\sqrt{2} \\ \hline 0 & 0 & 0 \\ 0 & 0 & 0 \end{array}\right],$$

and $P = I$ for $AP = QR$ in (6.22), and

$$T = \left[\begin{array}{cc|c} 1/\sqrt{10} & \sqrt{2}/6 & 1 \\ 0 & \sqrt{2}/3 & 1 \\ 0 & 0 & 1 \end{array}\right]$$

for $Q^\top AT = \begin{bmatrix} I & O \\ O & O \end{bmatrix}$ in (6.24). Putting $C = [\hat{c}_1, \hat{c}_2]$ and $D = [\hat{d}_1, \hat{d}_2]$ in (6.27), we can calculate $A^\wedge$ as

$$A^\wedge = \left[\begin{array}{cc|c} 1/\sqrt{10} & \sqrt{2}/6 & 1 \\ 0 & \sqrt{2}/3 & 1 \\ 0 & 0 & 1 \end{array}\right] \left[\begin{array}{cc|cc} 1 & 0 & 0 & 0 \\ 0 & 1 & 0 & 0 \\ \hline \hat{c}_1 & \hat{c}_2 & \hat{d}_1 & \hat{d}_2 \end{array}\right] \times \frac{1}{\sqrt{10}} \left[\begin{array}{cccc} -2 & 1 & 1 & -2 \\ 0 & -\sqrt{5} & \sqrt{5} & 0 \\ \hline \sqrt{5} & 0 & 0 & -\sqrt{5} \\ 1 & 2 & 2 & 1 \end{array}\right]$$

$$= \begin{bmatrix} -1/5 & -1/15 & 4/15 & -1/5 \\ 0 & -1/3 & 1/3 & 0 \\ 0 & 0 & 0 & 0 \end{bmatrix} + \frac{\hat{c}_1}{\sqrt{10}} \begin{bmatrix} -2 & 1 & 1 & -2 \\ -2 & 1 & 1 & -2 \\ -2 & 1 & 1 & -2 \end{bmatrix} + \frac{\hat{c}_2}{\sqrt{2}} \begin{bmatrix} 0 & -1 & 1 & 0 \\ 0 & -1 & 1 & 0 \\ 0 & -1 & 1 & 0 \end{bmatrix}$$

$$+ \frac{\hat{d}_1}{\sqrt{2}} \begin{bmatrix} 1 & 0 & 0 & -1 \\ 1 & 0 & 0 & -1 \\ 1 & 0 & 0 & -1 \end{bmatrix} + \frac{\hat{d}_2}{\sqrt{10}} \begin{bmatrix} 1 & 2 & 2 & 1 \\ 1 & 2 & 2 & 1 \\ 1 & 2 & 2 & 1 \end{bmatrix}.$$

Four free parameters $\hat{c}_1$, $\hat{c}_2$, $\hat{d}_1$, and $\hat{d}_2$ are contained in the least-square generalized inverse $A^\wedge$. ■

### 6.3.2 *Characterization*

Next, we give a characterization of the least-square generalized inverse $A^\wedge$ in an abstract form. Since $A^\wedge \boldsymbol{y}$ gives the minimum residual error for the equation $A\boldsymbol{x} = \boldsymbol{y}$, the inequality

$$\|AA^\wedge \boldsymbol{y} - \boldsymbol{y}\|_2{}^2 \leqq \|A\boldsymbol{x} - \boldsymbol{y}\|_2{}^2$$

holds for all $\boldsymbol{x} \in \mathbb{R}^n$. With the expression $\boldsymbol{x} = \boldsymbol{w} + A^\wedge \boldsymbol{y}$, this condition is rephrased to the statement that

$$\|(AA^\wedge - I)\boldsymbol{y}\|_2{}^2 \leqq \|(AA^\wedge - I)\boldsymbol{y} + A\boldsymbol{w}\|_2{}^2$$

holds for all $\boldsymbol{w} \in \mathbb{R}^n$. This means

$$A^\top (AA^\wedge - I)\boldsymbol{y} = \boldsymbol{0}, \tag{6.28}$$

as is explained in Remark 6.2. Since this should be true for all $\boldsymbol{y} \in \mathbb{R}^m$, it follows that

$$A^\top(AA^\wedge - I) = O,$$

*i.e.*,

$$(AA^\wedge)^\top A = A. \tag{6.29}$$

The condition (6.29) turns out (see Remark 6.5) to be equivalent to

$$AA^\wedge A = A, \qquad (AA^\wedge)^\top = AA^\wedge. \tag{6.30}$$

Thus, the least-square generalized inverse $A^\wedge$ is characterized by the condition (6.30).

As is expected from the above argument, the matrix $A^\wedge$ as defined in (6.27) satisfies the condition (6.30) and conversely, we can derive the expression (6.27) for $A^\wedge$ from the condition (6.30) by using (6.24).

**Remark 6.5.** We provide here a proof for the equivalence between (6.29) and (6.30). Obviously, (6.30) implies (6.29). To show the converse, we multiply (6.29) with $A^\wedge$ from the right to obtain $(AA^\wedge)^\top(AA^\wedge) = AA^\wedge$. This shows that $AA^\wedge$ is a symmetric matrix, *i.e.*, $(AA^\wedge)^\top = AA^\wedge$. By substituting this into (6.29), we obtain $AA^\wedge A = A$ in (6.30). In view of (6.9), it is quite natural that $AA^\wedge A = A$ follows from (6.29), since, for an equation $A\boldsymbol{x} = \boldsymbol{y}$ that has a solution, the least-square solution $A^\wedge \boldsymbol{y}$ is indeed a solution for the equation $A\boldsymbol{x} = \boldsymbol{y}$, and therefore, $A^\wedge$ is a generalized inverse, which satisfies the condition (6.9). ■

**Remark 6.6.** According to the general form (6.5) of a generalized inverse, the matrix $A^\wedge$ can be expressed as

$$A^\wedge = T \begin{bmatrix} I & B \\ C & D \end{bmatrix} S$$

for some nonsingular matrices $S$ and $T$ such that

$$A = S^{-1} \begin{bmatrix} I & O \\ O & O \end{bmatrix} T^{-1}.$$

Let $SS^\top$ be partitioned into blocks as

$$SS^\top = \begin{bmatrix} U_{11} & U_{12} \\ U_{21} & U_{22} \end{bmatrix}.$$

Then $A^\wedge$ satisfies the condition (6.30) if and only if $B = -U_{12}U_{22}^{-1}$. Note that $SS^\top$ is a positive-definite symmetric matrix, and therefore $U_{22}$ is nonsingular, $U_{22}^\top = U_{22}$, and $U_{21}^\top = U_{12}$. ■

**Example 6.5.** Let us obtain the least-square generalized inverse $A^\wedge$ of the $4 \times 3$ matrix $A$ of Example 6.1 by using the fact stated in Remark 6.6. For the matrix $S$ in (6.6) we have

$$SS^\top = \left[\begin{array}{cc|cc} 1 & 0 & 1 & -1 \\ 0 & 1 & 1 & 0 \\ \hline 1 & 1 & 3 & -1 \\ -1 & 0 & -1 & 2 \end{array}\right],$$

from which follows

$$\begin{bmatrix} b_{11} & b_{12} \\ b_{21} & b_{22} \end{bmatrix} = -\begin{bmatrix} 1 & -1 \\ 1 & 0 \end{bmatrix} \begin{bmatrix} 3 & -1 \\ -1 & 2 \end{bmatrix}^{-1} = \frac{1}{5}\begin{bmatrix} -1 & 2 \\ -2 & -1 \end{bmatrix}.$$

Therefore,

$$\begin{aligned} A^\wedge = &\begin{bmatrix} -1/5 & -1/15 & 4/15 & -1/5 \\ 0 & -1/3 & 1/3 & 0 \\ 0 & 0 & 0 & 0 \end{bmatrix} \\ &+ c_1 \begin{bmatrix} 1 & 0 & 0 & 0 \\ 1 & 0 & 0 & 0 \\ 1 & 0 & 0 & 0 \end{bmatrix} + c_2 \begin{bmatrix} 0 & 1 & 0 & 0 \\ 0 & 1 & 0 & 0 \\ 0 & 1 & 0 & 0 \end{bmatrix} \\ &+ d_1 \begin{bmatrix} 1 & 1 & 1 & 0 \\ 1 & 1 & 1 & 0 \\ 1 & 1 & 1 & 0 \end{bmatrix} + d_2 \begin{bmatrix} -1 & 0 & 0 & 1 \\ -1 & 0 & 0 & 1 \\ -1 & 0 & 0 & 1 \end{bmatrix}. \end{aligned}$$

There is a one-to-one correspondence, through a linear transformation, between these parameters $(c_1, c_2, d_1, d_2)$ and the parameters $(\hat{c}_1, \hat{c}_2, \hat{d}_1, \hat{d}_2)$ in Example 6.4. ■

**Remark 6.7.** The derivation of (6.28) is in fact a standard argument in the method of least squares, disguised here by the generalized inverse. Without reference to the generalized inverse, the argument goes as follows. For the least-square solution $\hat{\boldsymbol{x}}$ of the equation $A\boldsymbol{x} = \boldsymbol{y}$, we have

$$\|A\hat{\boldsymbol{x}} - \boldsymbol{y}\|_2{}^2 \leqq \|A\boldsymbol{x} - \boldsymbol{y}\|_2{}^2$$

for all $\boldsymbol{x} \in \mathbb{R}^n$. With the expression $\boldsymbol{x} = \boldsymbol{w} + \hat{\boldsymbol{x}}$, this condition is rephrased to the statement that

$$\|A\hat{\boldsymbol{x}} - \boldsymbol{y}\|_2{}^2 \leqq \|A\hat{\boldsymbol{x}} - \boldsymbol{y} + A\boldsymbol{w}\|_2{}^2$$

holds for all $\boldsymbol{w} \in \mathbb{R}^n$. This means

$$A^\top(A\hat{\boldsymbol{x}} - \boldsymbol{y}) = \mathbf{0}, \tag{6.31}$$

as is explained in Remark 6.2. The substitution of $\hat{\boldsymbol{x}} = A^{\wedge}\boldsymbol{y}$ into this equation yields (6.28). We add here that the equation (6.31) can be transformed to an equation in $\hat{\boldsymbol{x}}$:

$$A^{\top} A \hat{\boldsymbol{x}} = A^{\top} \boldsymbol{y}, \tag{6.32}$$

which is a fundamental equation, called the *normal equation*, in the method of least squares. ■

**Remark 6.8.** We mention here a formula

$$A^{\wedge} = (A^{\top} A)^{-} A^{\top} \tag{6.33}$$

for the least-square generalized inverse $A^{\wedge}$. Indeed we can show that the matrix $X = (A^{\top} A)^{-} A^{\top}$ on the right-hand side above satisfies the conditions $AXA = A$ and $(AX)^{\top} = AX$ in (6.30), as follows (in almost the same way as in Remark 6.4). Let $C$ be a matrix such that $A^{\top} AC = A^{\top}$, where the existence of such $C$ is guaranteed by $\mathrm{Im}(A^{\top} A) = \mathrm{Im}(A^{\top})$. Then

$$AXA = C^{\top} A^{\top} A(A^{\top} A)^{-} A^{\top} A = C^{\top} A^{\top} A = A,$$

$$AX = A(A^{\top} A)^{-} A^{\top} = C^{\top} A^{\top} A(A^{\top} A)^{-} A^{\top} AC = C^{\top} A^{\top} AC = (AX)^{\top}.$$

Note, however, that not every $A^{\wedge}$ can be expressed in the form of (6.33).[5] ■

### 6.3.3 *Expression of General Solutions*

The least-square generalized inverse $A^{\wedge}$ provides us with two different representations of least-square solutions:

$$\boldsymbol{x} = A^{\wedge}\boldsymbol{y} + \boldsymbol{z}, \qquad \boldsymbol{z} \in \mathrm{Ker}(A), \tag{6.34}$$

$$\boldsymbol{x} = A^{\wedge}\boldsymbol{y} + (I - A^{\wedge}A)\boldsymbol{s}, \qquad \boldsymbol{s} \in \mathbb{R}^n, \tag{6.35}$$

which are derived below.

First, we note that although the least-square solution $\boldsymbol{x}$ may not be uniquely determined, the vector $A\boldsymbol{x}$ is determined uniquely. This is indeed the case, since if both $\boldsymbol{x}^{(1)}$ and $\boldsymbol{x}^{(2)}$ attain the minimum residual error $d = \min_{\boldsymbol{x}} \|A\boldsymbol{x} - \boldsymbol{y}\|_2$, then by the parallelogram identity [6, Theorem 9.52] we have

$$\begin{aligned}
&\|A\boldsymbol{x}^{(1)} - A\boldsymbol{x}^{(2)}\|_2{}^2 \\
&= 2(\|A\boldsymbol{x}^{(1)} - \boldsymbol{y}\|_2{}^2 + \|A\boldsymbol{x}^{(2)} - \boldsymbol{y}\|_2{}^2) - \|A(\boldsymbol{x}^{(1)} + \boldsymbol{x}^{(2)}) - 2\boldsymbol{y}\|_2{}^2 \\
&= 2(\|A\boldsymbol{x}^{(1)} - \boldsymbol{y}\|_2{}^2 + \|A\boldsymbol{x}^{(2)} - \boldsymbol{y}\|_2{}^2) - 4\left\|A\left(\frac{\boldsymbol{x}^{(1)} + \boldsymbol{x}^{(2)}}{2}\right) - \boldsymbol{y}\right\|_2{}^2 \\
&\leqq 2(d^2 + d^2) - 4d^2 = 0,
\end{aligned}$$

[5] $\mathrm{rank}\,(A^{\top} A)^{-} A^{\top} = r$, but $\mathrm{rank}\, A^{\wedge}$ can be any integer ranging from $r$ to $\min(m, n)$.

which shows $A\boldsymbol{x}^{(1)} = A\boldsymbol{x}^{(2)}$.

With $\boldsymbol{x}^{(1)} = \boldsymbol{x}$ and $\boldsymbol{x}^{(2)} = A^{\wedge}\boldsymbol{y}$ in the above equation, we see $A(\boldsymbol{x} - A^{\wedge}\boldsymbol{y}) = \boldsymbol{0}$. Therefore, (6.34) is true for $\boldsymbol{z} = \boldsymbol{x} - A^{\wedge}\boldsymbol{y}$. Conversely, for $\boldsymbol{x}$ of the form of (6.34), we have

$$A\boldsymbol{x} = A(A^{\wedge}\boldsymbol{y} + \boldsymbol{z}) = A(A^{\wedge}\boldsymbol{y}),$$

which shows that $\boldsymbol{x}$ is a least-square solution. The other expression (6.35) follows immediately from (6.34) on the basis of the relation $\mathrm{Ker}(A) = \mathrm{Im}(I - A^{\wedge}A)$ shown in (6.11); see Sec. 6.1.3.

## 6.4 Moore–Penrose Generalized Inverse

### 6.4.1 *Definition and Construction*

In Sec. 6.3, we have considered the generalized inverse that gives a least-square solution for a system of equations $A\boldsymbol{x} = \boldsymbol{y}$ that does not necessarily possess a solution. In general, a least-square solution is not uniquely determined. With this in mind, let us consider the generalized inverse that attains the minimum norm among least-square solutions. Such a generalized inverse is called the *Moore–Penrose generalized inverse* and is denoted by $A^{+}$.

First, recall the equation (6.13) that is obtained from the QR decomposition of $A^{\top}$, and apply the QR decomposition to the right-hand side of this equation. Then we obtain

$$P^{\top}AQ = \tilde{Q}\begin{bmatrix} \hat{R} & O_{r,n-r} \\ O_{m-r,r} & O_{m-r,n-r} \end{bmatrix}, \tag{6.36}$$

where $\tilde{Q}$ is an orthogonal matrix of order $m$ and $\hat{R}$ is a nonsingular upper-triangular matrix of order $r$. Accordingly,

$$\hat{Q}^{\top}AQ = \begin{bmatrix} \hat{R} & O_{r,n-r} \\ O_{m-r,r} & O_{m-r,n-r} \end{bmatrix}, \tag{6.37}$$

where $\hat{Q} = P\tilde{Q}$ is an orthogonal matrix of order $m$.

With the change of variables $\tilde{\boldsymbol{x}} = Q^{\top}\boldsymbol{x}$ and $\tilde{\boldsymbol{y}} = \hat{Q}^{\top}\boldsymbol{y}$, the equation $A\boldsymbol{x} = \boldsymbol{y}$ is transformed to

$$(\hat{Q}^{\top}AQ)\tilde{\boldsymbol{x}} = \tilde{\boldsymbol{y}}. \tag{6.38}$$

Since

$$\|\tilde{\boldsymbol{x}}\|_2 = \|\boldsymbol{x}\|_2, \qquad \|(\hat{Q}^{\top}AQ)\tilde{\boldsymbol{x}} - \tilde{\boldsymbol{y}}\|_2 = \|A\boldsymbol{x} - \boldsymbol{y}\|_2,$$

finding the minimum norm solution $\boldsymbol{x}$ from among the least-square solutions for $A\boldsymbol{x} = \boldsymbol{y}$ is equivalent to finding the minimum norm solution $\tilde{\boldsymbol{x}}$ from among the least-square solutions for the equation (6.38).

For the transformed equation (6.38) with (6.37), the minimum norm solution among the least-square solutions is obviously given as

$$\begin{bmatrix} \tilde{x}_1 \\ \vdots \\ \tilde{x}_r \end{bmatrix} = \hat{R}^{-1} \begin{bmatrix} \tilde{y}_1 \\ \vdots \\ \tilde{y}_r \end{bmatrix}, \qquad \tilde{x}_{r+1} = \cdots = \tilde{x}_n = 0.$$

This relation can also be written as

$$\tilde{\boldsymbol{x}} = \begin{bmatrix} \hat{R}^{-1} & O \\ O & O \end{bmatrix} \tilde{\boldsymbol{y}}.$$

In the original variables we have

$$\boldsymbol{x} = Q \begin{bmatrix} \hat{R}^{-1} & O \\ O & O \end{bmatrix} \hat{Q}^\top \boldsymbol{y}.$$

Therefore, the Moore–Penrose generalized inverse $A^+$ is given as

$$A^+ = Q \begin{bmatrix} \hat{R}^{-1} & O \\ O & O \end{bmatrix} \hat{Q}^\top. \tag{6.39}$$

**Example 6.6.** Let us obtain the Moore–Penrose generalized inverse $A^+$ of the $4 \times 3$ matrix $A$ of Example 6.1. As the matrices $P$, $Q$, $\tilde{Q}$, and $\hat{R}$ for $P^\top A Q = \tilde{Q} \begin{bmatrix} \hat{R} & O \\ O & O \end{bmatrix}$ in (6.36), we can take $P = I$, $Q$ in Example 6.2,

$$\tilde{Q} = \frac{1}{\sqrt{10}} \left[\begin{array}{cccc} 2 & 0 & \sqrt{5} & 1 \\ -1 & \sqrt{5} & 0 & 2 \\ \hline -1 & -\sqrt{5} & 0 & 2 \\ 2 & 0 & -\sqrt{5} & 1 \end{array}\right], \qquad \hat{R} = \begin{bmatrix} \sqrt{15} & 0 \\ 0 & 3 \end{bmatrix}.$$

Then (6.39) gives

$$A^+ = \frac{1}{15} \begin{bmatrix} -2 & 1 & 1 & -2 \\ 1 & -3 & 2 & 1 \\ 1 & 2 & -3 & 1 \end{bmatrix}.$$

■

### 6.4.2 *Characterization*

Next, we give a characterization of the Moore–Penrose generalized inverse $A^+$ in an abstract form. Since $A^+$ is a least-square generalized inverse, we have

$$AA^+A = A, \qquad (AA^+)^\top = AA^+ \tag{6.40}$$

from (6.30). As shown in (6.35), the least-square solutions for the equation $A\boldsymbol{x} = \boldsymbol{y}$ can be written as

$$\boldsymbol{x} = A^+\boldsymbol{y} + (I - A^+A)\boldsymbol{s}, \qquad \boldsymbol{s} \in \mathbb{R}^n.$$

Therefore, $A^+\boldsymbol{y}$ has the minimum norm if and only if

$$\|A^+\boldsymbol{y}\|_2{}^2 \leqq \|A^+\boldsymbol{y} + (I - A^+A)\boldsymbol{s}\|_2{}^2$$

is true for all $\boldsymbol{s} \in \mathbb{R}^n$, and therefore (see Remark 6.2) we have

$$(I - A^+A)^\top A^+\boldsymbol{y} = \mathbf{0}.$$

Since this should be true for all $\boldsymbol{y} \in \mathbb{R}^m$, it follows that

$$(I - A^+A)^\top A^+ = O,$$

*i.e.*,

$$(A^+A)^\top A^+ = A^+. \tag{6.41}$$

The condition (6.41) is equivalent to

$$A^+AA^+ = A^+, \qquad (A^+A)^\top = A^+A, \tag{6.42}$$

which can be proved similarly as in Remark 6.5. Thus, the Moore–Penrose generalized inverse $A^+$ is characterized as a matrix that satisfies the conditions in (6.40) and (6.42).

As is expected from the above argument, the matrix $A^+$ as defined in (6.39) satisfies the conditions (6.40) and (6.42) and conversely, we can derive the expression (6.39) for $A^+$ from the conditions (6.40) and (6.42) by using (6.37).

For a given matrix $A$, the generalized inverses $A^-$, $A^\vee$, and $A^\wedge$ are not uniquely determined. In contrast, $A^+$ is uniquely determined, since the minimum norm solution among the least-square solutions is uniquely determined. This uniqueness can also be derived from the abstract characterizations (6.40) and (6.42) above. Indeed, for two matrices $X$ and $Y$ that satisfy (6.40) and (6.42), we have

$$\begin{aligned} X &= XAX = XAYAX = X(AY)^\top(AX)^\top = XY^\top A^\top X^\top A^\top \\ &= XY^\top A^\top = XAY = XAYAY = (XA)^\top(YA)^\top Y \\ &= A^\top X^\top A^\top Y^\top Y = A^\top Y^\top Y = YAY = Y. \end{aligned}$$

**Remark 6.9.** According to the general form (6.5) of a generalized inverse, the matrix $A^+$ can be expressed as

$$A^+ = T \begin{bmatrix} I_r & B \\ C & D \end{bmatrix} S$$

with some nonsingular matrices $S$ and $T$ such that

$$A = S^{-1} \begin{bmatrix} I_r & O \\ O & O \end{bmatrix} T^{-1}.$$

With this expression for $A^+$ we have

$$A^+AA^+ = T \begin{bmatrix} I_r & B \\ C & CB \end{bmatrix} S.$$

Therefore, $A^+AA^+ = A^+$ holds if and only if $D = CB$, in which case we have

$$A^+ = T \begin{bmatrix} I_r \\ C \end{bmatrix} \begin{bmatrix} I_r & B \end{bmatrix} S. \tag{6.43}$$

On the other hand, as discussed in Remarks 6.3 and 6.6, we have $(AA^+)^\top = AA^+$ and $(A^+A)^\top = A^+A$ if and only if

$$B = -U_{12}U_{22}^{-1}, \qquad C = -V_{22}^{-1}V_{21}, \tag{6.44}$$

where

$$SS^\top = \begin{bmatrix} U_{11} & U_{12} \\ U_{21} & U_{22} \end{bmatrix}, \qquad T^\top T = \begin{bmatrix} V_{11} & V_{12} \\ V_{21} & V_{22} \end{bmatrix}.$$

Therefore, the Moore–Penrose generalized inverse $A^+$ can also be given by (6.43) with (6.44). ■

**Example 6.7.** Let us obtain the Moore–Penrose generalized inverse $A^+$ of the $4 \times 3$ matrix $A$ of Example 6.1 by using the fact stated in Remark 6.9. Examples 6.5 and 6.3 show

$$B = \frac{1}{5} \begin{bmatrix} -1 & 2 \\ -2 & -1 \end{bmatrix}, \qquad C = \frac{1}{3} \begin{bmatrix} 1 & 1 \end{bmatrix}.$$

With $B$ and $C$ above and $S$ and $T$ in (6.6), the formula (6.43) gives

$$A^+ = \frac{1}{15} \begin{bmatrix} -2 & 1 & 1 & -2 \\ 1 & -3 & 2 & 1 \\ 1 & 2 & -3 & 1 \end{bmatrix},$$

in agreement with the result in Example 6.6. ■

**Remark 6.10.** For an $m \times n$ matrix $A$ of rank $r$ given in the form of

$$A = EF$$

with an $m \times r$ matrix $E$ and an $r \times n$ matrix $F$, the Moore–Penrose generalized inverse $A^+$ is known to be expressed as

$$A^+ = F^\top (FF^\top)^{-1}(E^\top E)^{-1}E^\top. \tag{6.45}$$

This can be proved as follows.

Let $X$ denote the matrix on the right-hand side of (6.45). Then $(AX)^\top = AX$ holds, since

$$AX = EF \cdot F^\top (FF^\top)^{-1}(E^\top E)^{-1}E^\top = E(E^\top E)^{-1}E^\top.$$

We also have

$$AXA = E(E^\top E)^{-1}E^\top \cdot EF = EF = A.$$

Thus, the condition (6.40) is satisfied. Furthermore, we have

$$XA = F^\top (FF^\top)^{-1}(E^\top E)^{-1}E^\top \cdot EF = F^\top (FF^\top)^{-1}F,$$

from which follows $(XA)^\top = XA$, and

$$\begin{aligned} XAX &= F^\top (FF^\top)^{-1}F \cdot F^\top (FF^\top)^{-1}(E^\top E)^{-1}E^\top \\ &= F^\top (FF^\top)^{-1}(E^\top E)^{-1}E^\top = X. \end{aligned}$$

Therefore, the condition (6.42) is also satisfied. Since $X$ satisfies both (6.40) and (6.42), it must be equal to the Moore–Penrose generalized inverse $A^+$, thereby proving (6.45).

Next, we discuss the consistency of the formula (6.45) with the expressions (6.43) and (6.44) in Remark 6.9. It follows from (6.2) that

$$A = S^{-1}\begin{bmatrix} I & O \\ O & O \end{bmatrix} T^{-1} = S^{-1}\begin{bmatrix} I \\ O \end{bmatrix}\begin{bmatrix} I & O \end{bmatrix} T^{-1},$$

which indicates a possible choice

$$E = S^{-1}\begin{bmatrix} I \\ O \end{bmatrix}, \qquad F = \begin{bmatrix} I & O \end{bmatrix} T^{-1}.$$

With the notation

$$(SS^\top)^{-1} = \begin{bmatrix} W_{11} & W_{12} \\ W_{21} & W_{22} \end{bmatrix}$$

we have $E^\top E = W_{11}$ and $E^\top = [W_{11}, W_{12}]S$, and therefore,

$$(E^\top E)^{-1}E^\top = \begin{bmatrix} I & W_{11}^{-1}W_{12} \end{bmatrix} S.$$

On the other hand, by a well-known formula [6, Theorem 1.5] for the inverse of a block matrix applied to

$$SS^\top = \begin{bmatrix} U_{11} & U_{12} \\ U_{21} & U_{22} \end{bmatrix},$$

we have

$$W_{11} = (U_{11} - U_{12}U_{22}^{-1}U_{21})^{-1}, \qquad W_{12} = -W_{11}U_{12}U_{22}^{-1}.$$

Then it follows that

$$(E^\top E)^{-1}E^\top = \begin{bmatrix} I & -U_{12}U_{22}^{-1} \end{bmatrix} S = \begin{bmatrix} I & B \end{bmatrix} S.$$

Similarly, we have

$$F^\top(FF^\top)^{-1} = T\begin{bmatrix} I \\ -V_{22}^{-1}V_{21} \end{bmatrix} = T\begin{bmatrix} I \\ C \end{bmatrix}.$$

By substituting these into (6.45) we obtain the expression (6.43) for $A^+$ in Remark 6.9. ■

## 6.5 Application

Generalized inverses have a variety of applications. For example, in statistics (in particular, regression analysis), generalized inverses (and projection matrices) serve as a fundamental theoretical tool [54, 55]. In addition, generalized inverses are widely used as useful tools in areas such as numerical computation, optimization, inverse problems, structural engineering, and robotics. In this section, we consider Newton's method for systems of equations with inconsistent degrees of freedom as an example of the application of generalized inverses to the field of numerical computation.

*Newton's method*, in its original form, is used to solve a system of equations involving as many equations as variables, say $n$ equations and $n$ variables:

$$\begin{cases} f_1(x_1, \ldots, x_n) = 0, \\ \qquad \vdots \\ f_n(x_1, \ldots, x_n) = 0, \end{cases}$$

which will be denoted as $\boldsymbol{f}(\boldsymbol{x}) = \boldsymbol{0}$. Starting with a suitable initial approximate solution $\boldsymbol{x}^{(0)}$, we compute a sequence of approximate solutions according to the iteration formula

$$\boldsymbol{x}^{(p+1)} = \boldsymbol{x}^{(p)} - J(\boldsymbol{x}^{(p)})^{-1}\boldsymbol{f}(\boldsymbol{x}^{(p)}) \qquad (p = 0, 1, 2, \ldots). \tag{6.46}$$

Here, $J(\boldsymbol{x})$ is the *Jacobian matrix* defined as

$$J(\boldsymbol{x}) = \begin{bmatrix} \dfrac{\partial f_1}{\partial x_1}(\boldsymbol{x}) & \cdots & \dfrac{\partial f_1}{\partial x_n}(\boldsymbol{x}) \\ \vdots & & \vdots \\ \dfrac{\partial f_n}{\partial x_1}(\boldsymbol{x}) & \cdots & \dfrac{\partial f_n}{\partial x_n}(\boldsymbol{x}) \end{bmatrix}. \tag{6.47}$$

In engineering we encounter a variety of situations in which we need to solve a system of equations with differing numbers of variables and equations:[6]

$$\begin{cases} f_1(x_1, \ldots, x_n) = 0, \\ \qquad \vdots \\ f_m(x_1, \ldots, x_n) = 0, \end{cases}$$

where $m \neq n$. In this case, the Jacobian matrix $J(\boldsymbol{x})$ is an $m \times n$ matrix, defined similarly as in (6.47). Being a rectangular matrix, the Jacobian matrix does not have the inverse that is required in Newton's method. In place of the iteration (6.46) of Newton's method, we can conceive of a similar iteration

$$\boldsymbol{x}^{(p+1)} = \boldsymbol{x}^{(p)} - J(\boldsymbol{x}^{(p)})^{+} \boldsymbol{f}(\boldsymbol{x}^{(p)}) \qquad (p = 0, 1, 2, \ldots) \tag{6.48}$$

using the Moore–Penrose generalized inverse $J(\boldsymbol{x}^{(p)})^{+}$ of $J(\boldsymbol{x}^{(p)})$. This method is referred to as the *generalized Newton's method*.[7]

If the approximate solutions generated by the generalized Newton's method (6.48) converge to $\boldsymbol{x}^*$, then $\boldsymbol{x}^*$ satisfies

$$J(\boldsymbol{x}^*)^{+} \boldsymbol{f}(\boldsymbol{x}^*) = \boldsymbol{0} \tag{6.49}$$

by (6.48). Since $\mathrm{Ker}(J(\boldsymbol{x}^*)^{+}) = \mathrm{Ker}(J(\boldsymbol{x}^*)^{\top})$ by (6.37) and (6.39), the condition (6.49) is equivalent to

$$J(\boldsymbol{x}^*)^{\top} \boldsymbol{f}(\boldsymbol{x}^*) = \boldsymbol{0}.$$

Since[8]

$$J(\boldsymbol{x}^*)^{\top} \boldsymbol{f}(\boldsymbol{x}^*) = \mathrm{grad} \left( \frac{1}{2} \sum_{i=1}^{m} f_i(\boldsymbol{x})^2 \right) \Bigg|_{\boldsymbol{x}=\boldsymbol{x}^*},$$

[6]This system of equations will also be denoted as $\boldsymbol{f}(\boldsymbol{x}) = \boldsymbol{0}$.

[7]The meaning of the "generalized Newton's method" varies with fields. For example, in the field of optimization, an extension of Newton's method to non-differentiable objective functions is referred to as "generalized Newton's method".

[8]"grad" stands for the *gradient vector*. That is, $\mathrm{grad}\, \varphi = (\partial\varphi/\partial x_1, \ldots, \partial\varphi/\partial x_n)^{\top}$ for a (differentiable) function $\varphi(x_1, \ldots, x_n)$.

we see further that $\boldsymbol{x}^*$ is a stationary point of $\frac{1}{2}\sum_{i=1}^{m} f_i(\boldsymbol{x})^2$. Although this does not necessarily mean $\sum_{i=1}^{m} f_i(\boldsymbol{x}^*)^2 = 0$ (*i.e.*, $\boldsymbol{f}(\boldsymbol{x}^*) = \boldsymbol{0}$), the generalized Newton's method is promising as an effective solution method.

**Example 6.8.** Consider the following system of equations with three variables and two equations:

$$f_1(x_1, x_2, x_3) = {x_1}^2 + {x_2}^2 + {x_3}^2 - 1 = 0,$$
$$f_2(x_1, x_2, x_3) = {x_1}^2 + {x_2}^2 - x_3 = 0.$$

The solutions $(x_1, x_2, x_3)$ are the intersection of a spherical surface and a paraboloidal surface, forming a circle expressed as ${x_1}^2 + {x_2}^2 = a$ and $x_3 = a$ with a positive real number $a = (\sqrt{5} - 1)/2$, which satisfies $a^2 + a - 1 = 0$ (Fig. 6.1). The generalized Newton's method (6.48) is applied with the initial solution $(-1, 2, 2)$. As is seen from Table 6.1, the convergence is fast enough, being quadratic convergence. ■

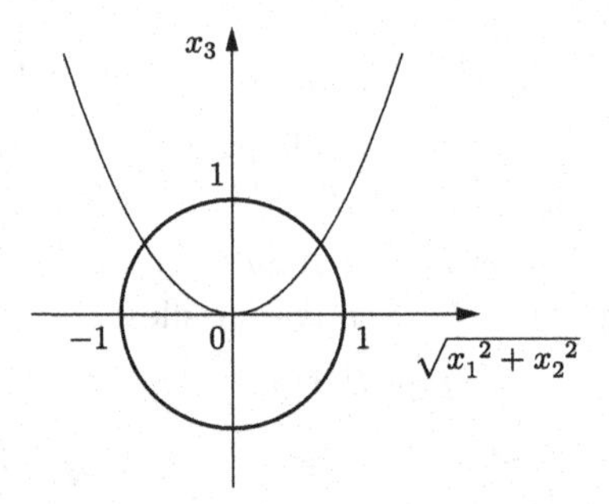


Fig. 6.1 A problem for the generalized Newton's method (Example 6.8).

As demonstrated in the above example, the generalized Newton's method using the Moore–Penrose generalized inverse copes successfully with inconsistent systems of equations with differing numbers of variables and equations, just as the ordinary Newton's method does for consistent systems of equations.

Table 6.1 Approximate solutions by the generalized Newton's method ($f_1 = {x_1}^2 + {x_2}^2 + {x_3}^2 - 1$, $f_2 = {x_1}^2 + {x_2}^2 - x_3$; Example 6.8 using double precision arithmetic).

| $p$ | $x_1^{(p)}$ | $x_2^{(p)}$ | $x_3^{(p)}$ | $f_1$ | $f_2$ | $({f_1}^2 + {f_2}^2)/2$ |
|---|---|---|---|---|---|---|
| 0 | −1.0000 | 2.0000 | 2.0000 | 8.00E+00 | 3.00E+00 | 3.65E+01 |
| 1 | −0.6000 | 1.2000 | 1.0000 | 1.80E+00 | 8.00E−01 | 1.94E+00 |
| 2 | −0.4111 | 0.8222 | 0.6667 | 2.90E−01 | 1.78E−01 | 5.78E−02 |
| 3 | −0.3561 | 0.7123 | 0.6190 | 1.74E−02 | 1.51E−02 | 2.65E−04 |
| 4 | −0.3516 | 0.7032 | 0.6180 | 1.04E−04 | 1.03E−04 | 1.06E−08 |
| 5 | −0.3516 | 0.7032 | 0.6180 | 4.29E−09 | 4.29E−09 | 1.84E−17 |

# Chapter 7

# Group Representation Theory

In this chapter we explain fundamental facts in group representation theory, which offers a mathematical methodology for expressing and utilizing the symmetry possessed by systems. The abstract structure of a group is described by concrete mathematical objects called representation matrices, and the symmetry possessed by engineering systems leads to block-diagonal decompositions of the matrices that describe the systems.

## 7.1 Systems with Symmetry

### 7.1.1 *Use of Symmetry*

As an example of a system that has symmetry, consider the *truss dome* (dome shaped truss structure) of Fig. 7.1. As is evident from the figure, this structure possesses a geometrical symmetry of regular hexagons. Suppose that the outermost nodes (at the bottom) are fixed. Then 217 free nodes remain. The system has the total of 651 ($= 3 \times 217$) degrees of freedom of deformation, since, for each node, there are three degrees of freedom of displacements in $x$, $y$, and $z$ directions. Suppose further that, as shown in the figure, a symmetric load is applied to the central part of the structure marked by the symbols ○ and △. By considering infinitesimal deformations and using a 651-dimensional vector $\boldsymbol{u}$ for the displacements of free nodes, we obtain a 651-dimensional system of linear equations:

$$K\boldsymbol{u} = \boldsymbol{f}.$$

The coefficient matrix $K$ here is a symmetric matrix ($K = K^{\top}$), called the *stiffness matrix*. The vector $\boldsymbol{f}$ on the right-hand side represents the external load.

Under the assumption that the member cross-sectional area and stiffness are all equal, this truss dome possesses regular hexagonal symmetry, not

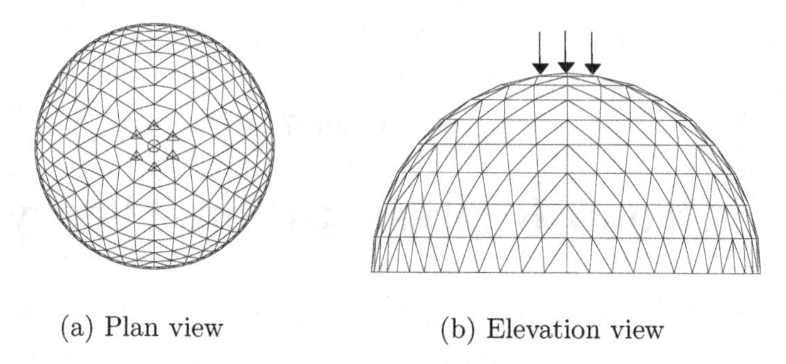

(a) Plan view (b) Elevation view

Fig. 7.1 A truss dome with regular hexagonal symmetry.

only geometrically but physically as well. Pertaining to this, we may be interested, for example, in the following questions:

- How many degrees of freedom are there for the deformation that possesses regular hexagonal symmetry?
- What structural characteristics does the stiffness matrix possess? Here, by "structural characteristics" we mean the properties that can be determined solely from symmetry.

In this chapter, we explain block-diagonalization of matrices based on group representation theory[1] as a systematic method of analysis for systems with symmetry. With this method we will be able to obtain the following facts for the truss dome above:[2]

- The deformation that keeps regular hexagonal symmetry has 61 degrees of freedom. If the applied external load possesses symmetry, the resulting deformation is also symmetric. Therefore, to find the deformation, we only have to solve a 61-dimensional system of linear equations, and need not solve the 651-dimensional full system.
- The stiffness matrix $K$ can be transformed to a block-diagonal form, as depicted in Fig. 7.2, by means of a transformation of the form $Q^{\top}KQ$ using an orthogonal matrix $Q$ that is determined independently of the values of physical parameters such as member cross-sectional area and stiffness. The block sizes in the block-diagonal form are 61, 48, 52, 56,

[1]Group representation theory is a comprehensive mathematical theory that covers a variety of topics other than block-diagonalization. See, *e.g.*, [56,57,59–65]. In this book, we restrict ourselves to fundamental issues about finite groups, but continuous groups are also important in applications.

[2]More details are provided in Remark 7.8 at the end of Sec. 7.4.4.

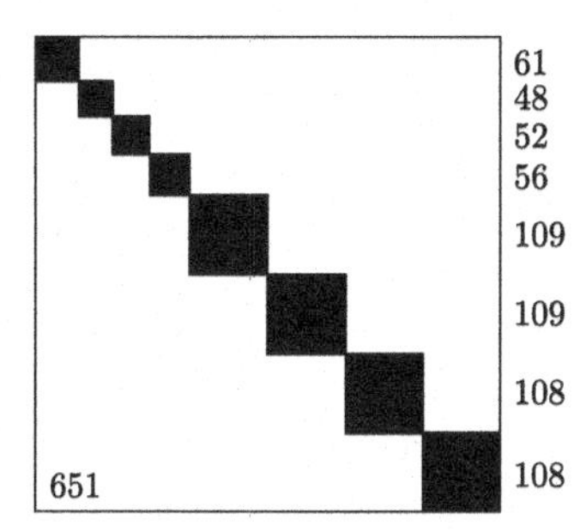


Fig. 7.2 Block-diagonal form of the stiffness matrix of the truss dome.

109, 109, 108, and 108; and the two blocks of size 109 and the two blocks of size 108 are the same matrices, respectively.

- As a consequence of the fact that the stiffness matrix $K$ can be block-diagonalized as in Fig. 7.2, the multiplicity structure of the eigenvalues of $K$ is obtained. As long as the parameter values such as member cross-sectional area and stiffness are not very special (or exceptional), the matrix $K$ has 217 $(= 61 + 48 + 52 + 56)$ simple eigenvalues and 217 $(= 109 + 108)$ eigenvalues of multiplicity 2.

### 7.1.2 *Mathematical Formulation of Symmetry*

How can we express symmetry in mathematical terms? Let us explain this using a simple example.

Consider a truss structure (*truss tent*) that possesses regular triangular symmetry, shown in Fig. 7.3. The three outer nodes 1, 2, and 3 are fixed with the only free node being the center node 0. Assume that the cross-sectional area and stiffness of the three members are the same, with the cross-sectional area $A$ and *Young's modulus* $E$. Also assume that force $f$ is applied to node 0 in the vertical direction downward.

Suppose that, prior to deformation, the node 0 is located at the origin $(0, 0, 0)$ and the nodes 1, 2, and 3 are at

$$\begin{aligned} (x_1, y_1, z_1) &= (1, 0, 2), \\ (x_2, y_2, z_2) &= \left(-\frac{1}{2}, \frac{\sqrt{3}}{2}, 2\right), \\ (x_3, y_3, z_3) &= \left(-\frac{1}{2}, -\frac{\sqrt{3}}{2}, 2\right). \end{aligned} \tag{7.1}$$

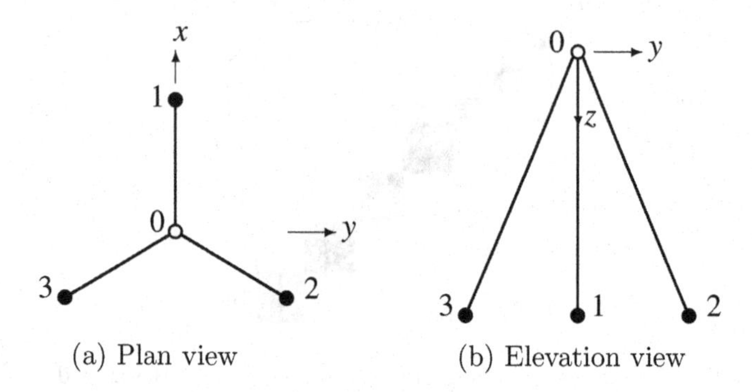


Fig. 7.3 A truss tent with regular triangular symmetry.

Then the lengths of the three members prior to deformation are all equal to $L = \sqrt{5}$. If the position of node 0 after deformation is denoted by $(x, y, z)$, the length of each member after deformation is given by

$$\hat{L}_i = ((x - x_i)^2 + (y - y_i)^2 + (z - z_i)^2)^{1/2} \qquad (i = 1, 2, 3).$$

Then the following equation expresses the balance of forces:

$$EA \sum_{i=1}^{3} \left( \frac{1}{L} - \frac{1}{\hat{L}_i} \right) \begin{bmatrix} x - x_i \\ y - y_i \\ z - z_i \end{bmatrix} = \begin{bmatrix} 0 \\ 0 \\ f \end{bmatrix}. \tag{7.2}$$

The position $(x, y, z)$ of node 0 at the equilibrium state is determined as the solution of this equation.

In considering infinitesimal deformation, we linearize the equation (7.2) in a neighborhood of $(x, y, z) = (0, 0, 0)$, to obtain a system of linear equations of the form

$$K\boldsymbol{u} = \boldsymbol{f}. \tag{7.3}$$

Here, $\boldsymbol{u} = (x, y, z)^\top$, $\boldsymbol{f} = (0, 0, f)^\top$, and the stiffness matrix

$$K = \sum_{i=1}^{3} \frac{EA}{L^3} \begin{bmatrix} x_i^2 & x_i y_i & x_i z_i \\ x_i y_i & y_i^2 & y_i z_i \\ x_i z_i & y_i z_i & z_i^2 \end{bmatrix} \tag{7.4}$$

is calculated as the Jacobian matrix[3] of the equation (7.2) evaluated at $(x, y, z) = (0, 0, 0)$. On substituting the numerical values in (7.1) into (7.4) we find that the matrix takes on the following very special form

$$K = \frac{EA}{L^3} \begin{bmatrix} 3/2 & 0 & 0 \\ 0 & 3/2 & 0 \\ 0 & 0 & 12 \end{bmatrix}, \tag{7.5}$$

[3]See (7.25) in Sec. 7.2.3 for the definition.

which is a diagonal matrix with repeated diagonal entries. Actually, this special form arises as a result of the geometrical symmetry of the structure, and the objective of this chapter is to explain the general mechanism thereof. As a preview, the general framework to be presented in this chapter is illustrated below for this example.

This truss tent has rotation symmetry around the $z$-axis at 120 degrees and reflection symmetry with respect to the $xz$-plane. The rotation and reflection are expressed through matrices

$$R = \begin{bmatrix} \cos(2\pi/3) & -\sin(2\pi/3) & 0 \\ \sin(2\pi/3) & \cos(2\pi/3) & 0 \\ 0 & 0 & 1 \end{bmatrix}, \qquad S = \begin{bmatrix} 1 & 0 & 0 \\ 0 & -1 & 0 \\ 0 & 0 & 1 \end{bmatrix}, \tag{7.6}$$

and the stiffness matrix $K$ in (7.5) satisfies

$$RK = KR, \qquad SK = KS, \tag{7.7}$$

which can be verified easily. Conversely, we can show (see Remark 7.1) that a matrix $K$ that satisfies (7.7) must take the following form:

$$K = \begin{bmatrix} \kappa & 0 & 0 \\ 0 & \kappa & 0 \\ 0 & 0 & \lambda \end{bmatrix} \tag{7.8}$$

for some $\kappa$ and $\lambda$.

The relation in (7.7) indicates the *commutativity* of the matrices $R$ and $S$ with the matrix $K$, the matrices $R$ and $S$ representing the transformations that keep the system invariant and the matrix $K$ describing the system in question. In general, the symmetry of a system is expressed in terms of the commutativity of the matrices for transformations that keep the system invariant with a matrix, say, $K$ for a system description; more details will be expounded in Sec. 7.2.3. From this commutativity follows the crucial fact that the matrix $K$ describing the system can be brought to a block-diagonal form by means of a transformation of the form $Q^\top KQ$ using an orthogonal matrix $Q$ (in the above example, $Q$ is the identity matrix).

**Remark 7.1.** Let us derive the particular form (7.8) of the matrix $K$ from the commutativity in (7.7). We put $K = (k_{ij} \mid i, j = 1, 2, 3)$ and substitute the concrete expressions of $R$ and $S$ in (7.6) into (7.7). From the relation $SK = KS$ in (7.7), we obtain

$$\begin{bmatrix} k_{11} & k_{12} & k_{13} \\ -k_{21} & -k_{22} & -k_{23} \\ k_{31} & k_{32} & k_{33} \end{bmatrix} = \begin{bmatrix} k_{11} & -k_{12} & k_{13} \\ k_{21} & -k_{22} & k_{23} \\ k_{31} & -k_{32} & k_{33} \end{bmatrix}.$$

Therefore, $k_{12} = k_{21} = k_{23} = k_{32} = 0$, that is,

$$K = \begin{bmatrix} k_{11} & 0 & k_{13} \\ 0 & k_{22} & 0 \\ k_{31} & 0 & k_{33} \end{bmatrix}.$$

Then, by using this in $RK = KR$, we obtain

$$\begin{bmatrix} \alpha \cdot k_{11} & -\beta \cdot k_{22} & \alpha \cdot k_{13} \\ \beta \cdot k_{11} & \alpha \cdot k_{22} & \beta \cdot k_{13} \\ k_{31} & 0 & k_{33} \end{bmatrix} = \begin{bmatrix} \alpha \cdot k_{11} & -\beta \cdot k_{11} & k_{13} \\ \beta \cdot k_{22} & \alpha \cdot k_{22} & 0 \\ \alpha \cdot k_{31} & -\beta \cdot k_{31} & k_{33} \end{bmatrix},$$

where $\alpha = \cos(2\pi/3)$ and $\beta = \sin(2\pi/3)$. This yields $k_{13} = k_{31} = 0$ and $k_{11} = k_{22}$. By setting $\kappa = k_{11} = k_{22}$ and $\lambda = k_{33}$, we obtain the expression in (7.8). ∎

In the above, we have illustrated the basic idea for a simple example. The objective of this chapter is to explain the general mechanism that underlies this kind of diagonalization (or block-diagonalization). In Sec. 7.2 we first explain how to describe symmetry in terms of groups, and then show mathematical properties of group representations in Sec. 7.3. A detailed account of block-diagonalization of matrices possessing group symmetry is given in Sec. 7.4.

## 7.2 Symmetry and Groups

The figures shown in Fig. 7.4 possess symmetry with respect to reflection and rotation. If we contemplate upon the meaning of symmetry, we would realize that symmetry can be captured as invariance to operations (transformations) such as rotation and reflection. The set of all operations that keep a given figure invariant forms an algebraic structure called a group. The concept of symmetry, as recognized by geometrical intuition, can be described mathematically by an algebraic object called a group.

### 7.2.1 *Groups*

Let us provide the definition and examples of a group.[4] Suppose that a *binary operation* "·" called *product* is defined on a set $G$. That is, we assume that, for every ordered pair $(g, h)$ of elements $g$ and $h$ of $G$, an element of $G$, denoted by $g \cdot h$, is specified as the product of $g$ and $h$. If this operation satisfies the following three conditions, then $G$ is called a *group*.

[4]See, *e.g.*, [60, 61] for details.

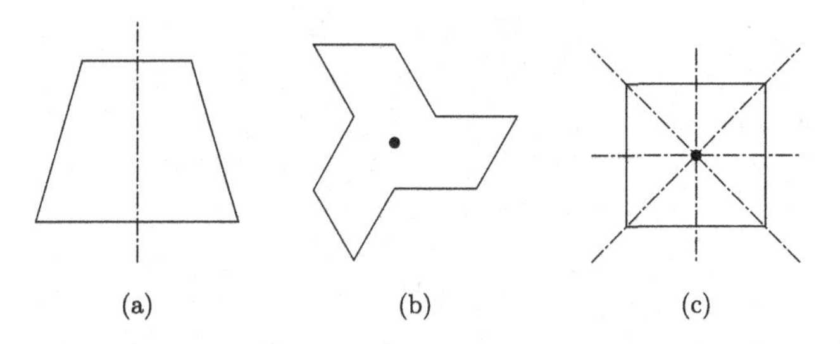


Fig. 7.4 Planar figures with symmetry (−·− : axis of reflection; ●: center of rotation).

(1) For any $g, h, k \in G$, it holds that

$$(g \cdot h) \cdot k = g \cdot (h \cdot k).$$

This is called the *associative law.*

(2) There exists an element $e \in G$ such that

$$e \cdot g = g \cdot e = g$$

for every $g \in G$. This element $e$ is called the *identity element.*

(3) For every $g \in G$ there exists an element $h \in G$ such that

$$g \cdot h = h \cdot g = e.$$

This element $h$ is called the *inverse* of $g$ and is denoted as $g^{-1}$.

From the above three conditions we can derive the facts that the identity element $e$ is unique, and that for each $g \in G$ the corresponding inverse element is uniquely determined by $g$. It is quite common to omit the symbol "$\cdot$" for product operation and write $g \cdot h$ simply as $gh$.

The set $G$ can be either a finite set or an infinite set, but in this book we will focus on the finite case. A group that is a finite set is called a *finite group.* The number of elements of a finite group $G$ is called the *order* of $G$, and is denoted as $|G|$.

The structure of a group $G$ can be captured easily if it is presented in the form of a *multiplication table.* A multiplication table is a table such as

| | $\cdots$ $h$ $\cdots$ |
|---|---|
| $\vdots$ | |
| $g$ | $gh$ |
| $\vdots$ | |

wherein the rows and columns correspond to the elements of $G$ and the value of the product $gh$ is given as the entry in the row of $g$ and the column of $h$.

It is often the case that, with some suitably selected elements $g_1, \ldots, g_k$ of group $G$, every element of $G$ can be obtained as a product formed by the selected elements $g_1, \ldots, g_k$ and their inverses $g_1^{-1}, \ldots, g_k^{-1}$. If this is the case, $g_1, \ldots, g_k$ are said to *generate* $G$. We also say that $g_1, \ldots, g_k$ are the *generators* of $G$ and express this as

$$G = \langle g_1, \ldots, g_k \rangle.$$

**Example 7.1.** Consider a group $G$ that is composed of two elements. One of the elements of $G$ is the identity element $e$. Let $s$ denote the other element. Then $s \cdot s$ is an element of $G$. If it is equal to $s$, then we can multiply the inverse of $s$ from the right of $s \cdot s = s$ to obtain $s = e$, which is a contradiction. Therefore, $s \cdot s = e$ and the multiplication table is as follows:

| | $e$ | $s$ |
|---|---|---|
| $e$ | $e$ | $s$ |
| $s$ | $s$ | $e$ |

We can interpret the element $s$ as representing a reflection operation in a plane. The identity element $e$ represents the operation that does nothing, and the relation $s \cdot s = e$ represents the fact that repeating the reflection operation twice is the same as doing nothing. With this interpretation, the group described by the above multiplication table is commonly denoted as $\mathrm{D}_1$ (see Example 7.3 below). The figure shown in Fig. 7.4(a) possesses the symmetry of the group $\mathrm{D}_1$.

Another interpretation is possible for the element $s$. If we make $s$ correspond to a 180 degree rotation operation centered at the origin in a plane, then the relation $s \cdot s = e$ remains true. With this interpretation, the group described by the above multiplication table is commonly denoted as $\mathrm{C}_2$ (see Example 7.2 below).

We have explained that two different geometrical interpretations, reflection and half-rotation, can be given to $s$. A different type of interpretation is also possible, that is, we may interpret $s$ as representing the permutation of two elements.

In this way, a group, which is defined by a multiplication table as an abstract mathematical structure, admits a variety of interpretations for concrete operations. In applications, groups are hardly ever given in an abstract form. Instead, we encounter a set of elements representing certain concrete operations and recognize the algebraic structure of a group by investigating the properties of those concrete operations. ∎

**Example 7.2.** Let $n$ be a natural number and $r$ be the $360/n$ degree ($2\pi/n$ radian) rotation operation around the origin in a plane. Then the operation $r^k$ represents a rotation of $2\pi k/n$ radian, where $k$ is the number of times the operation $r$ is performed. If we denote by $e = r^0$ the operation that does nothing, then $r^n = e$ holds. The set of these rotation operations forms a group by way of the product operation $r^i \cdot r^j = r^{i+j}$ and is expressed as

$$\mathrm{C}_n = \langle r \rangle = \{e, r, r^2, \ldots, r^{n-1}\}. \tag{7.9}$$

This group $\mathrm{C}_n$ is called the *cyclic group* of degree $n$. For example, the multiplication table of $\mathrm{C}_3$ is

| | $e$ | $r$ | $r^2$ |
|---|---|---|---|
| $e$ | $e$ | $r$ | $r^2$ |
| $r$ | $r$ | $r^2$ | $e$ |
| $r^2$ | $r^2$ | $e$ | $r$ |

(7.10)

and the figure shown in Fig. 7.4(b) possesses the symmetry of $\mathrm{C}_3$. ■

**Example 7.3.** As in Example 7.2, let $n$ be a natural number and $r$ be the $360/n$ degree ($2\pi/n$ radian) rotation operation around the origin in a plane, with $e$ expressing the operation that does nothing. In addition, let $s$ denote the reflection operation with respect to a line that passes through the origin. Rotation $r$ and reflection $s$ satisfy a relation

$$s \cdot r \cdot s \cdot r = e, \tag{7.11}$$

where the left-hand side represents the operation in which rotation $r$ is first executed, then reflection $s$, rotation $r$ and then lastly reflection $s$ is executed. In addition, as already stated, we have

$$r^i \cdot r^j = r^{i+j}, \qquad r^n = s \cdot s = e. \tag{7.12}$$

With these rules (or relations) we obtain a group

$$\mathrm{D}_n = \langle r, s \rangle = \{e, r, \ldots, r^{n-1}, s, sr, \ldots, sr^{n-1}\}, \tag{7.13}$$

which is called the *dihedral group* of degree $n$. For example, the multiplication table of $\mathrm{D}_3$ is given by

| | $e$ | $r$ | $r^2$ | $s$ | $sr$ | $sr^2$ |
|---|---|---|---|---|---|---|
| $e$ | $e$ | $r$ | $r^2$ | $s$ | $sr$ | $sr^2$ |
| $r$ | $r$ | $r^2$ | $e$ | $sr^2$ | $s$ | $sr$ |
| $r^2$ | $r^2$ | $e$ | $r$ | $sr$ | $sr^2$ | $s$ |
| $s$ | $s$ | $sr$ | $sr^2$ | $e$ | $r$ | $r^2$ |
| $sr$ | $sr$ | $sr^2$ | $s$ | $r^2$ | $e$ | $r$ |
| $sr^2$ | $sr^2$ | $s$ | $sr$ | $r$ | $r^2$ | $e$ |

. (7.14)

The upper left part of this multiplication table coincides with the multiplication table of $C_3$ shown in (7.10). The truss tent of Fig. 7.3 possesses the symmetry of $D_3$, where $r$ is interpreted as the 120 degree rotation around the $z$-axis and $s$ as the reflection with respect to the $xz$-plane. The figure shown in Fig. 7.4(c) possesses the symmetry of $D_4$. ∎

**Example 7.4.** For a natural number $n \geqq 1$, the set of all permutations of $\{1, 2, \ldots, n\}$ forms a group, with the product operation defined by the composition of permutations. This is called the *symmetric group* of degree $n$, and is commonly denoted as $S_n$. The order of $S_n$ is equal to $n!$. ∎

### 7.2.2 *Group Representation*

Suppose that, for each element $g$ of a group $G$, a nonsingular matrix $T(g)$ is associated, where the size of the matrix $T(g)$ is assumed to be the same for all $g$. The mapping (or correspondence) $T : g \mapsto T(g)$ is called a *representation* of the group $G$, if it satisfies the condition[5]

$$T(gh) = T(g)T(h) \qquad (g, h \in G). \tag{7.15}$$

The (common) size of the matrices $T(g)$, which we denote by $N$, is called the *degree* or *dimension* of the representation $T$. Thus, a representation of degree $N$ of a group $G$ is a collection $(T(g) \mid g \in G)$ of nonsingular matrices $T(g)$ of order $N$ that satisfy the condition (7.15). Each matrix $T(g)$ is called a *representation matrix.*

If we set $h = e$ (identity element) in (7.15), we obtain $T(g) = T(g)T(e)$, which implies

$$T(e) = I \quad \text{(unit matrix)} \tag{7.16}$$

since $T(g)$ is nonsingular. If we set $h = g^{-1}$ (inverse element of $g$) in (7.15), we obtain $T(e) = T(g)T(g^{-1})$, which implies

$$T(g^{-1}) = T(g)^{-1} \qquad (g \in G) \tag{7.17}$$

since $T(e) = I$.

If the group $G$ is generated by $g_1, \ldots, g_k \in G$, the matrices $T(g)$ for all $g \in G$ are determined from the matrices $T(g_1), \ldots, T(g_k)$ for the generators by the condition (7.15). For example, for an element $g$ of $G$ represented as $g = g_2^{-1} g_1{}^3$, we have $T(g) = T(g_2)^{-1} T(g_1)^3$. The condition (7.15) also

[5]In mathematical terminology, a representation of a group $G$ is a *homomorphism* from $G$ to $GL(N)$ for some $N$, where $GL(N)$ denotes the set of all nonsingular matrices of order $N$.

guarantees that value of $T(g)$ calculated in this way is determined uniquely by $g$ without depending on how the element $g$ is represented in terms of the generators.

When discussing group representations, it is important to clearly specify the set of numbers from which the entries of the representation matrices $T(g)$ are chosen. Usually we fix a *field*, say $F$, in which the four arithmetic operations (addition, subtraction, multiplication, and division) are defined, and assume that the entries of the matrices all belong to $F$. Part of the theory can possibly vary depending on the chosen field $F$. In this book we will explain the fundamental issues of the theory for $F = \mathbb{R}$ (field of real numbers) or $F = \mathbb{C}$ (field of complex numbers).

A representation $T$ is called a *unitary representation* if, for every $g \in G$, the representation matrix $T(g)$ is a unitary matrix, *i.e.*, if

$$T(g)^* T(g) = I \qquad (g \in G) \tag{7.18}$$

holds. Here, the symbol " * " represents the conjugate transpose of a matrix. As will be discussed later in Proposition 7.1 in Sec. 7.3.1, we can assume, without loss of generality, that $T$ is a unitary representation. When $F = \mathbb{R}$, the condition (7.18) takes the form of

$$T(g)^\top T(g) = I \qquad (g \in G). \tag{7.19}$$

In this case, $T$ is often referred to as an *orthogonal representation.*

**Example 7.5.** For every group $G$ we can obtain a one-dimensional representation $T$ by defining $T(g) = 1$ for all $g \in G$. This is called the *unit representation.* The unit representation is the same for all groups, and as such it does not reflect the characteristic structure of the group. But it plays an important role, *e.g.*, in considering irreducible decompositions, which we will explain later. ■

**Example 7.6.** For the group $\mathrm{D}_1 = \{e, s\}$ shown in Example 7.1, we can obtain a one-dimensional representation $T$ by defining

$$T(e) = 1, \qquad T(s) = -1.$$

The condition (7.15) is verified as follows:

$$\begin{aligned}
T(e)T(e) &= 1 \times 1 = 1 = T(e), \\
T(e)T(s) &= 1 \times (-1) = -1 = T(s) = T(es), \\
T(s)T(e) &= (-1) \times 1 = -1 = T(s) = T(se), \\
T(s)T(s) &= (-1) \times (-1) = 1 = T(e) = T(ss).
\end{aligned}$$

■

**Example 7.7.** For the group $\mathrm{D}_3 = \{e, r, r^2, s, sr, sr^2\}$ shown in Example 7.3, we can obtain a two-dimensional representation $T$ by defining

$$T(e) = \begin{bmatrix} 1 & 0 \\ 0 & 1 \end{bmatrix}, \quad T(r) = \begin{bmatrix} \alpha & -\beta \\ \beta & \alpha \end{bmatrix}, \quad T(r^2) = \begin{bmatrix} \alpha & \beta \\ -\beta & \alpha \end{bmatrix},$$

$$T(s) = \begin{bmatrix} 1 & 0 \\ 0 & -1 \end{bmatrix}, \quad T(sr) = \begin{bmatrix} \alpha & -\beta \\ -\beta & -\alpha \end{bmatrix}, \quad T(sr^2) = \begin{bmatrix} \alpha & \beta \\ \beta & -\alpha \end{bmatrix},$$

where $\alpha = \cos(2\pi/3)$ and $\beta = \sin(2\pi/3)$. We can easily verify that this $T$ satisfies the condition (7.15). Since the group $\mathrm{D}_3$ is generated by the two elements $r$ and $s$, all the matrices $T(g)$ for $g \in \mathrm{D}_3$ are determined by $T(r)$ and $T(s)$. For example, $T(sr^2) = T(s)T(r)^2$. ■

**Example 7.8.** Suppose that a permutation $\pi(g)$ of a set $P = \{1, 2, \ldots, N\}$ is associated with each element $g$ of a group $G$, and assume that

$$\pi(gh) = \pi(g)\pi(h) \qquad (g, h \in G) \tag{7.20}$$

holds true, where $\pi(g)\pi(h)$ denotes the permutation generated by first executing permutation $\pi(h)$ and then permutation $\pi(g)$. For each $g \in G$ let $T(g)$ be the *permutation matrix* that represents the permutation $\pi(g)$. That is, $T(g)$ is a matrix of order $N$ such that

$$(i,j) \text{ entry of } T(g) = \begin{cases} 1 & \text{if } \pi(g) \text{ moves } j \text{ to } i, \\ 0 & \text{otherwise.} \end{cases}$$

The family of matrices $T(g)$ defined in this manner satisfies the condition (7.15) and gives a representation of $G$. This is called a *permutation representation.*

In the truss tent of Fig. 7.3, for example, we have $G = \mathrm{D}_3 = \langle r, s \rangle$ with $r$ being the 120 degree rotation around the $z$-axis and $s$ the reflection with respect to the $xz$-plane. Permutations on the set $P = \{1, 2, 3\}$ of node numbers are caused by $r$ and $s$, and the corresponding permutation representation is given by

$$T(e) = \begin{bmatrix} 1 & 0 & 0 \\ 0 & 1 & 0 \\ 0 & 0 & 1 \end{bmatrix}, \quad T(r) = \begin{bmatrix} 0 & 0 & 1 \\ 1 & 0 & 0 \\ 0 & 1 & 0 \end{bmatrix}, \quad T(r^2) = \begin{bmatrix} 0 & 1 & 0 \\ 0 & 0 & 1 \\ 1 & 0 & 0 \end{bmatrix},$$

$$T(s) = \begin{bmatrix} 1 & 0 & 0 \\ 0 & 0 & 1 \\ 0 & 1 & 0 \end{bmatrix}, \quad T(sr) = \begin{bmatrix} 0 & 0 & 1 \\ 0 & 1 & 0 \\ 1 & 0 & 0 \end{bmatrix}, \quad T(sr^2) = \begin{bmatrix} 0 & 1 & 0 \\ 1 & 0 & 0 \\ 0 & 0 & 1 \end{bmatrix}.$$

This is a three-dimensional representation of $\mathrm{D}_3$. ■

**Example 7.9.** Let us enumerate the elements of a group $G$ in an arbitrary order, say, $g_1, g_2, \ldots, g_N$, where $N = |G|$. If we multiply $g_1, g_2, \ldots, g_N$ by $g \in G$ from the left, we obtain a permutation of $g_1, g_2, \ldots, g_N$. That is, each $g \in G$ determines a permutation $\pi(g)$ on $P = \{1, 2, \ldots, N\}$ according to $gg_j = g_{\pi(g)j}$ $(j = 1, 2, \ldots, N)$. The permutation representation $T$ induced from this permutation $\pi$ as in Example 7.8 is called the *regular representation* of $G$.

As an example, let us consider the regular representation $T$ of $\mathrm{D}_3$. If we enumerate the elements of $\mathrm{D}_3$ as

$$g_1 = e, \quad g_2 = r, \quad g_3 = r^2, \quad g_4 = s, \quad g_5 = sr, \quad g_6 = sr^2,$$

we have

$$\begin{aligned} &rg_1 = g_2, \quad rg_2 = g_3, \quad rg_3 = g_1, \quad rg_4 = g_6, \quad rg_5 = g_4, \quad rg_6 = g_5; \\ &sg_1 = g_4, \quad sg_2 = g_5, \quad sg_3 = g_6, \quad sg_4 = g_1, \quad sg_5 = g_2, \quad sg_6 = g_3 \end{aligned}$$

by the multiplication table (7.14). Therefore,

$$T(r) = \begin{bmatrix} 0 & 0 & 1 & 0 & 0 & 0 \\ 1 & 0 & 0 & 0 & 0 & 0 \\ 0 & 1 & 0 & 0 & 0 & 0 \\ 0 & 0 & 0 & 0 & 1 & 0 \\ 0 & 0 & 0 & 0 & 0 & 1 \\ 0 & 0 & 0 & 1 & 0 & 0 \end{bmatrix}, \qquad T(s) = \begin{bmatrix} 0 & 0 & 0 & 1 & 0 & 0 \\ 0 & 0 & 0 & 0 & 1 & 0 \\ 0 & 0 & 0 & 0 & 0 & 1 \\ 1 & 0 & 0 & 0 & 0 & 0 \\ 0 & 1 & 0 & 0 & 0 & 0 \\ 0 & 0 & 1 & 0 & 0 & 0 \end{bmatrix}.$$

Since $\mathrm{D}_3$ is generated by $r$ and $s$, the matrices $T(g)$ for all $g \in \mathrm{D}_3$ are determined by $T(r)$ and $T(s)$. For example, $T(sr^2) = T(s)T(r)^2$.

Regular representation is an important tool in theoretical development, and, in this book, it is used in the proof of Proposition 7.10 in Sec. 7.5 and in Example 7.29. ■

### 7.2.3 *Symmetry of Systems*

We now explain how to formulate the group symmetry possessed by a (physical or engineering) system as the properties of the equation describing the system. Suppose that a system in question is described by the following equation:

$$\boldsymbol{F}(\boldsymbol{u}) = \boldsymbol{0}. \tag{7.21}$$

Here, $\boldsymbol{u} \in \mathbb{R}^N$ is an $N$-dimensional vector that expresses the state of the system and $\boldsymbol{F} : \mathbb{R}^N \to \mathbb{R}^N$ is a (sufficiently smooth) function. For example, for the truss tent treated in Sec. 7.1.2 (see Fig. 7.3 and (7.2)), we have $\boldsymbol{u} = (x, y, z)^\top$ and

$$\boldsymbol{F}(\boldsymbol{u}) = \begin{bmatrix} F_1 \\ F_2 \\ F_3 \end{bmatrix} = \sum_{i=1}^{3} EA\left(\frac{1}{L} - \frac{1}{\hat{L}_i}\right) \begin{bmatrix} x - x_i \\ y - y_i \\ z - z_i \end{bmatrix} - \begin{bmatrix} 0 \\ 0 \\ f \end{bmatrix}. \tag{7.22}$$

The fact that the system described by (7.21) possesses the symmetry of a group $G$ can be formulated as the condition

$$T(g)\boldsymbol{F}(\boldsymbol{u}) = \boldsymbol{F}(T(g)\boldsymbol{u}) \qquad (g \in G) \tag{7.23}$$

on the function $\boldsymbol{F}(\boldsymbol{u})$. Here, $T(g)$ is an $N$-dimensional representation[6] of the group $G$. The property (7.23) is called the *equivariance* or *covariance* of the function $\boldsymbol{F}(\boldsymbol{u})$ to the group $G$. If the condition (7.23) is satisfied by the generators $g_1, \ldots, g_k$ of $G$, then[7] it is satisfied by all $g \in G$.

In the case of the truss tent, we have

$$G = \mathrm{D}_3 = \langle r, s \rangle = \{e, r, r^2, s, sr, sr^2\}.$$

Since $r$ represents a 120 degree rotation around the $z$-axis and $s$ is the reflection with respect to the $xz$-plane, we have

$$T(r) = \begin{bmatrix} \cos(2\pi/3) & -\sin(2\pi/3) & 0 \\ \sin(2\pi/3) & \cos(2\pi/3) & 0 \\ 0 & 0 & 1 \end{bmatrix}, \quad T(s) = \begin{bmatrix} 1 & 0 & 0 \\ 0 & -1 & 0 \\ 0 & 0 & 1 \end{bmatrix}. \tag{7.24}$$

For the function $\boldsymbol{F}(\boldsymbol{u})$ in (7.22), we have

$$T(r)\boldsymbol{F}(\boldsymbol{u}) = \boldsymbol{F}(T(r)\boldsymbol{u}), \qquad T(s)\boldsymbol{F}(\boldsymbol{u}) = \boldsymbol{F}(T(s)\boldsymbol{u}).$$

We note that, to verify this, we need to refer to the concrete values of the coordinates of the nodes given in (7.1).

The *Jacobian matrix* of equation $\boldsymbol{F}(\boldsymbol{u})$ is an $N \times N$ matrix defined by

$$J(\boldsymbol{u}) = \left(\frac{\partial F_i}{\partial u_j} \,\middle|\, i, j = 1, \ldots, N\right) = \begin{bmatrix} \dfrac{\partial F_1}{\partial u_1} & \cdots & \dfrac{\partial F_1}{\partial u_N} \\ \vdots & & \vdots \\ \dfrac{\partial F_N}{\partial u_1} & \cdots & \dfrac{\partial F_N}{\partial u_N} \end{bmatrix}. \tag{7.25}$$

[6] It should be clear that $T$ is a representation over $\mathbb{R}$. The complex case, with $\boldsymbol{u} \in \mathbb{C}^N$ and $\boldsymbol{F} : \mathbb{C}^N \to \mathbb{C}^N$, can be treated similarly with a representation $T$ over $\mathbb{C}$.

[7] If the condition (7.23) holds for $g = g_1, g_2 \in G$, then we obtain $T(g_1 g_2)\boldsymbol{F}(\boldsymbol{u}) = T(g_1)T(g_2)\boldsymbol{F}(\boldsymbol{u}) = T(g_1)\boldsymbol{F}(T(g_2)\boldsymbol{u}) = \boldsymbol{F}(T(g_1)T(g_2)\boldsymbol{u}) = \boldsymbol{F}(T(g_1 g_2)\boldsymbol{u})$ by using the property (7.15). By extending this argument we can show that the condition (7.23) for all generators $g_1, \ldots, g_k$ implies the condition (7.23) for every element $g$ of $G$.

By differentiating the equation (7.23) with respect to $\boldsymbol{u}$, we obtain

$$T(g)J(\boldsymbol{u}) = J(T(g)\boldsymbol{u})T(g) \qquad (g \in G), \tag{7.26}$$

which expresses the symmetry (equivariance) of the Jacobian matrix $J(\boldsymbol{u})$. In particular, if $\boldsymbol{u}$ possesses the symmetry of $G$, *i.e.*, if

$$T(g)\boldsymbol{u} = \boldsymbol{u} \qquad (g \in G), \tag{7.27}$$

then

$$T(g)J(\boldsymbol{u}) = J(\boldsymbol{u})T(g) \qquad (g \in G). \tag{7.28}$$

This equation shows that the Jacobian matrix $J(\boldsymbol{u})$ and each representation matrix $T(g)$ commute. In this way, the symmetry of a linearized system (or a linear system) can be expressed as the *commutativity* of the Jacobian matrix $J(\boldsymbol{u})$ with the representation matrices $T(g)$.

In the case of the truss tent, considering a linearized system corresponds to considering infinitesimal deformation, and the Jacobian matrix $J(\boldsymbol{u})$ evaluated at $\boldsymbol{u} = \boldsymbol{0}$ is the stiffness matrix $K$ in (7.5). The commutativity (7.28) for the generators $r$ and $s$ is written as

$$T(r)K = KT(r), \qquad T(s)K = KT(s) \tag{7.29}$$

using the representation matrices $T(r)$ and $T(s)$ in (7.24). Since $T(r)$ and $T(s)$ are equal to $R$ and $S$ in (7.6), respectively, (7.29) expresses the same conditions as (7.7) does.

Let us consider the symmetry of a potential system.[8] Recall that a system (7.21) of equations is called a *potential system*, if the function $\boldsymbol{F} : \mathbb{R}^N \to \mathbb{R}^N$ in (7.21) is given as

$$\boldsymbol{F} = \left( \frac{\partial U}{\partial u_1}, \ldots, \frac{\partial U}{\partial u_N} \right)^\top \tag{7.30}$$

in terms of a scalar function $U : \mathbb{R}^N \to \mathbb{R}$. The function $U(\boldsymbol{u})$ is called a *potential function*. The fact that a potential system possesses the symmetry of a group $G$ can be mathematically formulated as the condition that the potential function possesses *invariance* to $G$, which is expressed as

$$U(\boldsymbol{u}) = U(T(g)\boldsymbol{u}) \qquad (g \in G). \tag{7.31}$$

By differentiating this equation with respect to $\boldsymbol{u}$ and using (7.30), we obtain

$$\boldsymbol{F}(\boldsymbol{u}) = T(g)^\top \boldsymbol{F}(T(g)\boldsymbol{u}) \qquad (g \in G).$$

[8] All the truss structures appearing in this chapter are potential systems.

If the representation $T$ is orthogonal, we have $T(g)^\top = T(g)^{-1}$, and therefore the above equation can be rewritten as

$$\boldsymbol{F}(\boldsymbol{u}) = T(g)^{-1}\boldsymbol{F}(T(g)\boldsymbol{u}) \qquad (g \in G).$$

This is equivalent to the equivariance of function $\boldsymbol{F}(\boldsymbol{u})$ given in (7.23).

**Example 7.10.** Using the $\mathrm{D}_3$-symmetric truss in Fig. 7.5 as an example, we explain here how to derive the equivariance of the governing equation from the invariance of the potential function. This truss has three fixed nodes (indicated by • in the figure) and four free nodes 0, 1, 2, and 3 (indicated by ∘). Assume that the cross-sectional area and stiffness of all members are equal and that a $\mathrm{D}_3$-symmetric force is being exerted (for example, to node 0 in the vertical downward direction). Let $(x_i, y_i, z_i)$ denote the positions of nodes $i = 0, 1, 2, 3$ after deformation, and put

$$\boldsymbol{u}_0 = (x_0, y_0, z_0)^\top, \quad \boldsymbol{u}_1 = (x_1, y_1, z_1)^\top,$$
$$\boldsymbol{u}_2 = (x_2, y_2, z_2)^\top, \quad \boldsymbol{u}_3 = (x_3, y_3, z_3)^\top.$$

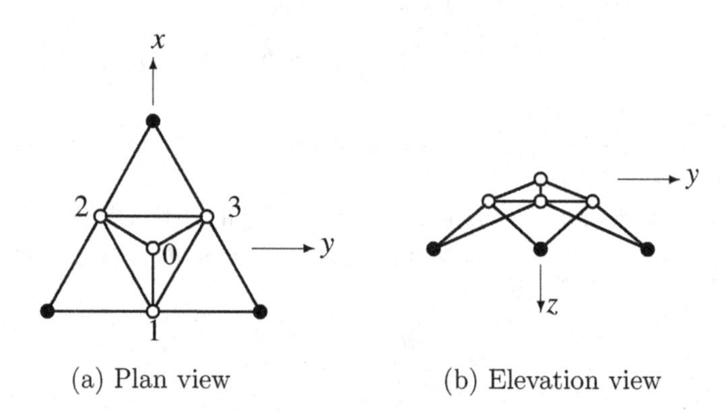


Fig. 7.5 A truss structure with regular triangular symmetry.

This truss structure is endowed with the symmetry of $r$, the 120 degree rotation around the $z$-axis, and $s$, the reflection with respect to the $xz$-plane. Let us express this fact as the properties of the potential function[9] $U(\boldsymbol{u}_0, \boldsymbol{u}_1, \boldsymbol{u}_2, \boldsymbol{u}_3)$.

The invariance of the potential function to rotation $r$ is expressed as

$$U(\boldsymbol{u}_0, \boldsymbol{u}_1, \boldsymbol{u}_2, \boldsymbol{u}_3) = U(R\boldsymbol{u}_0, R\boldsymbol{u}_3, R\boldsymbol{u}_1, R\boldsymbol{u}_2) \tag{7.32}$$

[9] If we define $\boldsymbol{u}$ as the 12-dimensional vector formed by vertically arranging vectors $\boldsymbol{u}_0$, $\boldsymbol{u}_1$, $\boldsymbol{u}_2$, and $\boldsymbol{u}_3$, then $U(\boldsymbol{u}_0, \boldsymbol{u}_1, \boldsymbol{u}_2, \boldsymbol{u}_3)$ is exactly in the form of $U(\boldsymbol{u})$.

with

$$R = \begin{bmatrix} \cos(2\pi/3) & -\sin(2\pi/3) & 0 \\ \sin(2\pi/3) & \cos(2\pi/3) & 0 \\ 0 & 0 & 1 \end{bmatrix}.$$

In regard to node 0, we have $U(\boldsymbol{u}_0, \cdots)$ on the left-hand side and $U(R\boldsymbol{u}_0, \cdots)$ on the right-hand side, where $R\boldsymbol{u}_0$ expresses the fact that $r$ causes a rotation in the $(x, y)$-coordinates without affecting the $z$-coordinate of node 0. For nodes 1, 2, and 3, we have $U(\cdot, \boldsymbol{u}_1, \boldsymbol{u}_2, \boldsymbol{u}_3)$ and $U(\cdot, R\boldsymbol{u}_3, R\boldsymbol{u}_1, R\boldsymbol{u}_2)$, the latter of which shows that a permutation of the node numbers is caused in addition to the rotation in the $(x, y)$-coordinates.

The invariance of the potential function to reflection $s$ is expressed as

$$U(\boldsymbol{u}_0, \boldsymbol{u}_1, \boldsymbol{u}_2, \boldsymbol{u}_3) = U(S\boldsymbol{u}_0, S\boldsymbol{u}_1, S\boldsymbol{u}_3, S\boldsymbol{u}_2) \tag{7.33}$$

with

$$S = \begin{bmatrix} 1 & 0 & 0 \\ 0 & -1 & 0 \\ 0 & 0 & 1 \end{bmatrix}.$$

In regard to nodes 0 and 1, we have $U(\boldsymbol{u}_0, \boldsymbol{u}_1, \cdots)$ on the left-hand side and $U(S\boldsymbol{u}_0, S\boldsymbol{u}_1, \cdots)$ on the right-hand side, since the inversion $y \mapsto -y$ of the $y$-coordinate is caused by $s$ whereas the $x$- and $z$-coordinates are not affected. For nodes 2 and 3, we have $U(\cdots, \boldsymbol{u}_2, \boldsymbol{u}_3)$ and $U(\cdots, S\boldsymbol{u}_3, S\boldsymbol{u}_2)$, since, in addition to the inversion of the $y$-coordinate, an exchange of the node numbers of 2 and 3 is caused.

Define $\boldsymbol{u}$ to be the 12-dimensional vector formed by vertically arranging vectors $\boldsymbol{u}_0$, $\boldsymbol{u}_1$, $\boldsymbol{u}_2$, and $\boldsymbol{u}_3$, and let

$$T(r) = \left[\begin{array}{c|c|c|c} R & & & \\ \hline & & & R \\ \hline & R & & \\ \hline & & R & \end{array}\right], \qquad T(s) = \left[\begin{array}{c|c|c|c} S & & & \\ \hline & S & & \\ \hline & & & S \\ \hline & & S & \end{array}\right]. \tag{7.34}$$

Then we can rewrite the equations (7.32) and (7.33) as

$$U(\boldsymbol{u}) = U(T(r)\boldsymbol{u}), \qquad U(\boldsymbol{u}) = U(T(s)\boldsymbol{u}). \tag{7.35}$$

A 12-dimensional representation $T$ of the group $\mathrm{D}_3$ is determined by (7.34), since the group $\mathrm{D}_3$ is generated by $r$ and $s$. The equation (7.35) gives a condition that is equivalent to the invariance of potential function $U(\boldsymbol{u})$ to $\mathrm{D}_3$, which reads

$$U(\boldsymbol{u}) = U(T(g)\boldsymbol{u}) \qquad (g \in \mathrm{D}_3).$$

■

**Remark 7.2.** In the equivariance condition (7.23) for the symmetry of a system, the action on $\boldsymbol{u}$ and the action on the value of $\boldsymbol{F}$ are described by the same representation $T$. Let us dwell on this point. Since the space of $\boldsymbol{u}$ and the space of the values of $\boldsymbol{F}$ are unrelated in general, it seems more natural to use different representations, $T_1$ and $T_2$, defined on the respective spaces, and to formulate the equivariance as

$$T_2(g)\boldsymbol{F}(\boldsymbol{u}) = \boldsymbol{F}(T_1(g)\boldsymbol{u}) \qquad (g \in G). \tag{7.36}$$

By differentiating (7.36) with respect to $\boldsymbol{u}$, we obtain

$$T_2(g)J(\boldsymbol{u}) = J(T_1(g)\boldsymbol{u})T_1(g) \qquad (g \in G). \tag{7.37}$$

This means that, for a state $\boldsymbol{u}$ satisfying the symmetry condition

$$T_1(g)\boldsymbol{u} = \boldsymbol{u} \qquad (g \in G), \tag{7.38}$$

we have

$$T_2(g)J(\boldsymbol{u}) = J(\boldsymbol{u})T_1(g) \qquad (g \in G). \tag{7.39}$$

Usually, in engineering systems, the Jacobian matrix $J(\boldsymbol{u})$ is nonsingular at some state $\boldsymbol{u} = \boldsymbol{u}_0$ with the symmetry (7.38) (in Example 7.10, for example, we can take the undeformed state as $\boldsymbol{u}_0$). With the notation $Q = J(\boldsymbol{u}_0)$ it follows from (7.39) that

$$T_2(g) = QT_1(g)Q^{-1} \qquad (g \in G).$$

Using this in (7.36) we obtain

$$T_1(g)Q^{-1}\boldsymbol{F}(\boldsymbol{u}) = Q^{-1}\boldsymbol{F}(T_1(g)\boldsymbol{u}) \qquad (g \in G).$$

This shows that the equivariance of the form of (7.23) holds for $\hat{\boldsymbol{F}}(\boldsymbol{u}) = Q^{-1}\boldsymbol{F}(\boldsymbol{u})$. Thus the formulation of equivariance in (7.23) is justified for mathematical arguments. ■

## 7.3 Properties of Group Representation

### 7.3.1 *Equivalence*

#### 7.3.1.1 *Definition*

Two representations $T_1$ and $T_2$ of a group $G$ are said to be *equivalent*, if there exists a nonsingular matrix $Q$ that satisfies the condition[10]

$$T_1(g) = Q^{-1}T_2(g)Q \qquad (g \in G). \tag{7.40}$$

[10]In this section, we consider representations over $F = \mathbb{R}$ and $F = \mathbb{C}$. In (7.40), $Q$ is a matrix over $F$. The equivalence defined by (7.40) is an equivalence relation (a binary relation that satisfies reflexive law, symmetric law, and transitive law) in the sense explained in Remark 1.3 in Sec. 1.1.2.

It is important that the nonsingular matrix $Q$ is independent of $g \in G$. Two representations that are not equivalent are said to be *inequivalent*.

The definition of equivalence (7.40) corresponds to a change of variables (basis change) in the space which $T(g)$ acts on. The meaning of (7.40) can be understood from a general fact on linear mappings and their matrix representations, but it can also be seen with reference to symmetry as follows. The symmetry of a state of a system is represented in the form of

$$T(g)\boldsymbol{u} = \boldsymbol{u} \qquad (g \in G)$$

as a property of the vector $\boldsymbol{u}$ which describes that state (see (7.27)). If the variable is transformed to $\tilde{\boldsymbol{u}} = Q^{-1}\boldsymbol{u}$, the expression of symmetry is changed to

$$(Q^{-1}T(g)Q)\tilde{\boldsymbol{u}} = \tilde{\boldsymbol{u}} \qquad (g \in G).$$

The representation matrix $Q^{-1}T(g)Q$ appearing in this equation is equivalent to $T(g)$ in the sense of (7.40).

We can regard equivalent representations as "essentially the same" and accordingly it is useful and effective to select a representation that has nice properties (in some appropriate sense) from the set of equivalent representations.

#### 7.3.1.2 *Unitarity*

Proposition 7.1 below allows us to restrict ourselves to unitary representations.

**Proposition 7.1.** *Each representation is equivalent to a unitary representation.*

**Proof.** Let $T$ be an arbitrary representation of a group $G$. Define a matrix

$$S = \sum_{h \in G} T(h)^* T(h).$$

Then, for all $g \in G$, we have

$$\begin{aligned} T(g)^* S T(g) &= T(g)^* \left( \sum_{h \in G} T(h)^* T(h) \right) T(g) \\ &= \sum_{h \in G} (T(h)T(g))^* (T(h)T(g)) \\ &= \sum_{h \in G} T(hg)^* T(hg) = \sum_{k \in G} T(k)^* T(k) = S, \end{aligned}$$

where we have used the fact that, when $h$ runs over all elements of $G$, $k = hg$ also runs over all elements of $G$. Since $S$ is a positive-definite Hermitian matrix, there exists a nonsingular matrix $Q$ that satisfies $QQ^* = S^{-1}$. If we define $\hat{T}(g) = Q^{-1}T(g)Q$ using this $Q$, then $\hat{T}$ and $T$ are mutually equivalent representations. Furthermore, $\hat{T}$ is a unitary representation, since

$$\hat{T}(g)^*\hat{T}(g) = Q^*T(g)^*(QQ^*)^{-1}T(g)Q = Q^*T(g)^*ST(g)Q = Q^*SQ = I.$$

□

Moreover, in considering equivalence between unitary representations, the transformation matrix $Q$ in (7.40) can be chosen to be a unitary matrix, as shown below (see also Remark 7.6 in Sec. 7.3.3).

**Proposition 7.2.** *For unitary representations $T_1$ and $T_2$ that are equivalent to each other, there exists a unitary matrix $Q$ that satisfies* (7.40).

**Proof.** First recall a fundamental fact (*polar decomposition*) that every nonsingular matrix $A$ can be uniquely represented as $A = UR$ with a unitary matrix $U$ and a positive-definite Hermitian matrix $R$. Since $T_1$ and $T_2$ are equivalent, there exists a nonsingular matrix $Q$ that satisfies $T_1(g) = Q^{-1}T_2(g)Q$ $(g \in G)$ as in (7.40). We represent this matrix $Q$ as $Q = UR$ with a unitary matrix $U$ and a positive-definite Hermitian matrix $R$. By $Q^*Q = R^*R = R^2$ and

$$T_1(g)T_1(g)^* = T_2(g)^*T_2(g) = I$$

we have

$$(T_1(g)^*RT_1(g))^2 = T_1(g)^*Q^*QT_1(g) = Q^*T_2(g)^*T_2(g)Q = Q^*Q = R^2.$$

This implies $T_1(g)^*RT_1(g) = R$, since, for positive-definite Hermitian matrices $X$ and $Y$, in general, $X^2 = Y^2$ implies $X = Y$. Then it follows that

$$T_2(g) = QT_1(g)Q^{-1} = URT_1(g)R^{-1}U^{-1} = UT_1(g)U^{-1}.$$

□

7.3.1.3 *Direct sum decomposition*

For two representations $T_1$ and $T_2$ of a group $G$, the representation defined by[11]

$$T_1(g) \oplus T_2(g) = \begin{bmatrix} T_1(g) & O \\ O & T_2(g) \end{bmatrix} \qquad (g \in G)$$

is called the *direct sum* of $T_1$ and $T_2$ and denoted by $T_1 \oplus T_2$.

**Example 7.11.** The representation matrices in (7.24) used to express $D_3$-symmetry of the truss tent are partitioned into blocks as follows:

$$T(r) = \left[\begin{array}{cc|c} \cos(2\pi/3) & -\sin(2\pi/3) & 0 \\ \sin(2\pi/3) & \cos(2\pi/3) & 0 \\ \hline 0 & 0 & 1 \end{array}\right], \qquad T(s) = \left[\begin{array}{cc|c} 1 & 0 & 0 \\ 0 & -1 & 0 \\ \hline 0 & 0 & 1 \end{array}\right].$$

Therefore, this $T$ is the direct sum of a two-dimensional representation (the one shown in Example 7.7) and the unit representation. ■

In Example 7.11 above, the given representation $T$ itself is already decomposed into a direct sum of two representations. Next we consider whether a given representation has an equivalent representation in the form of a direct sum. That is, we consider whether we can find a nonsingular matrix $Q$ that decomposes a given $T$ into two representations $T_1$ and $T_2$ as

$$Q^{-1}T(g)Q = \begin{bmatrix} T_1(g) & O \\ O & T_2(g) \end{bmatrix} \qquad (g \in G). \tag{7.41}$$

**Example 7.12.** In Example 7.8, a permutation representation of $D_3$:

$$T(e) = \begin{bmatrix} 1 & 0 & 0 \\ 0 & 1 & 0 \\ 0 & 0 & 1 \end{bmatrix}, \quad T(r) = \begin{bmatrix} 0 & 0 & 1 \\ 1 & 0 & 0 \\ 0 & 1 & 0 \end{bmatrix}, \quad T(r^2) = \begin{bmatrix} 0 & 1 & 0 \\ 0 & 0 & 1 \\ 1 & 0 & 0 \end{bmatrix},$$

$$T(s) = \begin{bmatrix} 1 & 0 & 0 \\ 0 & 0 & 1 \\ 0 & 1 & 0 \end{bmatrix}, \quad T(sr) = \begin{bmatrix} 0 & 0 & 1 \\ 0 & 1 & 0 \\ 1 & 0 & 0 \end{bmatrix}, \quad T(sr^2) = \begin{bmatrix} 0 & 1 & 0 \\ 1 & 0 & 0 \\ 0 & 0 & 1 \end{bmatrix}$$

has been shown for the truss tent in Fig. 7.3. If this $T$ is transformed to $\tilde{T}(g) = Q^{-1}T(g)Q$ with an orthogonal matrix

$$Q = \begin{bmatrix} 1/\sqrt{3} & 2/\sqrt{6} & 0 \\ 1/\sqrt{3} & -1/\sqrt{6} & 1/\sqrt{2} \\ 1/\sqrt{3} & -1/\sqrt{6} & -1/\sqrt{2} \end{bmatrix},$$

[11]To be more precise, for each $g \in G$, we associate $T_1(g) \oplus T_2(g)$, which is the direct sum of two matrices. This correspondence turns out to be a representation of $G$, which is called the direct sum (of representations) and denoted by $T_1 \oplus T_2$. By definition, we have $(T_1 \oplus T_2)(g) = T_1(g) \oplus T_2(g)$ for all $g \in G$.

it is decomposed as follows:

$$\tilde{T}(e) = \left[\begin{array}{c|cc} 1 & 0 & 0 \\ \hline 0 & 1 & 0 \\ 0 & 0 & 1 \end{array}\right], \quad \tilde{T}(r) = \left[\begin{array}{c|cc} 1 & 0 & 0 \\ \hline 0 & \alpha & -\beta \\ 0 & \beta & \alpha \end{array}\right], \quad \tilde{T}(r^2) = \left[\begin{array}{c|cc} 1 & 0 & 0 \\ \hline 0 & \alpha & \beta \\ 0 & -\beta & \alpha \end{array}\right],$$

$$\tilde{T}(s) = \left[\begin{array}{c|cc} 1 & 0 & 0 \\ \hline 0 & 1 & 0 \\ 0 & 0 & -1 \end{array}\right], \quad \tilde{T}(sr) = \left[\begin{array}{c|cc} 1 & 0 & 0 \\ \hline 0 & \alpha & -\beta \\ 0 & -\beta & -\alpha \end{array}\right], \quad \tilde{T}(sr^2) = \left[\begin{array}{c|cc} 1 & 0 & 0 \\ \hline 0 & \alpha & \beta \\ 0 & \beta & -\alpha \end{array}\right],$$

where $\alpha = \cos(2\pi/3)$ and $\beta = \sin(2\pi/3)$. The resulting representation $\tilde{T}$ is a direct sum of the unit representation and the two-dimensional representation shown in Example 7.7. ■

It is not always possible to decompose $T$ into a direct sum as in (7.41). If the family of matrices $Q^{-1}T(g)Q$ $(g \in G)$ does not decompose into a direct sum for any choice of a nonsingular matrix $Q$, then the representation $T$ is said to be *irreducible.* We will discuss properties of irreducible representations (and the precise definition) in Sec. 7.3.2.

Representations that are not irreducible are said to be *reducible.* If a representation is reducible, there is a possibility that $T_1$ and $T_2$ in (7.41) can be decomposed further. We will deal with this issue in Sec. 7.3.3.

### 7.3.2 *Irreducible Representations*

#### 7.3.2.1 *Definition and examples*

Let $T$ be a representation of degree $N$ of a group $G$. For each $g \in G$, the matrix $T(g)$ determines a linear transformation on the vector space $V = F^N$. A subspace $W$ of $V$ is called an *invariant subspace* (with respect to $G$), if it satisfies the condition:

$$T(g)\boldsymbol{w} \in W \text{ for all } \boldsymbol{w} \in W \text{ and all } g \in G. \tag{7.42}$$

The zero space $\{\mathbf{0}\}$ and the entire space $V$ are (trivial) invariant subspaces for any representation $T$. A representation $T$ that does not have an invariant subspace other than $\{\mathbf{0}\}$ and $V$ is called an *irreducible representation.* It is known as *Maschke's theorem* that this definition of irreducibility is equivalent to the condition that the family of matrices $Q^{-1}T(g)Q$ $(g \in G)$ does not decompose simultaneously into a direct sum for any choice of a nonsingular matrix $Q$. The latter condition is used in Sec. 7.3.1 as the definition of irreducibility.

**Remark 7.3.** The irreducibility of a representation depends on what field $F$ is assumed. As an example, consider the cyclic group $\mathrm{C}_3 = \{e, r, r^2\}$ of degree 3 and a representation thereof given as follows (see Example 7.2):

$$T(e) = \begin{bmatrix} 1 & 0 \\ 0 & 1 \end{bmatrix}, \quad T(r) = \begin{bmatrix} \alpha & -\beta \\ \beta & \alpha \end{bmatrix}, \quad T(r^2) = \begin{bmatrix} \alpha & \beta \\ -\beta & \alpha \end{bmatrix},$$

where $\alpha = \cos(2\pi/3)$ and $\beta = \sin(2\pi/3)$. This representation $T$ is irreducible as a representation over $\mathbb{R}$. This means that, for any choice of a real nonsingular matrix $Q$, the three matrices $Q^{-1}T(e)Q$, $Q^{-1}T(r)Q$, and $Q^{-1}T(r^2)Q$ cannot simultaneously be transformed into diagonal matrices. However, with a complex unitary matrix

$$Q = \frac{1}{\sqrt{2}} \begin{bmatrix} 1 & 1 \\ -\mathrm{i} & \mathrm{i} \end{bmatrix} \tag{7.43}$$

we can decompose $T$ as follows:

$$Q^{-1}T(e)Q = \begin{bmatrix} 1 & 0 \\ 0 & 1 \end{bmatrix}, \; Q^{-1}T(r)Q = \begin{bmatrix} \omega & 0 \\ 0 & \overline{\omega} \end{bmatrix}, \; Q^{-1}T(r^2)Q = \begin{bmatrix} \omega^2 & 0 \\ 0 & \overline{\omega}^2 \end{bmatrix},$$

where $\omega = \alpha + \mathrm{i}\beta = \exp(2\pi\mathrm{i}/3)$ and $\overline{\omega} = \alpha - \mathrm{i}\beta = \exp(-2\pi\mathrm{i}/3)$. Therefore, $T$ is not irreducible as a representation over $\mathbb{C}$. When we need to specify over which field the irreducibility is discussed, we use expressions like "irreducible over $\mathbb{R}$" (or "$\mathbb{R}$-irreducible") and "irreducible over $\mathbb{C}$" (or "$\mathbb{C}$-irreducible"). ∎

It is known that there are only finitely many irreducible representations of a (finite) group $G$ over a fixed field $F$, when mutually equivalent irreducible representations are identified with each other. Let us denote the family of all representatives from the equivalence classes of mutually equivalent irreducible representations by

$$(T^{\mu} \mid \mu \in R(G; F)). \tag{7.44}$$

Here, $\mu$ is an index for distinguishing the irreducible representations and $R(G; F)$ is the set of (the names of) the equivalence classes of the irreducible representations of group $G$ over $F$. According to Proposition 7.1, we can assume that $T^{\mu}$ are all unitary representations. Henceforth we will assume this. The degree of the irreducible representation $T^{\mu}$ is denoted as $N^{\mu}$.

The following facts are known for irreducible representations over the field $\mathbb{C}$ of complex numbers:

- The number of irreducible representations $|R(G; \mathbb{C})|$ is equal to the number of conjugacy classes[12] (Proposition 7.10).

[12] See Sec. 7.5.2 for the definition of conjugacy class.

- The sum of squares of the degree $N^\mu$ of irreducible representations is equal to the order $|G|$, *i.e.*,
$$\sum_{\mu \in R(G;\mathbb{C})} (N^\mu)^2 = |G|, \tag{7.45}$$
the proof of which is given in Example 7.29 in Sec. 7.5.
- The degree $N^\mu$ of an irreducible representation is a divisor of $|G|$.

**Example 7.13.** Let us enumerate the irreducible representations of the cyclic group $\mathrm{C}_3 = \{e, r, r^2\}$ of degree 3. For the field of real numbers $\mathbb{R}$, we have two irreducible representations, namely, the one-dimensional representation (unit representation):
$$T^{(+)}(e) = 1, \qquad T^{(+)}(r) = 1, \qquad T^{(+)}(r^2) = 1 \tag{7.46}$$
and the two-dimensional representation treated in Remark 7.3:
$$T^{(1)}(e) = \begin{bmatrix} 1 & 0 \\ 0 & 1 \end{bmatrix}, \quad T^{(1)}(r) = \begin{bmatrix} \alpha & -\beta \\ \beta & \alpha \end{bmatrix}, \quad T^{(1)}(r^2) = \begin{bmatrix} \alpha & \beta \\ -\beta & \alpha \end{bmatrix},$$
where $\alpha = \cos(2\pi/3)$ and $\beta = \sin(2\pi/3)$. We have
$$R(\mathrm{C}_3; \mathbb{R}) = \{(+), (1)\}$$
in the notation of (7.44). For the field of complex numbers $\mathbb{C}$, in contrast, all the irreducible representations are one-dimensional. There are three of those: the unit representation $\mu = (+)$ in (7.46) and $\mu = (1+), (1-)$ defined by
$$\begin{aligned} T^{(1+)}(e) &= 1, \quad T^{(1+)}(r) = \omega, \qquad T^{(1+)}(r^2) = \omega^2; \\ T^{(1-)}(e) &= 1, \quad T^{(1-)}(r) = \omega^{-1}, \quad T^{(1-)}(r^2) = \omega^{-2}, \end{aligned}$$
where $\omega = \exp(2\pi \mathrm{i}/3)$. That is,
$$R(\mathrm{C}_3; \mathbb{C}) = \{(+), (1+), (1-)\}.$$

For general $n$, the irreducible representations of $\mathrm{C}_n = \{e, r, \ldots, r^{n-1}\} = \langle r \rangle$ over $\mathbb{R}$ are given as follows (note that the representation is determined by the value at the generator $r$).

- If $n$ is odd: There is a one-dimensional representation and $(n-1)/2$ two-dimensional representations. The one-dimensional representation is given as $T^{(+)}(r) = 1$ (unit representation) and the two-dimensional irreducible representations are given as
$$T^{(j)}(r) = \begin{bmatrix} \cos(2\pi j/n) & -\sin(2\pi j/n) \\ \sin(2\pi j/n) & \cos(2\pi j/n) \end{bmatrix} \tag{7.47}$$
for $j = 1, \ldots, (n-1)/2$. Thus,
$$R(\mathrm{C}_n; \mathbb{R}) = \left\{ (+), (1), (2), \ldots, \left( \frac{n-1}{2} \right) \right\}.$$

- If $n$ is even: There are two one-dimensional representations and $(n-2)/2$ two-dimensional representations. The one-dimensional representations are given as $T^{(+)}(r) = 1$ (unit representation) and $T^{(-)}(r) = -1$. The two-dimensional irreducible representations are given by (7.47) for $j = 1, \ldots, (n-2)/2$. Thus,
$$R(\mathrm{C}_n; \mathbb{R}) = \left\{(+), (-), (1), (2), \ldots, \left(\frac{n-2}{2}\right)\right\}.$$

In the case of $\mathbb{C}$, all the irreducible representations are one-dimensional representations and they are given as follows.

- If $n$ is odd: In addition to the unit representation $(+)$, there are, for $j = 1, \ldots, (n-1)/2$, a pair of one-dimensional representations of $(j+)$ and $(j-)$ given as
$$T^{(j+)}(r) = \exp(\mathrm{i}2\pi j/n), \qquad T^{(j-)}(r) = \exp(-\mathrm{i}2\pi j/n). \tag{7.48}$$
Thus,
$$R(\mathrm{C}_n; \mathbb{C}) = \left\{(+), (1+), (1-), \ldots, \left(\frac{n-1}{2}+\right), \left(\frac{n-1}{2}-\right)\right\}.$$
There exist $n$ irreducible representations.
- If $n$ is even: In addition to the unit representation $(+)$, there is a one-dimensional representation given by $T^{(-)}(r) = -1$ and one-dimensional representations given by (7.48) for $j = 1, \ldots, (n-2)/2$. Thus,
$$R(\mathrm{C}_n; \mathbb{C}) = \left\{(+), (-), (1+), (1-), \ldots, \left(\frac{n-2}{2}+\right), \left(\frac{n-2}{2}-\right)\right\}.$$
There exist $n$ irreducible representations. ■

**Example 7.14.** Let us enumerate the irreducible representations of the dihedral group $\mathrm{D}_3 = \{e, r, r^2, s, sr, sr^2\}$ of degree 3. For both the field of real numbers $\mathbb{R}$ and that of complex numbers $\mathbb{C}$, this group has three irreducible representations: two one-dimensional representations given by
$$T^{(+,+)}(r) = 1, \qquad T^{(+,+)}(s) = 1; \tag{7.49}$$
$$T^{(+,-)}(r) = 1, \qquad T^{(+,-)}(s) = -1 \tag{7.50}$$
and a two-dimensional representation treated in Example 7.7:
$$T^{(1)}(r) = \begin{bmatrix} \alpha & -\beta \\ \beta & \alpha \end{bmatrix}, \qquad T^{(1)}(s) = \begin{bmatrix} 1 & 0 \\ 0 & -1 \end{bmatrix},$$
where $\alpha = \cos(2\pi/3)$ and $\beta = \sin(2\pi/3)$. Thus,
$$R(\mathrm{D}_3; \mathbb{R}) = R(\mathrm{D}_3; \mathbb{C}) = \{(+,+), (+,-), (1)\}.$$

For general $n$, the irreducible representations of $\mathrm{D}_n = \{e, r, \ldots, r^{n-1}, s, sr, \ldots, sr^{n-1}\} = \langle r, s\rangle$, over $\mathbb{R}$ or $\mathbb{C}$, are as follows.

- If $n$ is odd: There are two one-dimensional representations and $(n-1)/2$ two-dimensional representations. The one-dimensional representations are $(+,+)$ and $(+,-)$ in (7.49) and (7.50), and the two-dimensional irreducible representations are given as

$$T^{(j)}(r) = \begin{bmatrix} \cos(2\pi j/n) & -\sin(2\pi j/n) \\ \sin(2\pi j/n) & \cos(2\pi j/n) \end{bmatrix}, \quad T^{(j)}(s) = \begin{bmatrix} 1 & 0 \\ 0 & -1 \end{bmatrix} \tag{7.51}$$

for $j = 1, \ldots, (n-1)/2$. Thus,

$$R(\mathrm{D}_n; \mathbb{R}) = R(\mathrm{D}_n; \mathbb{C}) = \left\{ (+,+), (+,-), (1), (2), \ldots, \left(\frac{n-1}{2}\right) \right\}.$$

The relation in (7.45) is indeed true as follows:

$$2 \times 1^2 + \frac{n-1}{2} \times 2^2 = 2n = |\mathrm{D}_n|.$$

- If $n$ is even: There are four one-dimensional representations and $(n-2)/2$ two-dimensional representations. The one-dimensional representations are $(+,+)$ and $(+,-)$ in (7.49) and (7.50), and also $(-,+)$ and $(-,-)$ given by

$$T^{(-,+)}(r) = -1, \qquad T^{(-,+)}(s) = 1; \tag{7.52}$$

$$T^{(-,-)}(r) = -1, \qquad T^{(-,-)}(s) = -1. \tag{7.53}$$

The two-dimensional irreducible representations are those given by (7.51) for $j = 1, \ldots, (n-2)/2$. Thus,

$$\begin{aligned} &R(\mathrm{D}_n; \mathbb{R}) = R(\mathrm{D}_n; \mathbb{C}) \\ &= \left\{ (+,+), (+,-), (-,+), (-,-), (1), (2), \ldots, \left(\frac{n-1}{2}\right) \right\}. \end{aligned}$$

The relation in (7.45) holds as follows:

$$4 \times 1^2 + \frac{n-2}{2} \times 2^2 = 2n = |\mathrm{D}_n|.$$

■

**Example 7.15.** The irreducible representations of the symmetric group $\mathrm{S}_4$ of degree 4 are the same for the field of real numbers $\mathbb{R}$ and the field of complex numbers $\mathbb{C}$. There are two one-dimensional representations, one two-dimensional representation and two three-dimensional representations. The relation in (7.45) is satisfied as

$$1^2 + 1^2 + 2^2 + 3^2 + 3^2 = 24 = |\mathrm{S}_4|.$$

Also for general $n$, the irreducible representations of the symmetric group $\mathrm{S}_n$ have been investigated thoroughly, and are described in terms of a diagram called the *Young diagram*. ■

#### 7.3.2.2 *Schur's lemma*

In Sec. 7.2.3, we have seen that the symmetry of a system is expressed as the commutativity with the representation matrices (see (7.28) and (7.29)). In this section we will present propositions that unveil fundamental properties of a matrix that commutes with irreducible representations. Propositions 7.3 and 7.4 below are (both) referred to as *Schur's lemma.*

**Proposition 7.3.** *Let $F = \mathbb{R}$ or $F = \mathbb{C}$, and $A$ be an $m \times n$ matrix over $F$. Assume that $T_1$ and $T_2$ are irreducible matrix representations of a group $G$ over $F$ of dimensions $m$ and $n$, respectively, and that*

$$T_1(g)A = AT_2(g) \qquad (g \in G) \tag{7.54}$$

*holds. Then, we have the following.*
*(1) $A = O$ or else $A$ is square and nonsingular.*[13]
*(2) If $T_1$ and $T_2$ are inequivalent, then $A = O$.*

**Proof.** (1) For every $\boldsymbol{x} \in \mathrm{Ker}(A)$ we have $T_2(g)\boldsymbol{x} \in \mathrm{Ker}(A)$, since

$$A(T_2(g)\boldsymbol{x}) = T_1(g)(A\boldsymbol{x}) = \mathbf{0}$$

by (7.54). Hence, $\mathrm{Ker}(A)$ is an invariant subspace for $T_2$. Since $T_2$ is irreducible by the assumption, $\mathrm{Ker}(A)$ must be the zero space $\{\mathbf{0}\}$ or the entire space $F^n$. On the other hand, for every $\boldsymbol{y} \in \mathrm{Im}(A)$, we can take $\boldsymbol{x}$ that satisfies $\boldsymbol{y} = A\boldsymbol{x}$. Then

$$T_1(g)\boldsymbol{y} = T_1(g)(A\boldsymbol{x}) = A(T_2(g)\boldsymbol{x}) \in \mathrm{Im}(A),$$

which shows that $\mathrm{Im}(A)$ is an invariant subspace for $T_1$. Since $T_1$ is irreducible by the assumption, $\mathrm{Im}(A)$ must be the zero space $\{\mathbf{0}\}$ or the entire space $F^m$. Therefore, $A$ is either a zero matrix or a nonsingular matrix.

(2) If $A \neq O$, $A$ is nonsingular by (1). Then we have

$$T_2(g) = A^{-1}T_1(g)A \qquad (g \in G)$$

from (7.54). This shows that $T_1$ and $T_2$ are equivalent. Therefore, if $T_1$ and $T_2$ are inequivalent, then $A = O$. □

**Proposition 7.4.** *If*

$$T(g)A = AT(g) \qquad (g \in G) \tag{7.55}$$

*for a complex matrix $A$ and an irreducible representation $T$ over $\mathbb{C}$, then $A = \lambda I$ for some $\lambda \in \mathbb{C}$.*

---

[13] If $A$ is nonsingular, then, naturally, $m = n$. Therefore, if $m \neq n$, then $A = O$.

**Proof.** By virtue of a fundamental property of complex numbers $\mathbb{C}$, there exists an eigenvalue $\lambda \in \mathbb{C}$ of $A$, for which $A - \lambda I$ is singular. It follows from (7.55) that

$$T(g)(A - \lambda I) = (A - \lambda I)T(g) \qquad (g \in G),$$

where $T$ is irreducible. Then Proposition 7.3(1) shows that $A - \lambda I = O$. $\square$

**Example 7.16.** In Proposition 7.4 above, the assumption of $T$ being an irreducible representation over $\mathbb{C}$ is most important. The same statement is not true for an irreducible representation over $\mathbb{R}$. As an example, consider the two-dimensional representation of the cyclic group $\mathrm{C}_3 = \{e, r, r^2\}$ of degree 3 defined by

$$T(r) = \begin{bmatrix} \alpha & -\beta \\ \beta & \alpha \end{bmatrix}$$

with $\alpha = \cos(2\pi/3)$ and $\beta = \sin(2\pi/3)$. This representation $T$ is irreducible over $\mathbb{R}$ and reducible over $\mathbb{C}$ (see Remark 7.3). First, note that the condition of commutativity $T(g)A = AT(g)$ $(g \in \mathrm{C}_3)$ is equivalent to $T(r)A = AT(r)$. Next, a direct calculation shows that $T(r)A = AT(r)$ if and only if $A$ takes the following form:

$$A = \begin{bmatrix} a & -b \\ b & a \end{bmatrix} \qquad (a, b \in \mathbb{R}).$$

Therefore, $A$ is not necessarily of the form of $A = \lambda I$. Finally, if we work in $\mathbb{C}$, the matrix $A$ above is diagonalized as

$$Q^{-1} \begin{bmatrix} a & -b \\ b & a \end{bmatrix} Q = \begin{bmatrix} a + ib & 0 \\ 0 & a - ib \end{bmatrix}$$

using the complex unitary matrix $Q$ in (7.43). ■

**Remark 7.4.** As a variant of Proposition 7.4 we have the following statement. If the relation in (7.55) holds for a real matrix $A$ and an irreducible representation $T$ over $\mathbb{R}$, and, in addition, $T$ is irreducible as a representation over $\mathbb{C}$, then $A = \lambda I$ for some real number $\lambda \in \mathbb{R}$. The proof of this statement is almost the same as the proof of Proposition 7.4. First, we work over $\mathbb{C}$, thereby deriving the claim that $A = \lambda I$ for some complex number $\lambda \in \mathbb{C}$, and then observe that $\lambda$ must be a real number, since $A$ is a real matrix. ■

**Remark 7.5.** As an application of Schur's lemma, let us show that mutually equivalent irreducible unitary representations can be transformed to each other by unitary matrices. This provides an alternative proof of Proposition 7.2 for the special case of irreducible representations.

Let $F = \mathbb{R}$ or $F = \mathbb{C}$. Assume that two irreducible unitary representations $T_1$ and $T_2$ over $F$ are equivalent. Then there exists a nonsingular matrix $Q$ over $F$ that satisfies the condition $T_1(g) = Q^{-1}T_2(g)Q$ ($g \in G$) in (7.40). This condition, together with the unitarity of $T_1$, implies

$$I = T_1(g)T_1(g)^* = (Q^{-1}T_2(g)Q)(Q^{-1}T_2(g)Q)^*,$$

which yields

$$QQ^* = T_2(g)QQ^*T_2(g)^*.$$

By multiplying both sides with $T_2(g)$ from the right and using the unitarity $T_2(g)^*T_2(g) = I$, we see that

$$(QQ^*)T_2(g) = T_2(g)(QQ^*)$$

holds for all $g \in G$. The matrix $QQ^*$, which is a positive-definite Hermitian matrix, possesses a positive eigenvalue $\alpha \in \mathbb{R}$, for which we have

$$T_2(g)(QQ^* - \alpha I) = (QQ^* - \alpha I)T_2(g) \qquad (g \in G).$$

Since $QQ^* - \alpha I$ is singular and $T_2$ is irreducible, we must have $QQ^* - \alpha I = O$ by Proposition 7.3(1). Therefore, $\hat{Q} = Q/\sqrt{\alpha}$ is a unitary matrix that satisfies $T_1(g) = \hat{Q}^{-1}T_2(g)\hat{Q}$ ($g \in G$). It should be clear in the above argument that Schur's lemma for $\mathbb{C}$ (Proposition 7.4) is not used, and that an eigenvalue $\alpha$ exists in the case of $F = \mathbb{R}$ as well. ■

### 7.3.2.3 *Orthogonality*

Irreducible representations over the field of complex numbers $\mathbb{C}$ are endowed with a remarkable property (*orthogonality*) described in the next proposition. This property will be used in Sec. 7.5. We use the notation

$$\delta_{ij} = \begin{cases} 1 \ (i = j), \\ 0 \ (i \neq j), \end{cases} \tag{7.56}$$

which is referred to as the *Kronecker delta*.

**Proposition 7.5.** *For irreducible representations $T^\mu$ and $T^\nu$ ($\mu, \nu \in R(G;\mathbb{C})$) over $\mathbb{C}$ of a group $G$, we have:*[14]

$$\sum_{g \in G} T^\mu_{il}(g) T^\nu_{ms}(g^{-1}) = \begin{cases} \dfrac{|G|}{N^\mu} \delta_{is}\delta_{lm} & (\mu = \nu), \\ 0 & (\mu \neq \nu). \end{cases} \tag{7.57}$$

[14] $T^\mu_{il}(g)$ denotes the $(i, l)$ entry of the matrix $T^\mu(g)$, and $T^\nu_{ms}(g)$ denotes the $(m, s)$ entry of the matrix $T^\nu(g)$.

*If $T^\mu$ and $T^\nu$ are unitary representations, then*[15]

$$\sum_{g\in G} T^\mu_{il}(g)\overline{T^\nu_{sm}(g)} = \begin{cases} \dfrac{|G|}{N^\mu}\delta_{is}\delta_{lm} & (\mu=\nu), \\ 0 & (\mu\neq\nu). \end{cases} \tag{7.58}$$

**Proof.** Let $B$ be an arbitrary $N^\mu \times N^\nu$ matrix, where $N^\mu$ is the degree of $T^\mu$ and $N^\nu$ is the degree of $T^\nu$, and define

$$A = \frac{1}{|G|}\sum_{h\in G} T^\mu(h)BT^\nu(h^{-1}).$$

Then we have

$$T^\mu(g)A = AT^\nu(g) \qquad (g\in G). \tag{7.59}$$

Indeed, for each $g \in G$, we have

$$\begin{aligned} T^\mu(g)A &= T^\mu(g)\left(\frac{1}{|G|}\sum_{h\in G} T^\mu(h)BT^\nu(h^{-1})\right) \\ &= \frac{1}{|G|}\sum_{h\in G} T^\mu(g)T^\mu(h)BT^\nu(h^{-1}) \\ &= \frac{1}{|G|}\sum_{h\in G} T^\mu(gh)BT^\nu((gh)^{-1})T^\nu(g) \\ &= \left(\frac{1}{|G|}\sum_{k\in G} T^\mu(k)BT^\nu(k^{-1})\right)T^\nu(g) \\ &= AT^\nu(g), \end{aligned}$$

where we have used the fact that, when $h$ runs over all elements of $G$, $k = gh$ also runs over all elements of $G$. By (7.59) and Schur's lemma (Proposition 7.3(2) and Proposition 7.4) it follows that[16]

$$A = \begin{cases} \lambda(\mu, B)I_{N^\mu} & (\mu=\nu), \\ O & (\mu\neq\nu). \end{cases}$$

By choosing $B$ to be the matrix whose $(l,m)$ entry is 1 and all the other entries are 0, we obtain

$$\frac{1}{|G|}\sum_{g\in G} T^\mu_{il}(g)T^\nu_{ms}(g^{-1}) = \begin{cases} \lambda(\mu,l,m)\delta_{is} & (\mu=\nu), \\ O & (\mu\neq\nu) \end{cases} \tag{7.60}$$

[15] $\overline{T^\nu_{sm}(g)}$ means the complex conjugate of $T^\nu_{sm}(g)$.
[16] $I_{N^\mu}$ means the unit matrix of order $N^\mu$.

for each $(i, s)$. Furthermore, by setting $\mu = \nu$ and $i = s$ and taking the sum of the above expression for $i = 1, \ldots, N^\mu$ we obtain

$$N^\mu \lambda(\mu, l, m) = \frac{1}{|G|} \sum_{i=1}^{N^\mu} \sum_{g \in G} T^\mu_{mi}(g^{-1}) T^\mu_{il}(g) = \frac{1}{|G|} \sum_{g \in G} T^\mu_{ml}(e) = \delta_{ml},$$

which shows

$$\lambda(\mu, l, m) = \frac{1}{N^\mu} \delta_{ml}.$$

The substitution of this into (7.60) yields (7.57). In the unitary case we have

$$T^\nu_{ms}(g^{-1}) = \overline{T^\nu_{sm}(g)},$$

and therefore (7.58) follows from (7.57). □

### 7.3.3 *Irreducible Decomposition*

#### 7.3.3.1 *General case*

If a representation $T$ of a group $G$ is not irreducible,[17] it can be decomposed (see (7.41)) as

$$Q^{-1} T(g) Q = \begin{bmatrix} T_1(g) & O \\ O & T_2(g) \end{bmatrix} = T_1(g) \oplus T_2(g) \qquad (g \in G) \tag{7.61}$$

for some nonsingular matrix $Q$. If $T_1$ and/or $T_2$ are not irreducible, we can further perform the same decomposition. As a result of such a decomposition process, the given representation $T$ eventually decomposes into a finite number of irreducible representations. Each of the resulting irreducible representations is equivalent to one of the representative irreducible representations $T^\mu$ selected in (7.44). Let $T^\mu_i$ $(i = 1, \ldots, a^\mu)$ denote those irreducible representations which are contained in the decomposition of $T$ and are equivalent to $T^\mu$. Then

$$Q^{-1} T(g) Q = \bigoplus_{\mu \in R(G;F)} \bigoplus_{i=1}^{a^\mu} T^\mu_i(g) \qquad (g \in G) \tag{7.62}$$

for some nonsingular matrix $Q$. The nonnegative integer $a^\mu$ is referred to as the *multiplicity* of the irreducible representation $\mu$ in $T$, and is known to be determined independently of how the decomposition is constructed.[18]

[17] In this section, we continue to assume $F = \mathbb{R}$ or $F = \mathbb{C}$.
[18] In the case of $F = \mathbb{C}$, $a^\mu$ is expressed as in (7.106) in Sec. 7.5.

**Example 7.17.** The symbol $\bigoplus\bigoplus$ on the right-hand side of (7.62) means a nested block-diagonal structure. For example, if $R(G;F) = \{\mu, \nu\}$ for a group $G$ and the multiplicities of $\mu$ and $\nu$ in $T$ are $a^\mu = 2$ and $a^\nu = 3$, then the expression in (7.62) takes the form of

$$Q^{-1}T(g)Q = \begin{bmatrix} T_1^\mu(g) & & & & \\ & T_2^\mu(g) & & & \\ & & T_1^\nu(g) & & \\ & & & T_2^\nu(g) & \\ & & & & T_3^\nu(g) \end{bmatrix} \qquad (g \in G).$$

■

In the decomposition (7.62), we usually choose

$$T_i^\mu = T^\mu \qquad (i = 1, \ldots, a^\mu). \tag{7.63}$$

This choice brings the decomposition (7.62) to the following (nicer) form:

$$Q^{-1}T(g)Q = \bigoplus_{\mu \in R(G;F)} \bigoplus_{i=1}^{a^\mu} T^\mu(g) \qquad (g \in G). \tag{7.64}$$

The decomposition in (7.62) or (7.64) is called the *irreducible decomposition.*

Due to the fact that multiplicity $a^\mu$ is uniquely determined, the irreducible decomposition (7.64) is determined uniquely for $T$, once the representatives $T^\mu$ of the irreducible representations are fixed. Therefore, two representations are equivalent if and only if their irreducible decompositions (7.64) are the same.

To aggregate the $a^\mu$ irreducible representations that correspond to $\mu$ in (7.64), we define

$$\tilde{T}^\mu(g) = \bigoplus_{i=1}^{a^\mu} T^\mu(g) \qquad (g \in G) \tag{7.65}$$

and rewrite (7.64) as

$$Q^{-1}T(g)Q = \bigoplus_{\mu \in R(G;F)} \tilde{T}^\mu(g) \qquad (g \in G). \tag{7.66}$$

This form of decomposition is called the *isotypic decomposition.* Note that $\tilde{T}^\mu(g)$ is a matrix of order $a^\mu N^\mu$.

#### 7.3.3.2 *Unitary case*

In the previous section, we considered the decomposition into irreducible representations through a transformation of the form $Q^{-1}T(g)Q$ with a nonsingular matrix $Q$. For a unitary representation $T$, however, it is natural to choose as $Q$ a unitary matrix to preserve unitarity. This is indeed possible. If a unitary representation $T$ can be decomposed by using a nonsingular matrix as in (7.61), then it can be decomposed as

$$Q^{*}T(g)Q = \begin{bmatrix} T_1(g) & O \\ O & T_2(g) \end{bmatrix} = T_1(g) \oplus T_2(g) \qquad (g \in G) \tag{7.67}$$

by some unitary matrix $Q$.

**Proposition 7.6.** *A reducible unitary representation $T$ can be decomposed into the form of* (7.67) *with a unitary matrix $Q$.*

**Proof.** If a unitary representation $T$ is reducible, then there exists an invariant subspace $W$ that is neither the zero space nor the entire space; recall (7.42). The orthogonal complement of $W$, denoted as $W^{\perp}$, is an invariant subspace. Indeed, for any $\boldsymbol{v} \in W^{\perp}$ and $g \in G$, we have

$$(T(g)\boldsymbol{v}, \boldsymbol{w}) = (\boldsymbol{v}, T(g)^{*}\boldsymbol{w}) = (\boldsymbol{v}, T(g^{-1})\boldsymbol{w}) = 0 \qquad (\boldsymbol{w} \in W),$$

which shows that $T(g)\boldsymbol{v} \in W^{\perp}$. Let $Q$ be a matrix consisting of column vectors that form orthonormal bases of $W$ and $W^{\perp}$. Then $Q$ is a unitary matrix and $Q^{*}T(g)Q$ is decomposed as in (7.67). □

Proposition 7.6 above enables us to obtain a decomposition:

$$Q^{*}T(g)Q = \bigoplus_{\mu \in R(G;F)} \bigoplus_{i=1}^{a^{\mu}} T_i^{\mu}(g) \qquad (g \in G) \tag{7.68}$$

of the form of (7.62) using a unitary matrix $Q$. Here, each $T_i^{\mu}$ is an irreducible representation, which is actually a unitary representation due to the unitarity of $T$ and $Q$. It can be assumed further that $T_i^{\mu} = T^{\mu}$ for all $i$ as in (7.63), since mutually equivalent irreducible unitary representations can be transformed to each other using unitary matrices (see Proposition 7.2 as well as Remark 7.5).

To summarize, for a unitary representation $T$, both the irreducible decomposition (7.64) and the isotypic decomposition (7.66) are possible using a unitary transformation. That is,

$$Q^*T(g)Q = \bigoplus_{\mu \in R(G;F)} \bigoplus_{i=1}^{a^\mu} T^\mu(g) \qquad (g \in G), \tag{7.69}$$

$$Q^*T(g)Q = \bigoplus_{\mu \in R(G;F)} \tilde{T}^\mu(g) \qquad (g \in G) \tag{7.70}$$

for some unitary matrix $Q$.

**Example 7.18.** Suppose, for example, that $R(G;F) = \{\mu, \nu\}$ for a group $G$ and the multiplicities of $\mu$ and $\nu$ in a unitary representation $T$ are $a^\mu = 2$ and $a^\nu = 3$. Then the decompositions in (7.69) and (7.70) are given as

$$Q^*T(g)Q = \left[\begin{array}{cc|ccc} T^\mu(g) & & & & \\ & T^\mu(g) & & & \\ \hline & & T^\nu(g) & & \\ & & & T^\nu(g) & \\ & & & & T^\nu(g) \end{array}\right] = \left[\begin{array}{c|c} \tilde{T}^\mu(g) & \\ \hline & \tilde{T}^\nu(g) \end{array}\right],$$

where

$$\tilde{T}^\mu(g) = \left[\begin{array}{c|c} T^\mu(g) & \\ \hline & T^\mu(g) \end{array}\right], \qquad \tilde{T}^\nu(g) = \left[\begin{array}{c|c|c} T^\nu(g) & & \\ \hline & T^\nu(g) & \\ \hline & & T^\nu(g) \end{array}\right].$$

■

**Remark 7.6.** Since two mutually equivalent unitary representations $T_1$ and $T_2$ share the same irreducible decomposition (7.69), there exist unitary matrices $Q_1$ and $Q_2$ such that

$$Q_1^*T_1(g)Q_1 = \bigoplus_{\mu \in R(G;F)} \bigoplus_{i=1}^{a^\mu} T^\mu(g) = Q_2^*T_2(g)Q_2 \qquad (g \in G).$$

It then follows that $T_1(g) = Q^*T_2(g)Q$ for all $g \in G$ with $Q = Q_2Q_1^*$, which is a unitary matrix. Therefore, in discussing equivalence between unitary representations, it suffices to consider a transformation by a unitary matrix $Q$ in (7.40). We have thus arrived at an alternative proof of Proposition 7.2.

■

## 7.4 Block-Diagonalization under Group Symmetry

### 7.4.1 *Overview*

As previously mentioned, the objective of this chapter is to explain the mathematical principle of block-diagonalization such as the one shown in Fig. 7.2 (Sec. 7.1.1) for a truss dome with $D_6$-symmetry.

In Sec. 7.1.2, we have considered a truss tent that possesses $D_3$-symmetry as a very simple example. We have seen that the stiffness matrix $K$ becomes a diagonal matrix in (7.5) and verified, via concrete calculation in Remark 7.1, that this is a consequence of commutativity of $K$ with representative matrices formulated as

$$RK = KR, \qquad SK = KS. \tag{7.71}$$

In addition, in Sec. 7.2.3, we have explained that the symmetry of a linear system in general can be formulated as the commutativity

$$T(g)A = AT(g) \qquad (g \in G) \tag{7.72}$$

of the matrix $A$ describing the system and the representation matrix $T(g)$ describing the symmetry. The equation (7.71) above is an instance of (7.72); see also (7.29). Also in nonlinear systems, the Jacobian matrix has the commutativity of the form of (7.72); see (7.28).

In this section, we will make a full study of the block-diagonal structure of a matrix $A$ that satisfies the condition (7.72) for a group $G$. We assume that $F = \mathbb{R}$ or $F = \mathbb{C}$, that $T$ is an $N$-dimensional unitary representation, and that $A$ is an $N \times N$ matrix. Since every representation is equivalent to a unitary representation, the assumption of unitarity of $T$ does not impose any essential restrictions.

Let us review our notation. The family of all irreducible representations (the names thereof) over $F$ of a group $G$ is expressed as $R(G; F)$. The irreducible unitary representation corresponding to $\mu \in R(G; F)$ is denoted as $T^\mu$, the degree of which is $N^\mu$. In addition, the multiplicity of $\mu$ in the representation $T$ is denoted by $a^\mu$. The irreducible decomposition (7.69) and the isotypic decomposition (7.70) are given as follows:

$$Q^* T(g) Q = \bigoplus_{\mu \in R(G;F)} \bigoplus_{i=1}^{a^\mu} T^\mu(g) \qquad (g \in G), \tag{7.73}$$

$$Q^* T(g) Q = \bigoplus_{\mu \in R(G;F)} \tilde{T}^\mu(g) \qquad (g \in G), \tag{7.74}$$

where

$$\tilde{T}^{\mu}(g) = \bigoplus_{i=1}^{a^{\mu}} T^{\mu}(g) \qquad (g \in G) \tag{7.75}$$

and $Q$ is a unitary matrix.

The block-diagonalization of a matrix $A$ consists of the following two steps:

- Block-diagonalization I, which corresponds to the isotypic decomposition (7.74);
- Block-diagonalization II, which corresponds to the irreducible decomposition (7.73).

The first step, block-diagonalization I, works for both $F = \mathbb{R}$ and $F = \mathbb{C}$ in any group $G$. In contrast, the second step, block-diagonalization II, is basically valid for $F = \mathbb{C}$. However, it is worth noting that block-diagonalization II is also good for $F = \mathbb{R}$ in dihedral group $\mathrm{D}_n$ and symmetric group $\mathrm{S}_n$ which are important in applications (see Remark 7.8).

**Remark 7.7.** For a matrix $B$ that is not necessarily a square matrix, we can consider a (more general) commutativity condition of the form

$$T(g)B = BS(g) \qquad (g \in G) \tag{7.76}$$

involving two representations $T$ and $S$ of a group $G$. The generalization of this form is important in analyses of engineering systems arising naturally, for example, in sensitivity analyses. The method of block-diagonalization to be shown in this section can easily be extended to such cases [58]. ■

### 7.4.2 *Partition of Transformation Matrices*

The transformation matrix $Q$ for the irreducible decomposition (7.73) is a unitary matrix of order $N$, and its column set is partitioned into blocks in accordance with the irreducible decomposition. That is, we have

$$\begin{aligned} Q &= [\cdots | Q_1^{\mu}, Q_2^{\mu}, \ldots, Q_{a^{\mu}}^{\mu} | \cdots] \\ &= ((Q_i^{\mu} \mid i = 1, \ldots, a^{\mu}) \mid \ \mu \in R(G;F)), \end{aligned} \tag{7.77}$$

where $Q_i^{\mu}$ is an $N \times N^{\mu}$ matrix that corresponds to $(\mu, i)$. By (7.73) we have

$$\begin{aligned} (Q_i^{\mu})^* T(g) Q_i^{\mu} &= T^{\mu}(g) \qquad (g \in G), \\ T(g) Q_i^{\mu} &= Q_i^{\mu} T^{\mu}(g) \qquad (g \in G) \end{aligned}$$

for each $(\mu, i)$.

The partition of $Q$ that corresponds to the isotypic decomposition (7.74) is obtained by collecting $a^\mu$ blocks of $Q_i^\mu$ $(i = 1, \ldots, a^\mu)$ that correspond to the same $\mu$. That is, for each $\mu \in R(G; F)$, we define an $N \times a^\mu N^\mu$ matrix

$$Q^\mu = (Q_i^\mu \mid i = 1, \ldots, a^\mu) = [Q_1^\mu, Q_2^\mu, \ldots, Q_{a^\mu}^\mu], \tag{7.78}$$

to obtain

$$Q = \left[\cdots \middle| Q^\mu \middle| \cdots\right] = (Q^\mu \mid \mu \in R(G; F)). \tag{7.79}$$

By (7.74) we have

$$(Q^\mu)^* T(g) Q^\mu = \tilde{T}^\mu(g) \qquad (g \in G),$$
$$T(g) Q^\mu = Q^\mu \tilde{T}^\mu(g) \qquad (g \in G)$$

for each $\mu$.

**Example 7.19.** As in Example 7.18, let $T$ be a unitary representation of a group $G$ with $R(G; F) = \{\mu, \nu\}$, where the multiplicities of $\mu$ and $\nu$ are given by $a^\mu = 2$ and $a^\nu = 3$. Then the unitary matrix $Q$ for decomposing $T$ is partitioned as

$$Q = [Q_1^\mu, Q_2^\mu \mid Q_1^\nu, Q_2^\nu, Q_3^\nu] = [Q^\mu \mid Q^\nu]$$

(see Example 7.18), where

$$Q^\mu = [Q_1^\mu, Q_2^\mu], \qquad Q^\nu = [Q_1^\nu, Q_2^\nu, Q_3^\nu].$$

■

### 7.4.3 *Block-Diagonalization I*

Consider a matrix $A$ that commutes with the action of $G$ as in (7.72). We transform the matrix $A$ to[19]

$$\overline{A} = Q^* A Q \tag{7.80}$$

using the transformation matrix $Q$ in the irreducible decomposition (7.73). Then the matrix $\overline{A}$ is partitioned, in accordance with the partition of $Q = (Q_i^\mu)$ in (7.77), as

$$\overline{A} = (\overline{A}_{ij}^{\mu\nu} \mid i = 1, \ldots, a^\mu; j = 1, \ldots, a^\nu; \mu, \nu \in R(G; F)), \tag{7.81}$$

where

$$\overline{A}_{ij}^{\mu\nu} = (Q_i^\mu)^* A Q_j^\nu \quad (i = 1, \ldots, a^\mu; j = 1, \ldots, a^\nu; \mu, \nu \in R(G; F)).$$

The matrix $\overline{A}_{ij}^{\mu\nu}$ is an $N^\mu \times N^\nu$ matrix.

---

[19]This $\overline{\phantom{x}}$ has nothing to do with the complex conjugate. The equation (7.80) introduces a notation $\overline{A}$ for $Q^* A Q$.

**Example 7.20.** As in Example 7.18, let $T$ be a unitary representation with multiplicities $a^\mu = 2$ and $a^\nu = 3$. The transformation matrix $Q$ is partitioned as $Q = [Q_1^\mu, Q_2^\mu \mid Q_1^\nu, Q_2^\nu, Q_3^\nu]$ (see Example 7.19). Then the matrix $\overline{A}$ in (7.81) is as follows:

$$\overline{A} = \left[\begin{array}{cc|ccc} \overline{A}_{11}^{\mu\mu} & \overline{A}_{12}^{\mu\mu} & \overline{A}_{11}^{\mu\nu} & \overline{A}_{12}^{\mu\nu} & \overline{A}_{13}^{\mu\nu} \\ \overline{A}_{21}^{\mu\mu} & \overline{A}_{22}^{\mu\mu} & \overline{A}_{21}^{\mu\nu} & \overline{A}_{22}^{\mu\nu} & \overline{A}_{23}^{\mu\nu} \\ \hline \overline{A}_{11}^{\nu\mu} & \overline{A}_{12}^{\nu\mu} & \overline{A}_{11}^{\nu\nu} & \overline{A}_{12}^{\nu\nu} & \overline{A}_{13}^{\nu\nu} \\ \overline{A}_{21}^{\nu\mu} & \overline{A}_{22}^{\nu\mu} & \overline{A}_{21}^{\nu\nu} & \overline{A}_{22}^{\nu\nu} & \overline{A}_{23}^{\nu\nu} \\ \overline{A}_{31}^{\nu\mu} & \overline{A}_{32}^{\nu\mu} & \overline{A}_{31}^{\nu\nu} & \overline{A}_{32}^{\nu\nu} & \overline{A}_{33}^{\nu\nu} \end{array}\right].$$

■

We shall show that, as a consequence of the commutativity in (7.72), the matrix $\overline{A}$ in (7.81) is, in fact, a block-diagonal matrix. An example of this block-diagonal form is shown first, while the general expression of the block-diagonal form is given later in (7.88).

**Example 7.21.** According to the general result to be given in (7.88), the matrix $\overline{A}$ in Example 7.20 is a block-diagonal matrix of the following form:

$$\overline{A} = \left[\begin{array}{cc|ccc} \overline{A}_{11}^{\mu\mu} & \overline{A}_{12}^{\mu\mu} & O & O & O \\ \overline{A}_{21}^{\mu\mu} & \overline{A}_{22}^{\mu\mu} & O & O & O \\ \hline O & O & \overline{A}_{11}^{\nu\nu} & \overline{A}_{12}^{\nu\nu} & \overline{A}_{13}^{\nu\nu} \\ O & O & \overline{A}_{21}^{\nu\nu} & \overline{A}_{22}^{\nu\nu} & \overline{A}_{23}^{\nu\nu} \\ O & O & \overline{A}_{31}^{\nu\nu} & \overline{A}_{32}^{\nu\nu} & \overline{A}_{33}^{\nu\nu} \end{array}\right] = \left[\begin{array}{c|c} \overline{A}^{\mu} & O \\ \hline O & \overline{A}^{\nu} \end{array}\right].$$

■

We now derive the expression of the block-diagonal form. The commutativity condition (7.72) can be rewritten as

$$(Q^*T(g)Q)\,(Q^*AQ) = (Q^*AQ)\,(Q^*T(g)Q) \qquad (g \in G), \tag{7.82}$$

where $Q^*AQ = \overline{A} = (\overline{A}_{ij}^{\mu\nu})$ and $Q^*T(g)Q$ is a block-diagonal matrix in (7.73). A necessary and sufficient condition for (7.82) (or (7.72)) is that

$$T^\mu(g)\,\overline{A}_{ij}^{\mu\nu} = \overline{A}_{ij}^{\mu\nu}\,T^\nu(g) \qquad (g \in G) \tag{7.83}$$

holds for all $i = 1, \ldots, a^\mu$; $j = 1, \ldots, a^\nu$, and $\mu, \nu \in R(G; F)$.

If $\mu \neq \nu$, the irreducible representations $T^\mu$ and $T^\nu$ are inequivalent. Then, by Schur's lemma (Proposition 7.3(2)), it follows from (7.83) that

$$\overline{A}_{ij}^{\mu\nu} = O \qquad (i = 1, \ldots, a^\mu; j = 1, \ldots, a^\nu; \mu \neq \nu). \tag{7.84}$$

For each pair $(\mu, \nu)$, define a matrix

$$\overline{A}^{\mu\nu} = (\overline{A}_{ij}^{\mu\nu} \mid i = 1, \ldots, a^{\mu}; j = 1, \ldots, a^{\nu})$$

consisting of the blocks $\overline{A}_{ij}^{\mu\nu}$ with varying $i$ and $j$. Accordingly we may think of a coarser partition of $\overline{A}$, namely,

$$\overline{A} = (\overline{A}^{\mu\nu} \mid \mu, \nu \in R(G; F)). \tag{7.85}$$

The matrix $\overline{A}$ is a block-diagonal matrix with respect to this partition, since (7.84) is equivalent to

$$\overline{A}^{\mu\nu} = O \qquad (\mu \neq \nu). \tag{7.86}$$

The diagonal block $\overline{A}^{\mu} = \overline{A}^{\mu\mu}$ is a square matrix of order $a^{\mu} N^{\mu}$, which is given by

$$\overline{A}^{\mu} = (Q^{\mu})^* A Q^{\mu} \qquad (\mu \in R(G; F)). \tag{7.87}$$

In this way, we have obtained a block-diagonalization

$$\overline{A} = Q^* A Q = \bigoplus_{\mu \in R(G;F)} \overline{A}^{\mu} \tag{7.88}$$

corresponding to the isotypic decomposition (7.74) of $T$. Recall that (7.88) is illustrated in Example 7.21. We note again that $F = \mathbb{R}$ or $F = \mathbb{C}$.

### 7.4.4 *Block-Diagonalization II*

The decomposition of the form of (7.88) is valid over $\mathbb{R}$ and $\mathbb{C}$, but in the case of $\mathbb{C}$, the diagonal block $\overline{A}^{\mu}$ in (7.88) has an even more intricate block-diagonal structure. In the following discussion, we assume $F = \mathbb{C}$.

With $\mu = \nu$, the condition (7.83) for commutativity shows that

$$T^{\mu}(g)\, \overline{A}_{ij}^{\mu} = \overline{A}_{ij}^{\mu}\, T^{\mu}(g) \qquad (g \in G)$$

should hold for all $i, j = 1, \ldots, a^{\mu}$. Since $T^{\mu}$ is an irreducible representation, Schur's lemma for $\mathbb{C}$ (Proposition 7.4) shows that $\overline{A}_{ij}^{\mu}$ is a scalar multiple of the unit matrix, *i.e.*,

$$\overline{A}_{ij}^{\mu} = \alpha_{ij}^{\mu} I_{N^{\mu}} \tag{7.89}$$

for some complex number $\alpha_{ij}^{\mu} \in \mathbb{C}$. Here, $I_{N^{\mu}}$ stands for the unit matrix of order $N^{\mu}$.

**Example 7.22.** In Example 7.21, suppose that $N = 7$, $N^\mu = 2$, and $N^\nu = 1$. Then, still assuming multiplicities $a^\mu = 2$ and $a^\nu = 3$, we have $\overline{A}^\mu_{ij} = \alpha^\mu_{ij} I_2$ $(i,j=1,2)$ and $\overline{A}^\nu_{ij} = \alpha^\mu_{ij} I_1$ $(i,j=1,2,3)$, and hence

$$\overline{A} = \left[\begin{array}{cc|ccc} \alpha^\mu_{11}I_2 & \alpha^\mu_{12}I_2 & & & \\ \alpha^\mu_{21}I_2 & \alpha^\mu_{22}I_2 & & & \\ \hline & & \alpha^\nu_{11}I_1 & \alpha^\nu_{12}I_1 & \alpha^\nu_{13}I_1 \\ & & \alpha^\nu_{21}I_1 & \alpha^\nu_{22}I_1 & \alpha^\nu_{23}I_1 \\ & & \alpha^\nu_{31}I_1 & \alpha^\nu_{32}I_1 & \alpha^\nu_{33}I_1 \end{array}\right] = \left[\begin{array}{cccc|ccc} \alpha^\mu_{11} & & \alpha^\mu_{12} & & & & \\ & \alpha^\mu_{11} & & \alpha^\mu_{12} & & & \\ \alpha^\mu_{21} & & \alpha^\mu_{22} & & & & \\ & \alpha^\mu_{21} & & \alpha^\mu_{22} & & & \\ \hline & & & & \alpha^\nu_{11} & \alpha^\nu_{12} & \alpha^\nu_{13} \\ & & & & \alpha^\nu_{21} & \alpha^\nu_{22} & \alpha^\nu_{23} \\ & & & & \alpha^\nu_{31} & \alpha^\nu_{32} & \alpha^\nu_{33} \end{array}\right].$$

■

Each diagonal block $\overline{A}^\mu$ is a square matrix of order $a^\mu N^\mu$. By (7.89) we can decompose it to a block-diagonal form with $N^\mu$ diagonal blocks by suitably rearranging the rows and columns. Moreover, all the diagonal blocks will be the same $a^\mu \times a^\mu$ matrix

$$\tilde{A}^\mu = (\alpha^\mu_{ij} \mid i,j = 1,\ldots,a^\mu). \tag{7.90}$$

This means that, with a permutation matrix $\Pi^\mu$ expressing the rearrangement of the rows and columns, we have the decomposition

$$(\Pi^\mu)^* \overline{A}^\mu \Pi^\mu = \bigoplus_{k=1}^{N^\mu} \tilde{A}^\mu. \tag{7.91}$$

To combine (7.91) with (7.88), we introduce a permutation matrix

$$\Pi = \bigoplus_{\mu \in R(G;\mathbb{C})} \Pi^\mu, \tag{7.92}$$

which is of order $N$. Then we can obtain a block-diagonalization

$$\Pi^* \overline{A} \Pi = \bigoplus_{\mu \in R(G;\mathbb{C})} \bigoplus_{k=1}^{N^\mu} \tilde{A}^\mu \tag{7.93}$$

for the matrix $\overline{A}$. Here the diagonal blocks are given by $\tilde{A}^\mu$ in (7.90). This is the second step of our block-diagonalization (block-diagonalization II).

**Example 7.23.** For the matrix $\overline{A}$ of Example 7.22, we obtain

$$\Pi^* \overline{A} \Pi = \left[\begin{array}{cc|cc|ccc} \alpha^\mu_{11} & \alpha^\mu_{12} & & & & & \\ \alpha^\mu_{21} & \alpha^\mu_{22} & & & & & \\ \hline & & \alpha^\mu_{11} & \alpha^\mu_{12} & & & \\ & & \alpha^\mu_{21} & \alpha^\mu_{22} & & & \\ \hline & & & & \alpha^\nu_{11} & \alpha^\nu_{12} & \alpha^\nu_{13} \\ & & & & \alpha^\nu_{21} & \alpha^\nu_{22} & \alpha^\nu_{23} \\ & & & & \alpha^\nu_{31} & \alpha^\nu_{32} & \alpha^\nu_{33} \end{array}\right]$$

as the decomposition in (7.93). ■

**Remark 7.8.** The block-diagonalization II depends on Schur's lemma for $\mathbb{C}$ (Proposition 7.4), and, in general, is not valid for $\mathbb{R}$. Consider, for example, the following representation of $G = \mathrm{C}_3$:

$$T(r) = \left[\begin{array}{cc|cc} \alpha & -\beta & & \\ \beta & \alpha & & \\ \hline & & \alpha & -\beta \\ & & \beta & \alpha \end{array}\right]$$

with $\alpha = \cos(2\pi/3)$ and $\beta = \sin(2\pi/3)$. This representation contains the same two-dimensional representation with multiplicity 2. A matrix $A$ that satisfies the commutativity condition (7.72) takes the form of

$$A = \left[\begin{array}{cc|cc} a_{11} & -b_{11} & a_{12} & -b_{12} \\ b_{11} & a_{11} & b_{12} & a_{12} \\ \hline a_{21} & -b_{21} & a_{22} & -b_{22} \\ b_{21} & a_{21} & b_{22} & a_{22} \end{array}\right]$$

with arbitrary $a_{ij}, b_{ij} \in \mathbb{R}$ (see also Example 7.16). This is not in a block-diagonal form.

But dihedral group $\mathrm{D}_n$ and symmetric group $\mathrm{S}_n$ have a remarkable property that *every irreducible representation over* $\mathbb{R}$ *is also irreducible over* $\mathbb{C}$ (Examples 7.14 and 7.15). Based on this fact and the implication of Proposition 7.4 explained in Remark 7.4 in Sec. 7.3.2, we can carry out block-diagonalization II for the groups $\mathrm{D}_n$ and $\mathrm{S}_n$ without involving complex numbers. The block-diagonalization of the stiffness matrix of the $\mathrm{D}_6$-symmetric truss dome shown in Sec. 7.1.1 (Fig. 7.2) is a block-diagonalization II obtained according to this principle. The irreducible representations and the corresponding multiplicities for this truss dome are as follows:

| | One-dimensional representations | | | | Two-dimensional representations | |
|---|---|---|---|---|---|---|
| Irred. representation $\mu$ | $(+,+)$ | $(+,-)$ | $(-,+)$ | $(-,-)$ | (1) | (2) |
| Multiplicity $a^\mu$ | 61 | 48 | 52 | 56 | 109 | 108 |

The structure of the obtained block-diagonal matrix in Fig. 7.2 is consistent with the numerical data above. Since the size of the diagonal block corresponding to the unit representation $(+,+)$ is 61, we see that the $\mathrm{D}_6$-symmetric deformation for a $\mathrm{D}_6$-symmetric force can be determined by solving a 61-dimensional system of linear equations. ■

## 7.5 Characters

In this section we discuss characters. The concept of characters is not particularly necessary to describe the symmetry of engineering systems, but it is an important tool for developing the theory of group representation. In order to discuss orthogonality, which is a major property of characters, we need to work over $\mathbb{C}$. In this section, we assume $F = \mathbb{C}$.

### 7.5.1 *Definition*

For a representation $T$ over $\mathbb{C}$ of a group $G$, we define a function $\chi : G \to \mathbb{C}$ by

$$\chi(g) = \operatorname{Tr} T(g) \qquad (g \in G) \tag{7.94}$$

and call it the *character* of $T$. That is, the character of $T$ is the function that maps each $g \in G$ to the trace of the matrix $T(g)$. If $T$ is a representation of degree $N$, then

$$\chi(g) = \sum_{i=1}^{N} T_{ii}(g) \qquad (g \in G), \tag{7.95}$$

and, in particular,

$$\chi(e) = N,$$

where $e$ is the identity element.

If two representations are equivalent, then their characters are equal. This follows from $T_1(g) = Q^{-1}T_2(g)Q$ as

$$\operatorname{Tr} T_1(g) = \operatorname{Tr}(Q^{-1}T_2(g)Q) = \operatorname{Tr} T_2(g).$$

The converse is also true, that is, two representations with equal characters are equivalent.[20]

**Proposition 7.7.** *Two representations are equivalent if and only if their characters are equal.*

**Proof.** The necessity is explained above. The proof of the sufficiency will be given in Remark 7.9 in Sec. 7.5.4. □

For a unitary representation $T$, we have $T(g^{-1}) = T(g)^*$ $(g \in G)$ and hence[21]

$$\chi(g^{-1}) = \overline{\chi(g)} \qquad (g \in G). \tag{7.96}$$

[20] In this section, we are assuming $F = \mathbb{C}$, but Proposition 7.7 is true also for $F = \mathbb{R}$.
[21] $\overline{\phantom{x}}$ stands for the complex conjugate.

On the other hand, by Proposition 7.1, every representation is equivalent to a unitary representation. Therefore, the relation (7.96) above is true for any representation $T$ (whether unitary or not).

The character of an irreducible representation is called an *irreducible character.* The character of irreducible representation $T^{\mu}$ will be denoted as $\chi^{\mu}$, *i.e.*,

$$\chi^{\mu}(g) = \mathrm{Tr}\, T^{\mu}(g) \qquad (g \in G). \tag{7.97}$$

The following facts are important concerning irreducible characters:

- Irreducible characters are endowed with two types of orthogonality (Propositions 7.8 and 7.11).
- The number of irreducible characters is equal to the number of conjugacy classes (Proposition 7.10).
- The multiplicity $a^{\mu}$ in the irreducible decomposition (7.64) is expressed in terms of the inner product of the character $\chi$ of the given representation $T$ and the irreducible character $\chi^{\mu}$ (see (7.106)).

We will explain these facts in the following sections.

### 7.5.2 *Character Tables*

For two elements $g$ and $h$ of a group $G$, we say that $g$ is *conjugate*[22] to $h$ if there exists an element $k \in G$ such that $k^{-1}gk = h$. The equivalence classes induced from this conjugacy relation are called *conjugacy classes.* The size (number of elements) of the conjugacy class containing $g \in G$ is denoted as $c(g)$.

For example, the decompositions of $\mathrm{D}_3$ and $\mathrm{D}_6$ into conjugacy classes are as follows:

$$\begin{aligned} \mathrm{D}_3 &= \{e\} \cup \{r, r^2\} \cup \{s, sr, sr^2\}, \\ \mathrm{D}_6 &= \{e\} \cup \{r, r^5\} \cup \{r^2, r^4\} \cup \{r^3\} \cup \{s, sr^2, sr^4\} \cup \{sr, sr^3, sr^5\}. \end{aligned}$$

We have $c(e) = 1$, $c(r) = 2$ and $c(s) = 3$ in $\mathrm{D}_3$ and $\mathrm{D}_6$.

In general, a function $\psi : G \to \mathbb{C}$ is called a *class function* if

$$\psi(g) = \psi(k^{-1}gk) \qquad (g, k \in G). \tag{7.98}$$

That is, a class function is a function that takes a constant value on each conjugacy class. Every character $\chi$ of a group $G$ is a class function. Indeed,

$$\chi(k^{-1}gk) = \mathrm{Tr}\, T(k^{-1}gk) = \mathrm{Tr}\, [T(k)^{-1}T(g)T(k)] = \mathrm{Tr}\, T(g) = \chi(g)$$

[22]Conjugacy relation is an equivalence relation in the sense of Remark 1.3 (Sec. 1.1.2).

holds for all $g, k \in G$. In particular, irreducible characters are class functions, and conversely, every class function can be expressed as a linear combination of irreducible characters (Proposition 7.10).

The *character table* of a group $G$ is a table that lists the values of all irreducible characters for all conjugacy classes. For example, the character table of $\mathrm{D}_3$ is as follows:

| | $e$ | $r$ | $s$ |
|---|---|---|---|
| $\chi^{(+,+)}$ | 1 | 1 | 1 |
| $\chi^{(+,-)}$ | 1 | 1 | −1 |
| $\chi^{(1)}$ | 2 | −1 | 0 |
| $c$ | 1 | 2 | 3 |

(7.99)

where notations $(+,+)$, $(+,-)$, and $(1)$ are introduced in Example 7.14 in Sec. 7.3.2. The top row shows the representatives of the conjugacy classes, and the bottom row gives the values of $c$. The character table of $\mathrm{D}_6$ is as follows:

| | $e$ | $r$ | $r^2$ | $r^3$ | $s$ | $sr$ |
|---|---|---|---|---|---|---|
| $\chi^{(+,+)}$ | 1 | 1 | 1 | 1 | 1 | 1 |
| $\chi^{(+,-)}$ | 1 | 1 | 1 | 1 | −1 | −1 |
| $\chi^{(-,+)}$ | 1 | −1 | 1 | −1 | 1 | −1 |
| $\chi^{(-,-)}$ | 1 | −1 | 1 | −1 | −1 | 1 |
| $\chi^{(1)}$ | 2 | 1 | −1 | −2 | 0 | 0 |
| $\chi^{(2)}$ | 2 | −1 | −1 | 2 | 0 | 0 |
| $c$ | 1 | 2 | 2 | 1 | 3 | 3 |

. (7.100)

In the two examples above, the number of irreducible representations and the number of conjugacy classes are equal, and, as a result, the character tables are square (ignoring the $c$ row). This is indeed a very important fact that holds for arbitrary groups (Proposition 7.10).

### 7.5.3 *Orthogonality*

**Proposition 7.8 (First orthogonality relation).** *For $\mathbb{C}$-irreducible characters $\chi^\mu$ and $\chi^\nu$ of a group $G$, it holds that*

$$\sum_{g\in G} \chi^\mu(g)\overline{\chi^\nu(g)} = |G|\delta_{\mu\nu}, \tag{7.101}$$

*where $\delta_{\mu\nu}$ is the Kronecker delta introduced in* (7.56).

**Proof.** By (7.57) of Proposition 7.5 with $i = l$ and $m = s$ we have

$$\sum_{g\in G} T^\mu_{ii}(g)T^\nu_{mm}(g^{-1}) = \begin{cases} \dfrac{|G|}{N^\mu}\delta_{im} & (\mu = \nu), \\ 0 & (\mu \neq \nu). \end{cases}$$

For $\mu = \nu$, we take the sum over $i, m = 1, \ldots, N^\mu$ to obtain

$$\sum_{g \in G} \chi^\mu(g)\chi^\mu(g^{-1}) = |G|.$$

For $\mu \neq \nu$, we take the sum over $i = 1, \ldots, N^\mu$ and $m = 1, \ldots, N^\nu$ to obtain

$$\sum_{g \in G} \chi^\mu(g)\chi^\nu(g^{-1}) = 0.$$

We can combine the two into a single formula

$$\sum_{g \in G} \chi^\mu(g)\chi^\nu(g^{-1}) = |G|\delta_{\mu\nu}.$$

With the use of $\chi^\nu(g^{-1}) = \overline{\chi^\nu(g)}$ (see (7.96)), we obtain (7.101). □

**Example 7.24.** Let us verify the first orthogonality relation (7.101) of irreducible characters for dihedral group $D_6$. By the character table (7.100) we have

| | $e$ | $r$ | $r^2$ | $r^3$ | $r^4$ | $r^5$ | $s$ | $sr$ | $sr^2$ | $sr^3$ | $sr^4$ | $sr^5$ |
|---|---|---|---|---|---|---|---|---|---|---|---|---|
| $\chi^{(+,+)}$ | 1 | 1 | 1 | 1 | 1 | 1 | 1 | 1 | 1 | 1 | 1 | 1 |
| $\chi^{(+,-)}$ | 1 | 1 | 1 | 1 | 1 | 1 | $-1$ | $-1$ | $-1$ | $-1$ | $-1$ | $-1$ |
| $\chi^{(1)}$ | 2 | 1 | $-1$ | $-2$ | $-1$ | 1 | 0 | 0 | 0 | 0 | 0 | 0 |

and so forth. For $\mu = (+,+)$ and $\nu = (+,-)$, the sum of $\chi^\mu(g)\overline{\chi^\nu(g)}$ over all $g \in D_6$ on the left-hand side of (7.101) is equal to

$$1 \times 1 + \cdots + 1 \times 1 + 1 \times (-1) + \cdots + 1 \times (-1) = 0.$$

For $\mu = (+,-)$ and $\nu = (1)$, it is equal to

$$1 \times 2 + 1 \times 1 + 1 \times (-1) + 1 \times (-2) + 1 \times (-1) + 1 \times 1 + (-1) \times 0 + \cdots + (-1) \times 0 = 0.$$

For $\mu = \nu = (1)$, it is equal to

$$2^2 + 1^2 + (-1)^2 + (-2)^2 + (-1)^2 + 1^2 + 0^2 + \cdots + 0^2 = 12.$$

In either case, the relation (7.101) holds. ■

The orthogonality shown above can be interpreted as the orthogonality in a vector space. To see this, we first note that the set of all class functions forms a vector space over $\mathbb{C}$, which we denote by $V$. In the vector space $V$ we can define [6, Remark 9.19] an inner product $\langle \phi, \psi \rangle$ by

$$\langle \phi, \psi \rangle = \frac{1}{|G|} \sum_{g \in G} \overline{\phi(g)}\psi(g) \qquad (\phi, \psi \in V). \tag{7.102}$$

The equation (7.101) shows that the set of irreducible characters $(\chi^\mu \mid \mu \in R(G;\mathbb{C}))$ forms an orthonormal system with respect to this inner product.

Furthermore, the set of all irreducible characters forms an orthonormal basis; this fact is shown below. First we show the following proposition.

**Proposition 7.9.** *For any $\mathbb{C}$-irreducible representation $T$ of a group $G$ and any class function $\psi$, the following relation holds:*

$$\sum_{g\in G} \psi(g)T(g) = \lambda I, \tag{7.103}$$

*where $\lambda$ is a complex number given as*

$$\lambda = \frac{1}{N}\sum_{g\in G}\psi(g)\chi(g) = \frac{|G|}{N}\langle\overline{\psi},\chi\rangle.$$

*Here, $\chi$ is the character of $T$, $N$ is the degree of $T$, and $I$ represents the unit matrix of order $N$.*

**Proof.** Define $A = \sum_{g\in G}\psi(g)T(g)$. Then

$$\begin{aligned}
T(h)^{-1}AT(h) &= T(h)^{-1}\left(\sum_{g\in G}\psi(g)T(g)\right)T(h)\\
&= \sum_{g\in G}\psi(g)T(h^{-1})T(g)T(h) = \sum_{g\in G}\psi(g)T(h^{-1}gh)\\
&= \sum_{k\in G}\psi(hkh^{-1})T(k) = \sum_{k\in G}\psi(k)T(k) = A,
\end{aligned}$$

which shows $T(h)A = AT(h)$ for all $h \in G$. Therefore, by Schur's lemma (Proposition 7.4), we have $A = \lambda I$ for some $\lambda \in \mathbb{C}$. The value of $\lambda$ can be determined by calculating the trace of both sides of (7.103). □

**Proposition 7.10.** *The irreducible characters $(\chi^\mu \mid \mu \in R(G;\mathbb{C}))$ form a basis of the class functions. In particular, the number of irreducible characters $|R(G;\mathbb{C})|$ is equal to the number of conjugacy classes.*

**Proof.** Since we already know that irreducible characters $(\chi^\mu \mid \mu \in R(G;\mathbb{C}))$ form an orthonormal system, we only have to show that a class function $\phi$ that is orthogonal to all $\chi^\mu$ must be 0. We put $\psi = \overline{\phi}$ and assume that $\langle\overline{\psi},\chi^\mu\rangle = 0$ for all $\mu$.

For a representation $T$, in general, consider a matrix $A = \sum_{g\in G}\psi(g)T(g)$. If $T$ is irreducible, then $A = O$ by Proposition 7.9 and the assumption of $\langle \overline{\psi}, \chi^\mu \rangle = 0$. When $T$ is reducible, we can also conclude $A = O$ with the aid of the irreducible decomposition of $T$. Therefore, we have $A = O$ for any $T$. We now choose $T$ to be the regular representation (Example 7.9). Then each column of $A$ is a suitable arrangement of $\psi(g)$ for all $g \in G$, and therefore $\psi(g) = 0$ $(g \in G)$ follows from $A = O$.

In the above, we have proved that irreducible characters form a basis of the vector space $V$ of class functions. In particular, the number of irreducible characters is equal to the dimension of $V$. On the other hand, the dimension of $V$ is obviously equal to the number of conjugacy classes; the set of functions that are 1 in one conjugacy class and 0 elsewhere forms a basis of $V$. Therefore, the number of irreducible characters is equal to the number of conjugacy classes. □

**Example 7.25.** Let us verify Proposition 7.10 (the latter half) for dihedral groups $D_3$ and $D_6$. As shown in Sec. 7.5.2, $D_3$ has three conjugacy classes $\{e\}$, $\{r, r^2\}$, and $\{s, sr, sr^2\}$, and three irreducible representations $(+,+)$, $(+,-)$, and $(1)$. $D_6$ has six conjugacy classes $\{e\}$, $\{r, r^5\}$, $\{r^2, r^4\}$, $\{r^3\}$, $\{s, sr^2, sr^4\}$, and $\{sr, sr^3, sr^5\}$, and six irreducible representations $(+,+)$, $(+,-)$, $(-,+)$, $(-,-)$, $(1)$, and $(2)$. In either case, the number of irreducible characters is equal to the number of conjugacy classes, which is consistent with Proposition 7.10 (the latter half). ■

**Example 7.26.** Let us verify Proposition 7.10 (the latter half) for cyclic group $C_n$. In $C_n$ we have $k^{-1}gk = g$ for $k = r^i$ and $g = r^j$. Therefore, each element itself forms a conjugacy class, and there are $n$ conjugacy classes $\{e\}, \{r\}, \{r^2\}, \ldots, \{r^{n-1}\}$. On the other hand, as shown in Example 7.13, there are $n$ $\mathbb{C}$-irreducible representations. Thus Proposition 7.10 (the latter half) holds for $C_n$. ■

Lastly, let us show another orthogonality relation satisfied by irreducible characters. It should be clear that the summation on the left-hand side of (7.101) is taken over elements $g$ of $G$, whereas the summation in the equation (7.104) below is taken over irreducible representations $\mu \in R(G;\mathbb{C})$.

**Proposition 7.11 (Second orthogonality relation).**

$$\sum_{\mu \in R(G;\mathbb{C})} \chi^\mu(g)\overline{\chi^\mu(h)} = \begin{cases} \dfrac{|G|}{c(g)} & \text{if } g \text{ and } h \text{ belong to the same conjugacy class,} \\ 0 & \text{otherwise,} \end{cases} \tag{7.104}$$

*where $c(g)$ denotes the size (number of elements) of the conjugacy class to which $g$ belongs.*

**Proof.** By Proposition 7.10, we can regard the character table as a square matrix of order $|R(G;\mathbb{C})|$. Denote this matrix by $X$. The $(\mu, j)$ entry of matrix $X$ is equal to the value $\chi^\mu(g_j)$ of the $\mu$th irreducible character $\chi^\mu$ at the $j$th conjugacy class, where $g_j$ is an arbitrary representative of the $j$th conjugacy class. Let $C$ be the diagonal matrix with $c(g_j)/|G|$ ($j = 1, \ldots, |R(G;\mathbb{C})|$) as its diagonal entries. Then the first orthogonality relation (Proposition 7.8) can be recognized as the orthogonality among the row vectors of $X$ with respect to the weight $C$, which is expressed as

$$XCX^* = I.$$

This equation shows that $X$ and $CX^*$ are the inverse matrices to each other. Therefore, we have $CX^* \cdot X = I$, *i.e.*,

$$X^*X = C^{-1}.$$

This shows the orthogonality for the column vectors of matrix $X$, which is equivalent to the relation in (7.104). □

**Example 7.27.** Let us verify the second orthogonality relation (7.104) of irreducible characters for dihedral group $\mathrm{D}_6$. Note that the left-hand side of (7.104) is the inner product of column vectors of the character table (7.100). For $g = e$ and $h = r$, the equation (7.104) reads

$$1 \times 1 + 1 \times 1 + 1 \times (-1) + 1 \times (-1) + 2 \times 1 + 2 \times (-1) = 0.$$

For $g = r$ and $h = s$, it reads

$$1 \times 1 + 1 \times (-1) + (-1) \times 1 + (-1) \times (-1) + 1 \times 0 + (-1) \times 0 = 0.$$

For $g = h = s$, it reads

$$1^2 + (-1)^2 + 1^2 + (-1)^2 + 0^2 + 0^2 = 4 = \frac{12}{3} = \frac{|\mathrm{D}_6|}{c(s)}.$$

For $g = h = r^3$, it reads

$$1^2 + 1^2 + (-1)^2 + (-1)^2 + (-2)^2 + 2^2 = 12 = \frac{|\mathrm{D}_6|}{c(r^3)}.$$

In either case, the relation (7.104) holds. ■

### 7.5.4 *Multiplicity Formula*

We derive a formula for multiplicity $a^\mu$ in the irreducible decomposition (7.64) (or (7.69)) over $\mathbb{C}$ from the (first) orthogonality relation of irreducible characters.

By taking the trace of both sides of (7.64) (or (7.69)) we obtain

$$\chi(g) = \sum_{\mu \in R(G;\mathbb{C})} a^\mu \chi^\mu(g) \qquad (g \in G), \tag{7.105}$$

where $\chi$ stands for the character of the representation $T$. By multiplying this equation with $\overline{\chi^\nu(g)}$, taking the sum over $g \in G$, and using the first orthogonality relation (7.101) we obtain

$$\sum_{g \in G} \chi(g)\overline{\chi^\nu(g)} = \sum_{\mu \in R(G;\mathbb{C})} a^\mu \sum_{g \in G} \chi^\mu(g)\overline{\chi^\nu(g)} = |G| \sum_{\mu \in R(G;\mathbb{C})} a^\mu \delta_{\mu\nu} = |G| a^\nu.$$

Therefore,

$$a^\mu = \frac{1}{|G|} \sum_{g \in G} \chi(g)\overline{\chi^\mu(g)}. \tag{7.106}$$

It follows from (7.105) and (7.106) that

$$\begin{aligned}\sum_{g \in G} \chi(g)\overline{\chi(g)} &= \sum_{g \in G} \chi(g) \sum_{\mu \in R(G;\mathbb{C})} a^\mu \overline{\chi^\mu(g)} \\ &= \sum_{\mu \in R(G;\mathbb{C})} a^\mu \sum_{g \in G} \chi(g)\overline{\chi^\mu(g)} \\ &= |G| \sum_{\mu \in R(G;\mathbb{C})} (a^\mu)^2,\end{aligned}$$

which shows the relation

$$\frac{1}{|G|} \sum_{g \in G} |\chi(g)|^2 = \sum_{\mu \in R(G;\mathbb{C})} (a^\mu)^2. \tag{7.107}$$

In particular, a necessary and sufficient condition for $T$ to be irreducible is given by

$$\frac{1}{|G|} \sum_{g \in G} |\chi(g)|^2 = 1. \tag{7.108}$$

Since the left-hand side above can be calculated from the concrete values of $T$, this criterion is very useful in testing for the irreducibility of $T$.

**Example 7.28.** In Example 7.12 (Sec. 7.3.1), we have seen that the permutation representation $T$ of $D_3$ decomposes into a direct sum of the unit representation $(+,+)$ and the two-dimensional irreducible representation (1). The values of the irreducible characters $\chi^{(+,+)}$, $\chi^{(+,-)}$, and $\chi^{(1)}$ of $D_3$ and those of the character $\chi$ of $T$ are given by the character table (7.99) as follows:

| | $e$ | $r$ | $r^2$ | $s$ | $sr$ | $sr^2$ |
|---|---|---|---|---|---|---|
| $\chi^{(+,+)}$ | 1 | 1 | 1 | 1 | 1 | 1 |
| $\chi^{(+,-)}$ | 1 | 1 | 1 | −1 | −1 | −1 |
| $\chi^{(1)}$ | 2 | −1 | −1 | 0 | 0 | 0 |
| $\chi$ | 3 | 0 | 0 | 1 | 1 | 1 |

For $\mu = (+,+)$, $(+,-)$, and (1), the formula (7.106) yields

$$a^{(+,+)} = \frac{1}{6}(3+0+0+1+1+1) = 1,$$

$$a^{(+,-)} = \frac{1}{6}(3+0+0-1-1-1) = 0,$$

$$a^{(1)} = \frac{1}{6}(6+0+0+0+0+0) = 1.$$

These results coincide with the multiplicities we have had in Example 7.12. We can also verify the relation (7.107) as

$$\frac{1}{6}(3^2+0^2+0^2+1^2+1^2+1^2) = 2 = 1^2+0^2+1^2.$$

■

**Example 7.29.** Let us calculate the multiplicities in the irreducible decomposition of the regular representation of a group $G$; see Example 7.9 in Sec. 7.2.2 for the definition of the regular representation. The character $\chi$ of the regular representation is given as

$$\chi(g) = \begin{cases} |G| & (g = e), \\ 0 & (\text{otherwise}). \end{cases} \tag{7.109}$$

With this expression, the formula (7.106) yields

$$a^\mu = \frac{1}{|G|}\sum_{g\in G}\chi(g)\overline{\chi^\mu(g)} = \frac{1}{|G|}\chi(e)\overline{\chi^\mu(e)} = N^\mu.$$

That is, in the regular representation, the multiplicity $a^\mu$ of an irreducible representation $\mu$ is equal to its degree $N^\mu$. By substituting (7.109) and $a^\mu = N^\mu$ into (7.107), we can derive the relation

$$|G| = \sum_{\mu\in R(G;\mathbb{C})}(N^\mu)^2,$$

which is the formula introduced in (7.45) in Sec. 7.3.2. ■

**Remark 7.9.** Let us provide a proof for the sufficiency in Proposition 7.7 (the statement that two representations are equivalent if their characters are equal). As shown by the formula in (7.106), the multiplicities $a^{\mu}$ in the irreducible decomposition (7.64) are determined from the character of $T$. Hence, two representations with equal characters have the same irreducible decomposition, whereas two representations sharing the same irreducible decomposition are equivalent to each other. Therefore, two representations with equal characters are equivalent over $\mathbb{C}$. ■

# Bibliography

[**General (in English)**]
The following are textbooks for linear algebra in general written in English.

[1] D. S. Bernstein: *Matrix Mathematics: Theory, Facts, and Formulas*, 2nd ed., Princeton University Press, Princeton, 2009.
[2] F. R. Gantmacher: *The Theory of Matrices, Vol. I, Vol. II*, Chelsea, New York, 1959. Also: *Applications of the Theory of Matrices*, Interscience Publishers, New York, 1959; Dover, Mineola, New York, 2005.
[3] G. H. Golub and C. F. Van Loan: *Matrix Computations*, 4th ed., Johns Hopkins University Press, Baltimore, 2013.
[4] R. A. Horn and C. R. Johnson: *Matrix Analysis*, Cambridge University Press, Cambridge, 1985; 2nd ed., 2013.
[5] P. D. Lax: *Linear Algebra and Its Applications*, 2nd ed., John Wiley & Sons, Inc., Hoboken, NJ, 2007.
[6] K. Murota and M. Sugihara: *Linear Algebra I: Basic Concepts*, The University of Tokyo Engineering Course/Basic Mathematics, World Scientific and Maruzen Publishing, 2022.
[7] W.W. Sawyer: *An Engineering Approach to Linear Algebra*, Cambridge University Press, Cambridge, 1972.
[8] G. Strang: *Linear Algebra and Its Applications*, Academic Press, New York, 1976; 4th ed., Thomson Brooks/Cole, 2006.
[9] G. Strang: *Introduction to Linear Algebra*, 4th ed., Wellesley-Cambridge Press, Wellesley, MA, 2009.
[10] F. Zhang: *Matrix Theory: Basic Results and Techniques*, Springer, New York, 1999.

[**General (in Japanese)**]
The following are textbooks for linear algebra in general written in Japanese.

[11] H. Arai: *Linear Algebra—Fundamentals and Applications* (in Japanese), Nippon Hyoron Sha, Tokyo, 2006.
[12] T. Fujiwara: *Linear Algebra* (in Japanese), Iwanami Shoten, Tokyo, 1996.
[13] K. Hasegawa: *Linear Algebra* (Revised edition) (in Japanese), Nippon Hyoron Sha, Tokyo, 2015.

[14] Y. Ikebe, Y. Ikebe, S. Asai, and Y. Miyazaki: *Modern Linear Algebra Through Decomposition Theorems* (in Japanese), Kyoritsu Shuppan, Tokyo, 2009.
[15] M. Iri: *General Linear Algebra* (in Japanese), Asakura Publishing, Tokyo, 2009.
[16] M. Iri and T. S. Han: *Linear Algebra—Matrices and Their Normal Forms* (in Japanese), Kyouiku Shuppan, Tokyo, 1977.
[17] S. Kakei: *Engineering Linear Algebra* (New edition) (in Japanese), Suurikougaku-sha, Tokyo, 2014.
[18] A. Kaneko: *Lecture on Linear Algebra* (in Japanese), Saiensu-sha, Tokyo, 2004.
[19] H. Kimura: *Linear Algebra—Foundation of Mathematical Sciences* (in Japanese), University of Tokyo Press, Tokyo, 2003.
[20] T. Kusaba: *Linear Algebra* (Enlarged edition) (in Japanese), Asakura Publishing, Tokyo, 1988.
[21] M. Saito: *Introduction to Linear Algebra* (in Japanese), University of Tokyo Press, Tokyo, 1966.
[22] M. Saito: *Linear Algebra* (in Japanese), Tokyo Tosho, Tokyo, 2014.
[23] I. Satake: *Linear Algebra* (in Japanese), Shokabo Publishing, Tokyo, 1974.
[24] T. Tanino: *System Linear Algebra—Applications to Engineering Systems* (in Japanese), Asakura Publishing, Tokyo, 2013.
[25] T. Yamamoto: *Fundamentals of Matrix Analysis—Advanced Linear Algebra* (in Japanese), Saiensu-sha, Tokyo, 2010.

References for each chapter follow.

[**Chapter 1**] Ref. [16] above and

[26] R. Diestel: *Graph Theory*, 5th ed., Springer, Berlin, 2017.
[27] S. Fujishige: *Graph, Network, and Combinatorics* (in Japanese), Kyoritsu Shuppan, Tokyo, 2002.
[28] B. Korte and J. Vygen: *Combinatorial Optimization: Theory and Algorithms*, 6th ed., Springer, Berlin, 2018.
[29] K. Murota: *Matrices and Matroids for Systems Analysis*, Springer, Berlin, 2000.
[30] M. Sugihara and K. Murota: *Theoretical Numerical Linear Algebra* (in Japanese), Iwanami Shoten, Tokyo, 2009.
[31] R. S. Varga: *Matrix Iterative Analysis*, Prentice-Hall, Englewood Cliffs, New Jersey, 1962; 2nd ed., Springer, Berlin, 2000.
[32] R. J. Wilson: *Introduction to Graph Theory*, 5th ed., Prentice Hall, Harlow, 2010.

[**Chapter 2**] Refs. [25], [31] above and

[33] A. Berman and R. J. Plemmons: *Nonnegative Matrices in the Mathematical Sciences*, Academic Press, New York, 1979; SIAM, Philadelphia, 1994.
[34] W. Feller: *An Introduction to Probability Theory and Its Applications*, Vol. 1, 2nd ed., John Wiley and Sons, New York, 1957; 3rd ed., 1968.

[35] S. Karlin: *A First Course in Stochastic Processes*, Academic Press, New York, 1966; S. Karlin and H. M. Taylor: 2nd ed., 1975.
[36] H. Nikaido: *Linear Algebra for Economics* (in Japanese), Baifukan, Tokyo, 1961.

[**Chapter 3**] Ref. [36] above and

[37] V. Chvátal: *Linear Programming*, W. H. Freeman and Company, New York, 1983.
[38] G. B. Dantzig: *Linear Programming and Extensions*, Princeton University Press, Princeton, 1963.
[39] M. Fukushima: *Fundamentals of Nonlinear Optimization* (in Japanese), Asakura Publishing, Tokyo, 2001.
[40] M. Iri: *Linear Programming* (in Japanese), Kyoritsu Shuppan, Tokyo, 1986.
[41] H. Konno: *Linear Programming* (in Japanese), JUSE Press, Tokyo, 1987.
[42] A. Schrijver: *Theory of Linear and Integer Programming*, John Wiley and Sons, New York, 1986.
[43] S. J. Wright: *Primal-Dual Interior-Point Methods*, SIAM, Philadelphia, 1997.

[**Chapter 4**] Ref. [42] above and

[44] M. Newman: *Integral Matrices*, Academic Press, New York, 1972.
[45] A. Schrijver: *Combinatorial Optimization—Polyhedra and Efficiency*, Springer, Heidelberg, 2003.

[**Chapter 5**] Refs. [2], [44] above and

[46] F. E. Cellier: *Continuous System Modeling*, Springer, Berlin, 1991.
[47] I. Gohberg, P. Lancaster, and L. Rodman: *Matrix Polynomials*, Academic Press, New York, 1982; SIAM, Philadelphia, 2009.
[48] S. Kodama and N. Suda: *Matrix Theory for Systems Control* (in Japanese), 2nd ed., The Society of Instrument and Control Engineers, ed., Corona Publishing, Tokyo, 1981.
[49] P. Kunkel and V. Mehrmann: *Differential-Algebraic Equations: Analysis and Numerical Solution*, European Mathematical Society, Zürich, 2006.
[50] H. Maeda and T. Sugie: *Systems Control Theory for Advanced Control* (in Japanese), The Institute of Systems, Control and Information Engineers, ed., Asakura Publishing, Tokyo, 1990.
[51] N. Suda: *Linear System Theory* (in Japanese), The Institute of Systems, Control and Information Engineers, ed., Asakura Publishing, Tokyo, 1993.

[**Chapter 6**]

[52] A. Ben-Israel and T. N. E. Greville: *Generalized Inverses: Theory and Applications*, 2nd ed., Springer, New York, 2003.
[53] K. Kawaguchi: *Generalized Inverse and Its Applications to Structural Engineering* (in Japanese), Corona Publishing, Tokyo, 2011.
[54] C. R. Rao and S. K. Mitra: *Generalized Inverse of Matrices and Its Applications*, John Wiley and Sons, New York, 1970.

[55] H. Yanai and K. Takeuchi: *Projection Matrix, Generalized Inverse, and Singular Decomposition* (in Japanese), University of Tokyo Press, Tokyo, 1983.

**[Chapter 7]**

[56] A. Hattori: *Groups and Their Representations* (in Japanese), Kyoritsu Shuppan, Tokyo, 1967.
[57] T. Hirai: *Linear Algebra and Group Representation, Vol. I, Vol. II* (in Japanese), Asakura Publishing, Tokyo, 2001.
[58] K. Ikeda and K. Murota: *Imperfect Bifurcation in Structures and Materials—Engineering Use of Group-Theoretic Bifurcation Theory*, 3rd ed., Springer, New York, 2019.
[59] T. Inui, Y. Tanabe, and Y. Onodera: *Applied Group Theory: Group Representation and Physics* (in Japanese), Shokabo Publishing, Tokyo, 1976 (enlarged ed., 1980).
[60] S. Iyanaga and M. Sugiura: *Algebra for Applied Mathematicians* (in Japanese), Iwanami Shoten, Tokyo, 1960.
[61] T. Kondo: *Group Theory* (in Japanese), Iwanami Shoten, Tokyo, 1991.
[62] W. Miller, Jr.: *Symmetry Groups and Their Applications*, Academic Press, New York, 1972.
[63] J.-P. Serre: *Linear Representations of Finite Groups*, Springer, New York, 1977.
[64] I. Terada and K. Harada: *Group Theory* (in Japanese), Iwanami Shoten, Tokyo, 2006.
[65] T. Yamanouchi and M. Sugiura: *Introduction to Continuous Groups* (in Japanese), Baifukan, Tokyo, 1960 (newly-formatted, 2010).

# Index

Printed in the United States
by Baker & Taylor Publisher Services